Biotic Regulation of the Environment

Key Issue of Global Change

Springer
London
Berlin
Heidelberg
New York
Barcelona
Hong Kong
Milan
Paris
Santa Clara
Singapore
Tokyo

Victor G. Gorshkov, Vadim V. Gorshkov and
Anastassia M. Makarieva

Biotic Regulation of the Environment

Key Issue of Global Change

Springer

Published in association with
Praxis Publishing
Chichester, UK

PRAXIS

Professor V. G. Gorshkov
Leading Scientific Researcher
Department of Theoretical Physics
Petersburg Nuclear Physics Institute
St Petersburg
Russia

Dr V. V. Gorshkov
Senior Scientific Researcher
Laboratory of Ecology of Plant Communities
Komarov Botanical Institute
St Petersburg
Russia

Dr A. M. Makarieva
Junior Scientific Researcher
Department of Theoretical Physics
Petersburg Nuclear Physics Institute
St Petersburg
Russia

SPRINGER–PRAXIS BOOKS IN ENVIRONMENTAL SCIENCES
SUBJECT *ADVISORY EDITOR*: John Mason B.Sc., Ph.D.

ISBN 1-85233-181-X Springer-Verlag Berlin Heidelberg New York

British Library Cataloguing in Publication Data
Gorshkov, V. G. (Victor G.), 1935–
Biotic regulation of the environment : key issue of global change. — (Springer-Praxis books in environmental sciences)
1. Biotic communities 2. Sustainable development 3. Environmental protection
I. Title II. Gorshkov, Vadim V. III. Makarieva, Anastassia M.
577.8′2 ISBN 1-85233-181-X

Library of Congress Cataloging-in-Publication Data
Gorshkov, V. G., 1935–
Biotic regulation of the environment : key issue of global change/Victor G. Gorshkov, Vadim V. Gorshkov, and Anastassia M. Makarieva.
p. cm. — (Springer-Praxis books in environmental sciences)
Includes bibliographical references (p.).
ISBN 1-85233-181-X (alk. paper)
1. Biotic communities. 2. Ecosystem management. 3. Nature—Effect of human beings on. 4. Environmental policy.
I. Makar'eva, Anastasiia Mikhailovna. II. Gorshkov, Vadim V. III. Title.
IV. Series.
QH541.G64 2000
577.2′2—dc21 00-037164

Printed by MPG Books Ltd, Bodmin, Cornwall, UK

Cover design: Jim Wilkie
Copy editor: Rachael Wilkie
Typesetting: Originator, Gt. Yarmouth, Norfolk, UK

Printed on acid-free paper supplied by Precision Publishing Papers Ltd, UK

Contents

Preface

At the beginning of the third millennium, amidst the consequences of the explosive scientific and technological progress of the twentieth century, humanity is faced with a vague perspective of further development of civilisation. On the one hand, the inherent human claim for improving living standards urges further acceleration of economic growth and exploitation of biospheric resources. On the other hand, it becomes evident that uncontrolled spontaneous development of modern civilisation, determined by catering for immediate human demands, goes hand in hand with global environmental devastation. This, in its turn, inevitably impairs the quality of life and undermines the security of human existence.

Humanity is searching for a compromise between the two trends. Hopes for sustainable development are most often associated with creation of technologies capable of imposing a stabilising impact upon the global environment. This way of solving the problem necessitates further enhancement of civilisation's power, growth of human population and the inevitable cultivation of the remaining natural biota and other biospheric resources. The fact that many difficult issues have been successfully settled so far with the help of science and technology, works to convince people that current environmental problems of humanity can be similarly resolved. However, the statement that a technological solution to the problem of global environmental security is in principle possible, is not self-evident and requires a detailed scientific investigation.

A different path of development compatible with long-term environmental safety lies through conservation and restoration of a substantial part of the Earth's biosphere in its natural, nonperturbed state bearing in mind the stabilising potential of the natural biota of Earth with respect to the global environment. Stabilising properties of the natural nonperturbed biota become evident from the fact that the global environment has been supported in a stable state suitable for life for the last four billion years in spite of external destabilising factors. Restoration of the stabilising biotic potential would mean relaxation of anthropogenic pressure on perturbed territories and complete abandonment of further cultivation of the natural biota. This strategy sets a ceiling to exploitation of biospheric resources, economic growth and global human population numbers.

There is a very essential difference between the two alternative strategies of civilisation's development. Ample scientific evidence accumulated up to now proves the reality of achieving a sustainable state of civilisation, were the second path of development associated with conservation of the stabilising biotic potential the one to be followed.

In contrast, sustainability of the first strategy of development based on complete cultivation of the biosphere and technological means of ensuring environmental stability remains entirely unproved. It may not be considered as real until a serious scientific investigation of this question yields a positive answer. Only then will humanity be free to choose between the alternative strategies of development, which may be done on the basis of the democratic choice of the majority of people of Earth.

Until such research is conducted, there is no free choice but only one safe strategy of civilisation development, namely to rely on the conservation of the natural biota of Earth. The technological way of development along which modern civilisation is now spontaneously moving is burdened with the risk of global ecological catastrophe and places the very existence of humanity under threat.

This book presents an attempt to perform scientific analysis of possible strategies for the development of civilisation and to work out quantitative recommendations that could be used by people when choosing their future. The obtained results show that it is impossible to substitute the existing natural biotic mechanism of environmental stabilisation by any technological means, whatever the advanced stage of technological progress of civilisation.

Acknowledgements

We are particularly grateful to Professor K.Ya. Kondratyev who persuaded us of the necessity of writing the book. We express our deep appreciation to Dr Brendan Mackey for most fruitful discussions of the major issues of the book. Special thanks are due to the Administration of the Department of Theoretical Physics of Petersburg Nuclear Physics Institute for providing Victor Gorshkov and Anastassia Makarieva with excellent working conditions. Our work was in part supported by the Russian State Committee of Ecology.

We dedicate this book to Ella Gorshkova

1

General overview

This chapter presents an outline of all the problems to be covered in the book.

The natural biota (i.e. flora and fauna) of Earth undisturbed by human activities regulates the environment on both local and global scales and compensates for any deviations from the optimal environmental characteristics suitable for life as a whole and humans in particular. Natural biotic regulation of the environment is based on genetic information encoded in genomes of natural biological species combined into ecological communities. When the degree of anthropogenic cultivation of the biota goes beyond the ecologically-permissible level, the remaining natural biota of Earth loses its ability to stabilise the global environment. As a result, the global environment degrades. Namely the human-induced perturbation of the natural biota rather than direct anthropogenic forcing appears to be the primary cause of the observed global change. We quantify the area that has to be occupied by natural biota unperturbed by modern industrial society in order to ensure long-term environmental stability on Earth.

1.1 EXTERNAL ENVIRONMENT AND INTERNAL MILIEU

Each organism is characterised by an internal milieu and exists in an external environment, which drastically differ from one another. Such a situation is possible due to the fact that every organism has a natural envelope that protects and separates it from the external environment. Trees have bark, mammals have skin and hair, birds have feathers, living cells have special protecting membranes, etc. The internal milieu of an organism is maintained by the well coordinated work of its internal organs. Failure of any of the internal organs leads to deterioration of the internal milieu and impairment of the general condition (health) of the individual.

Functioning of the internal organs of the organism relies on the continuous consumption of nutrients and energy from the external environment. Within the living body, organic nutrients are decomposed and excreted from the organism together with thermal energy that is released in the course of metabolic processes. As a result, the external environment changes. Each organism is characterised by a rather narrow interval of environmental conditions where it can live (temperature, humidity, air pressure, availability of nutrients, low concentration of toxins, etc.). If the external environment is not continuously supplied with necessary nutrients and not cleared of the excreta, it soon becomes unfit for life of any organism. Thus, it is evident that in any environment, species cannot exist sustainably in isolation from other species.

Life of individuals of any species is only possible provided there is correlated interaction with other species of the biota. Consequently, natural biota consists of ecological communities of species. Excreta of one species become nutrients for another and vice versa. Only then does there open a theoretical possibility for the environment to remain stable. However, long-term environmental stability may be only guaranteed if the functioning of all the species in the community is rigidly correlated, similar to the work of internal organs within a living body.

The statement that the natural biota may really *work* to stabilise the environment may seem somewhat odd to a human being. Modern humans live in artificial conditions that emerged as the main product of civilisation. Each normal human being is obliged to perform a certain amount of work. Most everyday actions of modern people are aimed at the maintenance of stability of various elements of civilisation. The majority of these actions are in conflict with the inherent desires of people. This is manifested by the fact that most people would prefer to work much less than they actually do. In the modern world work takes on average 6–8 hours per day. There is a constant demand imposed by trade unions on employers all over the world to reduce working hours. In other words, most people prefer rest to work, all other factors remaining the same.

During their free time, people act in accordance with their genetically-coded behavioural programme of positive and negative emotions. They engage in various types of social activities, enjoy sports, fishing, hiking and other forms of recreation. In contrast, animals do that all the time. However, by doing so, animals at the same time perform strictly specified work on the stabilisation of the environment. Everything the animals do has a meaning and contributes to the stability of their ecological community. One might say that people first do what they have to do, then what they wish to do, while animals do both things at the same time. This drastic difference in behaviour of humans and the rest of the biosphere sometimes precludes people from a correct understanding of the observed ecological phenomena.

An ecological community taken together with its environment forms a *local ecosystem*. The main difference between an organism and a local ecosystem is that the latter does not have an envelope that would delineate its internal milieu (living area of all organisms of the community) from its external environment (area where there are no living beings). A partial diffusive envelope can be exemplified by lower soil horizons that separate the internal milieu of terrestrial ecological communities from the lifeless lithosphere. Concentrations of various elements in soil differ drastically from corresponding concentrations in the Earth's crust in very much the same manner as, for example, body temperature of endothermic (warm-blooded) animals differs from that of the external environment.

The scientific question of whether the natural biota adapts to the external environment that changes arbitrarily due to random physical, chemical and biological processes or whether the natural biota forms and maintains its environment itself, is, as we show below, of critical importance to modern humanity.

1.2 ADAPTATION TO, OR REGULATION OF, THE ENVIRONMENT?

The concept of adaptation to changing environment forms the basis of the Darwinian theory of evolution. In the first half of the 18th century, a Swedish naturalist, Karl Linnaeus, created his famous classification of biological species on the basis of morphological similarities and differences (Linnaeus, 1789). Linnaeus thought that species did not change with time. Charles Darwin analysed paleodata and the extant species to put forward the statement that similar species had a common origin in the global process of biological evolution. Darwin further assumed that evolutionary process represents continuous accumulation of hereditary changes in each individual, which is followed by natural selection of those individuals that are best adapted to the existing environment, i.e. leave in this environment the greatest number of offspring. Darwin thought that natural selection is absolutely analogous to the artificial selection that is performed by people to create new breeds of animals and sorts of plants. The primary driving force of evolution was, consequently, a spontaneous change of the environment that brought about changes in the natural selection priorities.

According to the adaptation concept, gradual accumulation of hereditary changes finally leads to the fact that individuals of a given species acquire specific traits of a new species. That continuous process is perceived as a succession of extinctions of old species and beginnings of new ones. Accumulation of hereditary changes may proceed along several different ways. Some species give rise to two or more new species, while others (evolutionary dead-ends) become extinct without giving rise to a new species altogether. Extinction of such species counteracts possible exponential growth of the total number of species in the biosphere. It is assumed that all those processes combine into the evolutionary pattern known from paleodata.

In the 20th century, when the genetic nature of hereditary changes became evident, the Darwinian approach was modified to form the basis of the so-called paradigm of neo-darwinism (Dobzhansky, 1951; Mayr, 1963; Ayala and Fitch, 1997). According to that paradigm, the global environment of our planet appears to be suitable for life owing to an exclusively lucky orbital position occupied by the Earth in the solar system. Within a broad corridor of physical environmental conditions that are possible given that orbital position, the natural biota of Earth is capable of adapting to practically any changes of the environment. In any case, during the nearly four billion years' period of life existence, there had been no catastrophic global environmental changes to which the biota could not adapt.

Within the paradigm it is well admitted that substantial changes in the environment may be initiated by the biota itself. To these changes the biota is also able to adapt, so that a circular process entails (biotic modification of the environment → genetic adaptation to it → appearance of new species → new biotic modifications of the environment, etc.). A classic example of drastic biotic impact imposed on the environment is the transition from the oxygen-free to the present-day atmosphere, that occurred more than a billion years ago and was presumably triggered by some major evolutionary changes in the biota (Cloud, 1972; Kasting, 1987).

According to the neo-darwinism paradigm, evolution and continuous genetic adaptation to a changing environment are the principal properties of life in general. There are no specific environmental conditions that would be optimal for life as a whole. Any environment becomes optimal provided the biota is given sufficient time to adapt to it.

Genetic basis of the adaptation is provided by the observed genetic polymorphism (i.e. genetic non-identity of individuals in a population) and mutability (i.e. appearance of new genetic options, not found in parental lines). The most adapted individuals are, by definition, those producing the maximum number of offspring. Genetic variants imparting the highest reproductive capacity to their carriers propagate in the population. Biosphere is composed of chaotically interacting species continuously adapting to changing environment. All the observed evolutionary changes known from paleodata are explained by continuous genetic adaptation and natural selection of individuals.

The neo-darwinistic paradigm completely excludes any possibility of biotic regulation of the environment aimed at conservation of a particular set of optimal environmental conditions. Genetic adaptation necessarily implies a rigid correlation between the new genetic information of a species and the new environment where such adaptation takes place. The genetic programme of an adapting species loses information about the former environment. As a result, the genetic programme of an adapting species cannot, in principle, include a programme of actions aimed at relaxation of the environment to the initial state. Thus, genetic adaptation and long-term environmental stability are, in principle, incompatible (see also Section. 1.7).

Within the conventional biological paradigm, the observed radical change of the global environment caused by large-scale transformation of natural ecological communities into agro-, silvy- and maricultures designed to cater for growing human needs, is often envisaged as but one of the stages of the conventional evolutionary process. The struggle with industrial pollution that acts to change the environment in a way unfavourable for humans is put forward as the major ecological task to be solved by humanity. The on-going cultivation of the global biota by humans and transformation of the biosphere into a global biosystem—noosphere, that solely provides for the needs of a single species, *Homo sapiens*—is thought to be a natural process as well.

Natural biodiversity is thus treated as the genetic resource of humankind, that can be used in the future at some advanced stage of development of biotechnology and gene engineering. It is assumed that biodiversity is a common name for such types of diversity as species diversity (i.e. the variety of the extant species in the biosphere) and intraspecific genetic variability, which serves as the necessary material basis for the genetic adaptation and evolution of new species. To conserve biodiversity, people create gene banks and zoos, along with national parks that account for less than one per cent of the total Earth's surface and do not impede the free extensive as well as intensive development of modern civilisation.

Unlimited economic growth, necessarily based on a continuous increase in rates of exploitation of biospheric resources, is envisaged as the only possibility to provide for the escalating needs of the growing global population of humans.

It may seem that some of the above conclusions, especially those concerning human strategy in the modern world, are not related to the purely biological paradigm of genetic adaptation. However, these conclusions logically follow unambiguously from that paradigm, which has been dominating biological science during the last hundred years and has become common, at least implicitly, to most mentality patterns that proliferate in the modern world. Another possibility is that this paradigm appeared itself as a product of the European social mentality at the end of the 19th century, which was celebrated by a remarkable development of industry and technology and favoured the anthropocentric view of the world.

1.3 MAJOR INCONSISTENCIES IN THE GENETIC ADAPTATION PARADIGM

The paradigm outlined above of genetic adaptation is unable to provide explanations of some widespread phenomena observed in the biosphere. Among others are ecological restrictions apparently imposed on population densities of most natural species in the biosphere (Chapter 3), formation of stable ecological communities with rigid internal correlation of species (Chapters 4 and 5), approximately equal evolutionary species lifespans in different biological kingdoms (Chapter 11), etc. All these problems will be discussed in detail in the chapters that follow.

One may name, however, two major contradictions that immediately catch the eye in the genetic adaptation paradigm:

1. Why, in spite of continuous adaptation to an ever-changing environment, do all species retain discreteness both in space and time? In other words, why are there no transient forms between the extant as well as between the extinct species known from paleodata?
2. Why, in spite of uncontrollable changes of the environment, especially those due to the biotic impact, have the global environmental conditions remained within the life-compatible interval during the whole period of life existence?

We now discuss these two questions in more detail.

1.4 DISCRETENESS AND STABILITY OF BIOLOGICAL SPECIES

The available paleodata testify for morphologic and, consequently, genetic (Jackson, 1990) constancy of all species-specific characteristics during the whole period of existence of most (at least 90%) of the species studied (Gould and Eldridge, 1993). No transient forms that would bridge two discrete acts of speciation are observed. The classical example of gradual speciation, family *Equidae*, has been recently shown to be but a misinterpretation of the paleodata (MacFadden, 1993). A closer analysis showed that what had been traditionally conceived as a succession of transient forms that had finally shaped into the modern horse, often proved to be a set of discrete contemporary species that coexisted in both space and time.

The extant species also demonstrate strict discreteness. There are neither intermediate forms between related species, nor processes of their formation being observed. Hybrid zones that are sometimes considered as a possible seeding of evolutionary process, are strongly restricted and occupy a few per cent of the biosphere at most (Raven and Johnson, 1988). All these facts suggest that the period of speciation (i.e. appearance and spread of a new species) takes a much shorter time compared to the average time of species existence (a few million years) (Gould and Eldridge, 1993). During the speciation 'burst', organisms undergo rapid morphologic and genetic change to remain further unchanged for the rest of the new species' existence.

One may conclude, therefore, that the evolution of a species is not a result of the gradual accumulation of relatively frequent minor modifications of the hereditary programme of a species. Rather, evolutionary process is discontinuous and is due to infrequent but radical changes of the genetic programme of a species. The continuous 'daily' process of mutations that supports intraspecific genetic polymorphism bears no relation to the evolutionary process. Rather it can be conceived as random deviations from a normal hereditary programme that is coded genetically in DNA molecules and is known as the species' genome. Random genetic deviations accumulate as a result of the mutational process and erase the genetic information of the species. The number of such deviations cannot increase infinitely, but should be limited by selection that, due to its function, may be called stabilising. In the process of stabilising selection, individuals with too many genetic deviations are forced out from the population. Thus, under natural conditions, the genetic programme of a species is prevented from decay.

However, under distorted conditions individuals with genetic defects can accumulate. Artificial selection uses this fact to create new breeds of economically-important plants and animals. Unlike evolutionary changes, many artificially-created genetic changes are reversible. When placed under natural conditions, many domestic species recover their normal (wild-type) genetic programme which assures maximum competitiveness of individuals in their natural ecological niche. For example, doves (*Columba livia*) living free in cities have rather uniform morphology and are practically identical with wild doves, though urban doves descend from various domesticated breeds that differ drastically from one another and from the wild doves. As soon as the press of artificial selection was relaxed, the most competitive wild phenotype was restored.

In some cases, when two genetically different populations of the same species live in different environmental conditions, individuals taken from one population and placed on the territory of the other appear poorly fitted to the alien environment and lose competition with aboriginal individuals. Such facts are interpreted as important empirical evidence in favour of the existence of genetic adaptation.

However, the possibility of the existence of normal and distorted environment is completely ignored in such considerations. Suppose that in the normal environment individuals of a certain species need both to swim and to walk. In one distorted environment they only need to walk, in the other, only to swim. In both distorted environments individuals will lose one of the two abilities, because competitive interaction in the distorted environments will not be able to support both. Individuals that are able both to walk and to swim in the environment where swimming is the only thing needed will have no advantage over those capable only of swimming, so the ability to walk will finally vanish. So, in the first environment individuals will only be able to walk, in the second one, only to swim. After changing environment, in both cases they will die. But, evidently, their genetic differences are not an example of adaptation, i.e. acquiring new information about changed environment. On the contrary, genetic differences between the two hypothetical populations clearly represent erosion of the original genetic information (see also Section 9.8).

The above consideration may be illustrated by the example of domesticated plants and animals. Under natural conditions these plants and animals cannot compete with normal individuals of the corresponding wild species, because those hereditary properties that make them useful for humans (high productivity of milk, high degree of fat, extremely large size of edible parts of plants) prove to be disadvantageous and do not correspond to the maximum competitiveness under natural conditions.

This alternative explanation of the observed pattern relies on the notions of normal (optimal) and distorted environment, which are by definition absent from the genetic adaptation concept. The genetic adaptation concept states as an axiom that since species adapt to changing environment, any environment may become optimal and there are no favourites among possible environments. After such a statement is made, the observed pattern is unambiguously interpreted as independent evidence in favour of the genetic adaptation. In reality, however, this explanation is dependent on that critical statement about environments, which in itself remains absolutely unproved. As shown above, if we replace this statement by the opposite one, the observed pattern can be consistently explained not only without involving genetic adaptation but from an opposite point of view. Thus the fact that genetically-different conspecific organisms sometimes behave differently in different environments is not in itself a testimony for genetic adaptation (see also Section 9.8).

On the above basis, the following explanation of the observed discreteness of species appears justified. Due to the random character of the mutation process, different individuals in a population have different locations of genome sites with erased genetic information. The total amount of such sites, i.e. the total amount of erased genetic information, is limited by the stabilising selection and should be approximately equal in all competitive individuals. All the meaningful genetic information is the same in all individuals of all populations of the same species. Namely this fact determines the observed constancy of species-specific characteristics in space and time.

We have seen, therefore, that the available empirical data on species discreteness are inconsistent with the assumption about continuous genetic adaptation to changing environment. This suggests that there exists a certain optimal for life environment, which is maintained and controlled by the biota itself, information about the characteristics of that environment being genetically encoded in biological species and kept intact during the most time of the species' existence.

1.5 GLOBAL ENVIRONMENT FORMED BY THE NATURAL BIOTA

The notion *global environment* comprises physical and chemical conditions encountered on the planet's surface and in the planet's climate. A critical characteristic of the latter is the mean global surface temperature. The temperature of Earth's surface is determined by the balance of incoming solar radiation and thermal radiation that is emitted by the planet into space. This balance is totally determined by the amount of solar radiation that is reflected by the planet back to space (the planetary albedo) and absorption of thermal radiation by the so-called greenhouse gases. At constant flux of solar radiation incident upon Earth, surface temperature can assume almost arbitrary (including life-incompatible) values depending on values of planetary albedo and greenhouse effect (see Table 8.1), which are completely determined by inherent environmental characteristics of the planet.

The hydrosphere of Earth is sufficient for total glaciation of the Earth's surface, which corresponds to the global average surface temperature of about −100°C. On the other hand, complete evaporation of Earth's hydrosphere would lead to a catastrophic greenhouse effect with the global surface temperature rising to several hundred degrees Celsius, similar to the situation on Venus. In both cases life would be impossible. No *physical* barriers are known that would prevent the modern climate of Earth from spontaneous transition to any of the two extreme life-incompatible stable states (Chapter 8).

Any stable state is characterised not by the absolute constancy of its characteristics, but by the absence of irreversible changes in their values. Certain non-zero fluctuations of environmental characteristics are always present in any stable state. To evaluate their importance, their values should be compared to the threshold fluctuations that undermine the system's stability and elicit irreversible changes of the initially stable state. On this basis, the most remarkable property of Earth's climate is not its variability due to a succession of glacial and inter-glacial periods, volcanic activities, drift of the continents, meteorite falls, etc., but the fact that in spite of all those perturbations the climate has remained suitable for life during the last four billion years. In this sense Earth's climate is indeed stable.

The observed peculiarities of the Earth's climate provide the empirical basis for the concept, where the natural biota of Earth is regarded as the only mechanism ensuring maintenance of life-compatible environmental conditions on both global and local scales, i.e. the concept of the biotic regulation of the environment.

1.6 BIOTIC REGULATION OF THE ENVIRONMENT

According to this concept, the main property of life is the ability of biological species to perform certain specific work aimed at maintenance of a particular set of environmental conditions favourable to the biota itself. Complex interaction of living organisms with their environment necessitates the formation of internally correlated ecological communities of species. Within the community, individuals of different species interact with each other in a correlated manner. Such correlation can be compared to correlated functioning of cells and organs within a multicellular organism.

Only those species that are able to perform necessary work on the regulation of the environment have a chance to persist in the biosphere and enter a certain community. Such species maintain optimal population density and produce an optimal rather than maximal number of offspring. By doing so, they ensure stationary distribution of all biotic characteristics of the environment and, in particular, prevent population explosions that may lead to complete degradation of the corresponding community.

Such a situation is often interpreted as paradoxical. It is argued that natural selection would never favour genes beneficial to the community as a whole but detrimental to the individuals possessing them (Baerlocher, 1990). In that context, the property of an individual to restrict itself to producing an optimal number of progeny (instead of the maximum possible) is conceived as detrimental. However, in such a consideration the question of the long term environmental stability where individuals of a given species are to exist is completely neglected. A species unable to control its population density and tending to increase its population number infinitely would be prosperous at first, but in the end will inevitably undermine the ecosystem's resilience and degrade together with its environment. This gives an evolutionary advantage to those species (and, consequently, individuals) who possess a genetic control of their population density coupled to the existing environmental conditions.

To give a vivid example of the impertinence of the above objection regarding natural selection, one may rightly ask how the multicellular organism evolved from unicellular ones, if selection could not in principle favour biological objects acting for the common (instead of their own) good? There is no doubt that cells within a multicellular organism apparently function for the benefit of the organism as a whole, individual proliferation of many types of cells being strongly prohibited. When such coordinated interaction of cells is broken, and cells of a particular tissue begin to grow at the expense of the others (cancer), the individual becomes ill, loses competitiveness and is forced out from a population by normal individuals. Such an individual further dies together with the 'most prosperous' cancer cells. The answer that is in the course of evolution association of separate cells into a multicellular organism became possible due to the appearance of selection at a higher level, i.e. at the level of organisms instead of separate cells. Thus, cells that were able to suppress their own ambitions for the common good ensured stability of the internal milieu of the organism and, by doing so, imparted to it a high competitive capacity. As a result, organisms with coordinated (instead of competitive) functioning of cells do dominate in the biosphere. Similarly, coordinated interaction of individuals of different species within ecological communities is prevented from disintegration by selection at the community level (see Chapter 2, in particular Section 2.11).

The optimal number of offspring is determined by the condition of the maximum efficiency of the biotic regulation power achieved by the community. The maximum possible number of offspring is determined by the food and territory resources available at the moment. A spontaneous transition of any species to producing the maximum possible number of offspring testifies to an impaired genetic programme of that species. Such a transition, similar in its effect to a cancer tumour, disintegrates the normal work of the community and impairs its regulatory potential. As a result, such a community loses its competitiveness and is forced out by another community where the same species retains its normal genetic programme and produces the optimum (instead of the maximum) number of offspring.

Species that work to maintain a stable environment should, apparently, prevent their genetic programme from spontaneous changes. In other words, spontaneous genetic adaptation to random fluctuations of the environment should be strongly prohibited. There should be a mechanism operating in populations of biological species that would stabilise the genetic programme of species and prevent accumulation of random genetic changes that erase meaningful genetic information of species. Thus, according to the biotic regulation concept, biological species should retain their genetic stability over the geological time-scale. Species must not only be adapted to a given environment (i.e. be possible to exist within it) but also be able to perform correlated interaction with other species in the community. These restrictions explain the observed discreteness and morphological constancy of both the extinct and extant species.

The observed genetic polymorphism of individuals of natural species corresponds to random deviations from the normal genetic programme that cannot be eliminated by the stabilising selection due to the limited sensitivity of the latter (Chapters 9 and 10). Stabilising selection operates most efficiently under natural environmental conditions of a corresponding ecological niche. Under perturbed environmental conditions stabilising selection becomes weaker. This leads to increased level of genetic polymorphism in the population and accumulation of individuals with genetic deviations. When the natural environment is restored, process of stabilising selection regains its efficiency and individuals with genetic deviations are expelled from the population. As a result, the level of genetic polymorphism drops to its initial value.

Evolutionary transitions to new species may only occur provided that the species' ability to stabilise the environment is conserved. The mechanism providing for the appearance of such species is that of competitive interaction of ecological communities. New species that are not able to correlate with the others to keep the environment stable, adversely change the competitive capacity of their communities. As a result, communities patronising such species are forced out by other communities, where all species (both evolutionary old and new) perform correlated interaction.

Over larger time-scales (of the order of billion years) evolutionary process may be accompanied by considerable environmental changes. The transition of environment from one favourable for life state to another is brought about by the restructuring of the biota itself. In other words, during the long-term evolutionary process there may be formed new communities of species that are more competitive than the old ones and for which quite another environment may become optimal. Such communities force out the previously-dominating ones and transform the former environment into a favourable state for themselves to maintain further unchanged over another billion of years. Such large-scale evolutionary change may be exemplified by the widely-cited transition from the oxygen-free atmosphere to the present-day one. This is believed to be due to the appearance of photosynthesising organisms that both produced oxygen and used it for breathing. Communities with photosynthesis appeared to be more competitive than those composed of anaerobic organisms, the result being the major restructuring of the biosphere as a whole. Note, however, that during that large-scale evolutionary process the environment all the time remained under the control of the biota, be that biota composed of either evolutionary new (photosynthesising) or old (anaerobic) organisms.

Thus, long-term environmental changes in the biosphere are explained by the fact that, despite the universal biochemical character of life, different types of environment appear to be optimal for different types of organisation of living beings. The inherent ability of life to control its environment being conserved, evolutionary changes of dominant organisms in the biosphere may lead to even significant environmental modifications at no risk of spontaneous adverse changes of the environment that could threaten the existence of life as a whole. As far as these modifications are due to the fact that the newly arisen biota appears to be more competitive than the old one, these modifications are inherently irreversible.

1.7 CONCEPTS OF GENETIC ADAPTATION AND BIOTIC REGULATION ARE MUTUALLY EXCLUSIVE

We have so far considered two possible hereditary patterns of reactions of living beings to environmental changes.

According to the biotic regulation concept, any deviation from the optimal state of the environment necessarily elicits a correlated reaction of all species of the community directed at compensation of that deviation. In other words, any random fluctuations of the environment are counteracted by a biotically-driven negative feedback, so that the favourable for life optimal environment is maintained in a stable state.

Note that the very existence of an optimal environment implies a correlation between morphological and behavioural properties of living beings on the one hand, and characteristics of their optimal environment on the other hand. Such correlation may be called adaptedness or adaption. The ability of leaves to absorb solar radiation; the ability of roots to take in water solutions of nutrients from soil; the ability of animals to run, swim, fly and climb the trees; the ability of endothermic animals to keep constant body temperature using feather and hair etc. are but a few examples of how the organisms are adapted to their optimal environment. Such adaption (sometimes called also adaptation) is a hereditary property of the organisms, characterising the *state* of their correlation with the environment.

According to the concept of genetic adaptation briefly outlined in Section 1.2, individuals of any species genetically adapt to the changing environment. The new environment becomes optimal to those organisms that have adapted to it, their genetic programme being altered accordingly. Individuals of certain species adapt to the presence of individuals of other species in very much the same manner as they adapt to abiotic environmental conditions. As a result, a concordant coexistence of species emerges. Such reaction of the biota to environmental perturbations corresponds to the process of genetic adaptation to changing environment.

It should be emphasised that the *process* of genetic adaptation consists in a directional change of the genetic programme of a species, the ultimate goal being to ensure correlation of the genetically-modified species with a new arbitrary environment. This process is opposed to the *state* of genetic adaption to the specific optimal environment, that is kept constant by the biota.

These two reactions to environmental changes—biotic regulation and genetic adaptation—are incompatible in the sense that only one of them is actually realised. Individuals of biological species may either change the environment to the initial optimal state, or change themselves by genetically adapting to the new environment. There is no compromise between the two possibilities.

If the biotic regulation strategy is realised, biological species should retain unchanged the information about characteristics of the optimal environment. This information, being genetically encoded into species, governs their work on the maintenance of the optimal environment. Thus, biological species cannot change genetically when adapting to a new environment. Otherwise the information about how to maintain the optimal environment will be lost.

If the natural biota as a whole follows the strategy of genetic adaptation, no biotic regulation of the environment is possible. Biotic regulation is not a random coincidence of properties of randomly coexisting species, but a complex genetic programme of correlated interaction of individuals of different species within the community. This programme is aimed at maintenance of a specific environment optimal for the whole community. The complexity of such a programme by far exceeds the complexity of the genetic programme of a single species, that ensures stable internal milieu of organisms through correlated interactions of various types of cells. During genetic adaptation to a changing environment species change their genetic programme and the new environment becomes optimal for those who survive best, i.e. produce the maximum number of progeny. It is highly improbable that those who produce the maximum progeny will randomly invent a new programme of regulation of a new environment. Such a situation can be compared to a hypothetical case when cell lines from different tissues of a single organism are selected *in vitro* for a sufficiently long period of time for their ability to reproduce at the maximum possible rate. It is evident that genetically-modified cells would then no longer be able to function in an organism without disintegrating its internal milieu.

Another argument against coexistence of the processes of genetic adaptation and biotic regulation is that when you are able to regulate the environment, keeping it at a certain optimum, you have no need to change genetically. Also, when you are able to adapt to any environment, there is no need to spend efforts to regulate something. Thus we conclude that biotic regulation and genetic adaptation are mutually exclusive.

Which of the two strategies dominates in the natural biota is to be determined from the available empirical data that allow unambiguous interpretation. As shown above (see also Chapters 9, 10 and 11), the available data on species discreteness and physical instability of the Earth's climate testify in favour of the biotic regulation concept.

1.8 EMPIRICAL EVIDENCE FOR THE BIOTIC REGULATION OF THE ENVIRONMENT

There is direct evidence suggesting that the global environment of Earth is formed and maintained by the natural biota. We list these arguments below.

1. There is a constant, though small, net influx of inorganic carbon entering the biosphere due to filtration from the Earth's mantle. In a billion years such flux could increase the atmospheric CO_2 concentration by a factor of ten thousand (!) as compared to its modern value. (For comparison, at present humanity is seriously concerned by a 30% increase in the atmospheric CO_2.) This would increase the average global surface temperature by several hundred degrees Celsius. The fact that the global surface temperature has remained suitable for life during four billion years suggests that the atmospheric concentration of carbon dioxide retained its order of magnitude during that period (i.e. it did not change more than tenfold in either direction). One may therefore conclude that a compensating mechanism should be operating removing the excessive inorganic carbon from the atmosphere. Such a mechanism is indeed discovered. It is the biotic depositing of inactive organic carbon in sediments. As follows from the paleodata, the biotic sedimentation compensates the net emission of inorganic carbon very precisely. As soon as a random precise coincidence of two independent fluxes appears improbable, this fact points unambiguously to the biotic regulation mechanism operating in the biosphere (see Section 5.4 for more details).
2. Molar ratios of certain important inorganic nutrients dissolved in the ocean coincide with stoichiometric ratios of those elements observed in biochemical reactions of synthesis and destruction of organic matter by the oceanic biota. This is an indication that oceanic concentrations of nutrients are formed and maintained by the oceanic biota itself (Redfield, 1958; Chen *et al.*, 1996; see Section 6.1 for more details).
3. River run-off is equal to the amount of water that evaporates from the oceanic surface but precipitates on land. The observed global river run-off is three times lower than precipitation on land. It means that two-thirds of the precipitated water evaporates from land, and only one-third is brought from

the ocean. Most of the water that evaporates from land comes from the vegetation cover. Plants spend the most part of the absorbed energy of solar radiation on transpiration. Thus, the precipitation régime on land is also under biotic control (see Section 6.6 for more details).

4. The atmospheric CO_2 concentration coincides with the average global concentration of the dissolved CO_2 in the surface oceanic layer (CO_2 solubility equals unity at 15°C) and is three times lower than the CO_2 concentration at oceanic depths. Such difference is maintained by the biological pump. Diffusion of inorganic carbon from the depths to the surface is compensated by photosynthesis of organic carbon at the surface and its sinking down to depth where it is decomposed by heterotrophic organisms. Thus the oceanic biota maintains atmospheric CO_2 concentration four times lower than it would have been in the absence of the biota (see Section 6.1 for more details).

Note that arbitrary changes of marine ecosystems as a result of their possible cultivation by humans may lead to disintegration of the biological pump which is ensured by strictly specified correlated interactions of synthesisers and reducers in natural marine communities. As a result of world-wide cultivation of the oceanic biota, atmospheric concentration of CO_2 would increase by a factor of four bringing about a catastrophic greenhouse effect and, possibly, other adverse environmental changes that at present are difficult even to be outlined. Thus large-scale cultivation of marine ecosystems presents a major threat to the long-term stability of the environment of Earth. Those optimists who plan to feed the growing population of humans by marine products are in fact preparing a global ecological catastrophe. The same statement holds true for people who plan to introduce large-scale perturbations into marine systems in order to increase their productivity and make them absorb carbon from the atmosphere (see, e.g. Martin *et al.*, 1990 for discussion of such a possibility and Section 6.1 for more details). The effect of breaking the harmonic balance of the nonperturbed biota of the ocean[1] is likely not only to be the opposite to what is naively expected, but catastrophic in its irremediability.

5. The available data on changes of oxygen and carbon content in the atmosphere indicate that the nonperturbed biota of the ocean absorbs excessive atmospheric CO_2 and thus partially compensates anthropogenic carbon emissions at a rate comparable to that of fossil fuel burning. The terrestrial biota substantially perturbed by anthropogenic activities has lost its stabilising ability and at present adds to the anthropogenic perturbation of the environment (see Section 6.3 for more details).
6. Any large-scale external perturbation of forest communities (fire, windfall, cutting) brings about biological processes of recovery known as succession. Direct observations show that in the course of succession the community changes various parameters of its environment (pH, humidity, light and

[1] The oceanic biota as a whole may be considered nonperturbed due to the fact that people consume only a small part of its primary production in contrast to the situation with terrestrial ecosystems.

temperature régimes, etc.) by orders of magnitude. When the initial stationary stable state is recovered, the forest community is able to further maintain its own environment unchanged for infinitely long periods of time in the absence of large-scale external perturbations. This provides unambiguous evidence for biotic regulation in the forest ecosystems (see Sections 6.7 and 6.8 for more details).

1.9 STABILITY OF LIFE ORGANISATION

Let us now discuss major properties of life that enable living beings to combine into ecological communities capable of maintaining long-term environmental stability (see Chapter 2 for detailed coverage of this issue).

Any living organism represents a super-organised, internally-correlated system that exists due to the external fluxes energy. The level of organisation (complexity) of living beings by far exceeds that of physical fluxes of energy used by life. This means that no living object will ever arise spontaneously in any fluxes of external energy. This also means that the external fluxes of energy are unable to support the level of organisation of living objects for a long time. Due to this fact, all living objects inevitably decay (die) in spite of the constant presence of external energy fluxes. The decay of living objects is manifested as a decreasing level of their organisation that tends to physical equilibrium with the inanimate environment. This process ends with death of the organism well before that physical equilibrium is reached. Note that the ability to reproduce, inherent to biological objects, does not help to maintain the long-term stability of organisation of living beings. Such ability is itself subject to decay.

To maintain the level of organisation achieved in the course of evolution, life has invented the mechanism of selection that we call stabilising, in accordance with its function. This way of maintaining stability is unique to life and is never encountered in inanimate nature. It consists of the following. All biological species necessarily exist in the form of populations, i.e. sets of uncorrelated individuals with approximately the same physiologically-meaningful hereditary programme. Within a population, individuals compete with each other in an aggressive manner. This means that competitive interaction does not depend on the availability/unavailability of resources such as food, territory, etc., but is an inherent property of living individuals.

In the course of competitive interaction, non-competitive individuals (i.e. those with an impaired hereditary programme) are forced out from the population, even though such individuals may remain quite viable. The vacancies appearing in the population are filled by the progeny of normal individuals. As a result, the population sustains its status of a set of equally competitive individuals with the initial high level of organisation. It is natural to call such a mechanism of maintaining stability *stabilising selection*, though in biological textbooks this term usually refers to stabilisation of the phenotype (i.e. morphological and behavioural properties of individuals), rather than to stabilisation of the genetic hereditary programme as a whole (which, naturally, includes stabilisation of the phenotype as well).

For stabilising selection to operate, individuals within the population should not be correlated with each other, so that exclusion of any individual from the population should not affect the well-being of the others. If individuals form an internally-correlated association (e.g. a family of animals), elimination of any of its members would impair the correlated organisation of the association in very much the same manner as ablation of a certain organ would impair the correlated functioning of the organism. Thus, maintenance of internal correlation of associations of individuals may be ensured only by stabilising selection operating at a higher level than individuals, namely in a population of associations with competitive interaction between them (see Fig. 2.3).

Internally-correlated associations of biological objects are encountered at all levels of life organisation, e.g. association of cells in a multicellular organism, association of social insects in structures such as bee-hives and ant hills and, finally, association of individuals of different species in ecological communities. Similarly to other levels of organisation, internal correlation of species within an ecological community can only be maintained as a result of competitive interaction of homologous communities within a non-correlated set of communities. A population of homologous (uniform) communities together with their environment is known as ecosystem. One can speak about a forest ecosystem, coral reef ecosystem, etc.

Internal correlation of human societies is also maintained in the same manner. Sociality (i.e. the ability of humans to form a stable sustainable society) is a complex hereditary species-specific property genetically encoded into the *Homo sapiens* species. It may be maintained only in the course of competitive interaction of different societies (countries). Thus, if the global population of people was united into a single, globally-correlated society (this may be done by forming a culturally uniform global nation), the ability of sociality would undergo genetic decay and finally would be lost.

The aggressive mode of competitive interaction inherent to biological objects at any level of organisation explains the incessant political and economical competition between different countries, which often have tragic consequences and at present may even threaten the very existence of the humankind (e.g. nuclear arms race). Here two points need to be stressed. First, even though the competitive interaction of nations is the only mechanism that prevents sociality of humans from genetic decay, the process of genetic decay, as we show in Section 9.2, is a very slow one and takes very long periods of time. Thus, if we humans weakened the intensity of present-day international confrontations for several decades or even hundreds and thousands of years, that would not threaten our hereditary genetic programme with regard to sociality, but would only help to establish a more secure world for all people. Secondly, as far as the ability to compete is inherent to any living being including humans, it is useless to try to suppress it altogether. The only thing that can be done is to enhance peaceful forms of competitive interaction at the expense of political and economical confrontations, which actually present the most danger to the fragile balance of the modern world. This may be done by imparting increasing importance to such forms of peaceful international competition as sports championships, musical festivals, scientific achievements, etc.

1.10 MECHANISM OF BIOTIC REGULATION

To regulate the environment, ecological communities use processes of synthesis (production) and decomposition (destruction) of organic matter. In the absence of external physical fluxes of biogens to and from the local ecosystem, their concentrations inside the local ecosystem will only remain stable if biological synthesis of organic matter is precisely compensated by biological destruction. Thus, in the absence of external perturbations, communities tend to maintain close biochemical cycles of all biogens in order not to disturb the optimal characteristics of the environment.

If external physical fluxes of certain biogens are smaller than biological productivity (and, consequently, destructivity) of the community, the latter is able to form and easily maintain concentrations of these biogens inside the local ecosystem at a level that can differ significantly from that in the external milieu. For example, concentrations of various elements in soil differ drastically from corresponding concentrations in the Earth's crust or the atmosphere. It means that in natural ecosystems the rate of physical and chemical degradation of soil (soil erosion) is substantially lower than the rate of the compensating process of soil recovery performed by ecological communities. In a situation when physical fluxes of biogens are negligible compared to biotic ones, even a single ecological community is able to maintain concentrations of these biogens at the optimal level in the local ecosystem.

In many cases, external fluxes of biogens are considerably larger than the community's productivity. For example, physical mixing in the atmosphere and ocean is so large that it is not possible to discriminate between the biotic and abiotic environment. In such a situation, optimal concentrations of biogens can be maintained only by a large number of uniform biological communities occupying large territories of the Earth's surface. Such biogens, for example atmospheric CO_2, may be called globally regulated. The process of their regulation is organised as follows.

If the external concentration of a certain globally-regulated biogen differs from the community's optimum, the community activates processes aimed to compensate for that difference. Direction and rates of these processes are the same in communities of equal competitiveness. Compensating processes can be based on increasing productivity as compared to destructivity, or vice versa. For example, if the global atmospheric concentration of CO_2 exceeds the community's optimum, the community can try to decrease the internal CO_2 concentration of the local ecosystem depositing excessive CO_2 in organic form. This will induce a physical influx of CO_2 into the local ecosystem.

If such minor local change gives the community advantage, i.e. it makes it more competitive, such a community can force out other communities that cannot perform such change. As a result, there appears a large set of equally competitive communities all ensuring the same flux. Thus there will be a global flux of CO_2 to biota until the global atmospheric CO_2 concentration equals the community's optimum. The excessive atmospheric CO_2 will be removed from the atmosphere and deposited in organic form in humus or other organic stores. So, small relative changes of concentration of biogens performed by single communities may lead to large absolute changes in global environment (see Section 5.6 for more details).

Internal correlation of individuals inside an ecological community is characterised by a certain radius, i.e. it becomes weaker with distance and dies out at a certain critical value of it. In other words, biological communities, as all other internally-correlated biological objects (e.g. bodies of living organisms), are characterised by finite size. This property of ecological communities follows from the necessity to form populations consisting of a large number of communities in order to support their stability. Thus, a single community cannot occupy a very large territory. However, due to the absence of visible boundaries delimiting the adjacent communities, the characteristic size of ecological communities is difficult to evaluate directly. Visible boundaries are only observed in the simplest communities encountered in the biosphere—epilithic lichens. A lichen community consists of only two species, an alga (synthesiser) and a fungus (reducer). Individuals of both species are so tightly correlated that lichens are formally classified as species. The picture on the book cover shows different communities of *Lecidea spp.* that are delimited from each other by visible boundaries.

Indirect estimates show that ecological communities never exceed several tens of metres in size (Section 3.5). In a non-perturbed forest, large trees are likely to perform the function of the community centre, around which the other community components (bacteria, fungi, small invertebrates) are tightly combined.

Such consideration renders 'homeless' large animals (reptiles, birds, mammals) in the sense that they apparently do not belong to any particular ecological community due to the small size of the latter as compared to the feeding territories and home ranges of most animals. Rather, large animals may be regarded as a certain component of the environment, which is regulated by the communities in the same manner as the concentrations of globally-regulated biogens. Large animals are encountered in the overwhelming majority of ecosystems, which indicates that their presence in the community contributes to the competitive capacity of the latter (see Sections 4.5 and 6.7 for discussion of ecological functions of large animals). Correlated interaction of large animals with other species in the community mostly consists in a correct choice of food and a strictly-specified share of consumption allocated to large animals within the community. The community is able to control the share of consumption of every particular species of large animal via optimisation of their population density. If the population density of a given species of large animal deviates from the optimum, the ecological community may react by changing the environmental conditions favourable for that species. For example, when too many large herbivores are present, the community may reduce production of edible parts of plants and, at the same time, increase production of poisonous mushrooms or thorny parts of plants. As a result, the average time spent by a large animal in that community will be reduced and, accordingly, their birth rate and death rate will be affected as well. All normal communities proceed with such impact until the population density of large animals relaxes to its initial value optimal for all normal communities.

1.11 NATURAL DISTRIBUTION OF ENERGY CONSUMPTION OVER INDIVIDUALS OF DIFFERENT BODY SIZE

The complex programme of biotic regulation, genetically encoded into natural biological species, comprises information about peculiarities of functioning and optimal population densities of species, that uniquely determine the amount of energy consumed by individuals of a particular species.

Figure 1.1 gives distribution of consumption of the primary production of plants over heterotrophs of different body sizes. This distribution is based on published data for different natural terrestrial ecosystems and is universal for all ecological communities of the terrestrial biota. Notice that the most part of primary production (more than 90%) is consumed by the smallest organisms, mostly bacteria and fungi. One may say that bacteria and fungi together with photosynthesising plants constitute the core of any ecological community. Medium-sized organisms, invertebrates mostly, consume less than 10% of the primary production leaving less than one per cent of it to be consumed by large vertebrates.

The observed distribution of energy consumption is not random. It is arranged in such a way that the smallest absolute amount of energy is to be consumed by the largest organisms that are characterised by the largest relative fluctuations of consumption. Large animals consume plant production by the confiscation of substantial amounts of biomass accumulated in leaves, branches, trunks of living trees, etc. This inevitably leads to sharp fluctuations of a community's biomass. In other words, environmental impact of a single large animal is very substantial and sometimes can even lead to complete degradation of the community on a local scale. To prevent such situations on a larger scale, population densities of large animals are kept low in natural communities, so that the cumulative energy consumption of large animals and its absolute fluctuations appear to be small.

Each bacterial cell consumes but a tiny part of the net primary production, the total flux of consumption being ensured by a huge population of bacteria. Relative fluctuations of consumption of the smallest organisms appear to be very low due to the operation of the law of great numbers (see Section 3.5 for more details). The smallest organisms are therefore allocated the largest absolute part of consumption (more than 90%, Figure 1.1). Such distribution allows the community to keep absolute fluctuations of consumption introduced by any species below the natural level of fluctuations of plant productivity. Thus the distribution in Figure 1.1 corresponds to the maximum possible stability of the community's organisation.

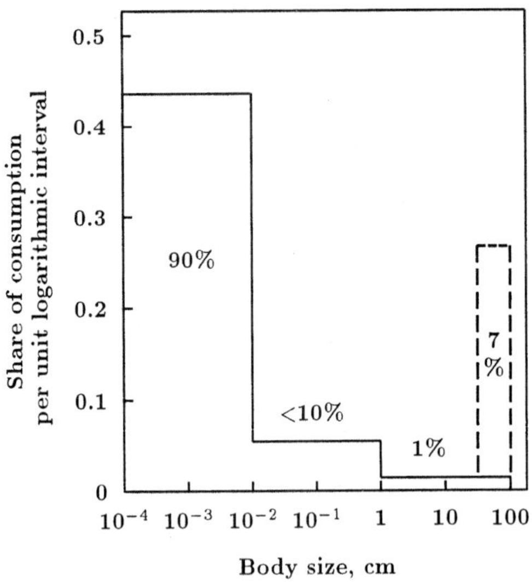

Figure 1.1. Share of consumption of the net primary production allocated to heterotrophic individuals (bacteria, fungi, animals) of different body size in natural terrestrial ecosystems (Gorshkov, 1981). The solid curve gives the universal distribution found in all nonperturbed terrestrial ecosystems. The area enveloped by the solid curve is equal to unity. Numbers in per cent give the relative input of different parts of the histogram. The dashed line describes violation of the natural distribution caused by the present-day anthropogenic perturbation of the biosphere. Area under the anthropogenic peak (7%) corresponds to food of humans, cattle fodder and anthropogenic consumption of wood (Gorshkov, 1995).

Each organism is characterised by a strictly-specified set of internal organs and a strictly-specified distribution of fluxes of nutrients and energy going through different organs. Similarly, each community is characterised by 1) a strictly-specified species composition and 2) a strictly-specified distribution of fluxes of matter and energy going through different species in the community.

Each biological species is characterised by its hereditary genetic information. Random changes of genetic information of natural species as well as violation of the natural distribution of energy fluxes in the community disintegrate the internal community structure, reduce its regulatory potential and, consequently, elicit unfavourable environmental changes. Evidently, most environmental risks are associated with reduction of energy fluxes ensuring functioning of the dominant species of the community—plants, bacteria and fungi. Disturbance of these community components has an immediate adverse impact on environmental quality. In contrast, even complete elimination of species of large animals affects the community's well-being only after a lapse of time.

Significant danger is also associated with increase in population density of large animals, when their consumption goes beyond the limits prescribed by the natural distribution (Figure 1.1). Such an increase occurs at the expense of the smaller-sized species that perform the largest amount of work on regulation of the environment. The fact that the distribution presented in Figure 1.1 is universal for all terrestrial ecosystems suggests that communities where large animals are allowed to dominate energetically lose their ability to regulate the environment and cannot compete with normal communities where the share of energy consumption by large animals is limited. Thus, biotic regulation of the environment on a global scale corresponds to the condition when the biosphere as a whole conforms to the natural energy distribution represented in Figure 1.1. It means, in particular, that large mammals altogether may consume not more than one per cent of the net primary production of the global biota.

Natural ecological communities are completely destroyed on territories occupied by arable lands, where people cultivate monocultures and consume more than half of their primary production. Arable lands account for less than 10% of the continental area of Earth. Taking into account that people do not consume primary production of the oceanic biota, we conclude that the anthropogenic consumption of primary production associated with arable lands does not significantly exceed the ecological limit outlined by the natural distribution (Figure 1.1). However, with allowance made for pastures and, especially, exploitation of forests, the global consumption of the net primary production of the biosphere by humankind appears to exceed the ecological limit by nearly an order of magnitude (dotted line in Figure 1.1). This results in violation of the global biotic regulation mechanism and entails the observed global environmental changes, in particular the unprecedented growth of atmospheric CO_2 concentration, acid rains, ozone depletion, etc.

The modern practice of forest exploitation is based on so-called forest management. Forest management essentially represents a series of anthropogenic impacts imposed on the forest community in order to maximise biological production of economically-valuable wood. The regulatory potential of forest communities is completely neglected and destroyed during such management. For example, fires (often referred to as *natural* disturbances[2]) are commonly used in many countries to speed up growth of valuable tree sorts. Such policy completely ignores the fact that fires eliminate the most part of the organic matter accumulated in the community and, by doing so, deprive of nutrition the dominant species of the community (bacteria and fungi) that under natural conditions perform, together with plants, the most important work on environmental regulation. Burning of organic matter, or in any other way its confiscation from the forest community, can be compared to extermination of food resources that support a working population of the society.

It is often stated that anthropogenic management may *improve* some ecological function of forests that at the moment seems to people to be the most important, e.g. to enhance sequestration of carbon in forests by restructuring their species composition (see, e.g. Shvidenko *et al.*, 1997). It is clear to everybody that only

[2] Fires are indeed encountered in nature. However, indirect estimates show that more than 99% of modern spontaneous fires are due to anthropogenic activities (see Section 6.8.3). Under natural environmental conditions, fires represent a natural calamity and are encountered at least one hundred times less often than at present. To cope with fires, the natural biota has invented a special mechanism of the quickest possible recovery of ecosystems disturbed by fires, i.e. the mechanism of forest succession (Sections 6.7 and 6.8).

those who know how a mechanism works may improve it. As shown in Chapter 7 (see also Section 1.13 below), ecological communities are characterised by unprecedented complexity and interrelation of all their components. Fluxes of information going through natural communities are such that people will never be able to account for all the characteristics controlled by the biota. Thus, people can in no way improve the functioning of the natural biota. In very much the same way that any medicine given to healthy people may only impair their condition, any anthropogenic impact imposed on natural communities may only disintegrate their functioning and weaken their regulatory potential.

Absolutely all medicines are carefully tested for possible dangerous indirect effects before they are offered to people. Chemists selling medicines are legally responsible for their quality. Medicines are tested on short-lived animals, while the characteristic lifetime of trees is equal to 100–200 years. Thus, it is impossible to test the proposed 'improving' ecological measures for their possible indirect effects, because such testing will take at least several decades. This fact allows people to speak eloquently about how the natural biota can be restructured in favour of people. When the dangerous, possibly irreversible, environmental effects of the implemented measures come up in a hundred years, there will be nobody to punish. Thus, all ecological decisions connected with restructuring of natural communities are loaded with environmental danger.

Meanwhile, estimates based on the analysis of the global carbon cycle (Section 6.5) show that if forest-rich states reduced the rate of forest exploitation by approximately 40% and allowed natural processes of recovery of forest communities to dominate in the protected areas, the present-day growth of atmospheric CO_2 could be stopped. That effect would be due not to fixation of a certain amount of carbon in wood, as it is often stated, but due to restoration of the regulatory potential of the whole of the natural biota in the protected areas. Note that most developed countries which have completely exhausted their natural forests remain perfectly prosperous in the economic sense without forest industry incorporated into their economy. In countries where forest industry does exist, it does not play a major role in the economy. For example, according to the World Bank data (World Bank, 1997) the contribution of forest industry into the Russian Gross National Product does not exceed one per cent.

All this means that a complete permanent ban on commercial logging in all the remaining global forest estates will by no means lead to collapse of the global economic system. If such a radical measure is taken, the present-day global changes will not only be stopped but even reversed. As a result, the nonperturbed preindustrial global environment will be restored. Meanwhile, the present-day situation remains paradoxical, when the natural regulatory mechanism of forests communities, which is vitally important to the whole of humanity, is being destroyed by about one per cent of the world's working population. Future generations of humans are likely to categorise modern forest exploitation together with such antihumanitarian crimes of the past as, for example, the Inquisition.

1.12 CONSERVING BIODIVERSITY OR BIOTIC REGULATION?

For a very long period of civilisation development, natural species of the biosphere have been classified as useful for people, useless for people and harmful for people. Useful species were those few that were actively used in agriculture, cattle-breeding, medicine etc.; harmful species were those that interfered with anthropogenic activities. Useful species were protected, harmful species were struggled with. The rest of the biosphere species, of no particular interest to the general public, were not protected and could be done away with if their conservation could stop the progress of civilisation.

At present, there exist many scientific as well as public movements for the conservation of biodiversity. But still there are no commonly-accepted scientific arguments for the conservation of biodiversity. Arguments either are based on aesthetics or represent an appeal for the unique nature of species genomes and the impossibility of restoring them in case of species extinction. The weakness of such arguments is often evident even to those people who put them forward. The overwhelming majority of people cannot afford to watch biological species within their natural ecological niche. Most people have to content themselves with infrequent visits to zoos, where a random selection of species (mostly large animals) is usually represented. Nobody has ever tried to conserve species of bacteria, though they, as we have seen, play a most important role in the biosphere. Regarding the genetic aspect of species conservation, one should note that, until now, people have been using about several hundred species from the existing ten million (Thomas, 1990) and the number of the widely used species has no tendency to grow. There are no reasonable grounds to expect that all the ten million species of the biosphere will ever be used by humankind. It remains unclear why people should protect them, often at the expense of economic growth, the benefits of which, in contrast to that of conserving biodiversity, are very transparent. It is not surprising that the decision-makers in all countries are very reluctant to sacrifice economic progress to a poorly-argued task of biodiversity conservation.

The concept of the biotic regulation of the environment provides a very clear scientific base to the necessity of biodiversity conservation. Natural species of the biosphere are not aesthetic objects to be looked at by humans, they are not hypothetical genetic resources of humankind or potential forest pests. They are indispensable parts of the working mechanism to maintain an environment favourable for people on both local and global scales. Each species inside the community performs strictly specified work on the stabilisation of the environment. The programme of this work is determined by the species genome. And, due to the unique nature of species genomes, this work cannot be done by any other species.

To sum up, there is an urgent need for a new conservation strategy aimed at ensuring the continued biotic regulation of global environmental conditions, as the present focus on the conservation of biodiversity *per se* is incapable of securing the Earth's environment. This new strategy must recognise the need that sufficiently large areas of terrestrial landscapes be allowed to continue without human interference.

1.13 BIOTIC REGULATION CANNOT BE REPLACED BY TECHNOLOGY

Natural biota has been keeping the global environment in a favourable state for life during the four billion years of life existence. There is no doubt that, provided the necessary limitations on anthropogenic activities are imposed, the natural biota will be able to further perform that important regulatory function.

On the other hand, the on-going development of civilisation yields considerable progress in industrial technology. Many technological processes of today, though expensive, ensure almost total purification of wastes and do not cause local pollution. It seems that further progress in science and technology may result in the creation of a technological system which would be able to regulate the environment on a global scale and could substitute the existing natural mechanism of the biotic regulation.

At a closer look, however, such a possibility proves to be improbable (see Chapter 7 for a detailed treatment of this problem). This statement can be easily proved when comparing information-processing capacities of the natural biota and civilisation.

The main energy flux in any natural community goes through plants, fungi and unicellular organisms. Cellular metabolism results in strictly-specified ordered chemical reactions that pertain to all molecules of the cell. It means that the flux of information goes through memory cells of molecular size. Using the known value of cellular metabolism, it is possible to estimate information flux going through one bacterial cell. It turns out to be of the order of information flux processed in a modern PC. In the biosphere, each square micron of the Earth's surface is covered by about one hundred cells of various organisms. Surface area of the biosphere coincides with that of Earth and constitutes $5 \cdot 10^{26}$ square microns. Thus, the total number of cells in the biosphere is of the order of 10^{28}–10^{29} cells, i.e. about 20 orders of magnitude more than the world population of humans. Even if every person on Earth each had one PC, our civilisation would be able to process an information flux that is 20 orders of magnitude smaller than is processed by the natural biota.

No computer technology of future generations can cover the information flux gap of 20 orders of magnitude between the biota and civilisation. To do that, it would be necessary to build computers with memory cells of molecular size and cover the Earth's surface with a continuous network of such computers, i.e. to recreate the natural biota in its preindustrial state. Thus, we conclude that natural biota is the only possible mechanism of formation and stabilisation of a suitable-for-life global environment and climate.

1.14 ECOLOGICAL PROBLEMS OF HUMANKIND

Ecological problems of humankind are unprecedented in the biosphere. No other species has ever come so close to complete extermination of its own ecological niche. What is more, no other species has ever threatened the integrity of the biosphere as a whole as people do today. We will now briefly discuss the peculiarities of our own species that have caused such a situation.

All life-important knowledge of the environment in immobile organisms—plants, fungi—is confined to their genetic hereditary information. Meanwhile, in order to move, locomotive animals need to have concrete information about the place where they live and about other components of their environment that are encountered during the lifespan of a certain organism. Such information changes from one generation to another during the whole time of species existence and therefore cannot be written in genomes of locomotive animals. Once written, it may become useless for the next generation. Thus, during their lifespan, locomotive animals need to process additional non-genetic information about their environment. Such information accumulates in the memory of each locomotive organism (which in most animals is localised in the brain) and vanishes with the death of the organism.

Transmission of non-genetic information from one generation to another is biologically forbidden. Such a situation guarantees against transmission of wrong information about environmental conditions that may change during the lifespan of the next generation. Mountains become hills, brooks turn into rivers, lakes become bogs, old trees are replaced by young ones, etc. Information that correctly characterises the environment at one moment of time, may become wrong at the next moment. Transmission of such information may lead to wrong behaviour of the organism in the next generation and disintegration or even degradation of the biological community. That is why transmission of this information is genetically forbidden in all biological species except *Homo sapiens*.

Homo sapiens proved to be able to accumulate cultural information that, as well as the genetic information of a species, can be transmitted from generation to generation. But, unlike genetic information, the amount of cultural information increases from generation to generation. *Homo sapiens* is a unique biological species that can learn and use knowledge of the environment accumulated by former generations.

At present, the cultural information of the whole of humankind is comparable to, or even exceeds, the genetic information of *Homo sapiens* as a species (Chapter 7). The major part of the present-day cultural information of humankind is represented by scientific information about the surrounding phenomena. Fundamental studies of physical, chemical and biological laws of nature gave people an opportunity to work out technology-based applications of the acquired knowledge. People were able to inhabit all continental areas of Earth and began actively to explore outer space.

Development of technology enabled people to adapt to almost any environmental conditions encountered in the biosphere. When inhabiting new territories, people transformed nature into a most favourable state for them. People cut down forests and dried out bogs to build houses, turned large territories into fields and pastures to get food. Mines, roads and factories were built. These activities were necessarily accompanied by industrial and agricultural wastes being emitted into the transformed environment.

In the meantime, people changed the structure of natural biological communities, reducing population numbers of those species that they were not dependent upon, and increasing population numbers of useful species, that gave necessary products like meat, crops, timber etc. People also changed genetic programmes of natural species, creating new breeds of cattle and new sorts of agricultural plants. As a result, ecological communities began to lose their ability to compensate environmental perturbations, while the human-induced degradation of the environment became the most important of the destabilising factors.

People began to pollute the environment in the ancient past. But until the beginning of the 20th century, natural biota had been able to counteract the anthropogenic pollution (Section 6.4). Pollution was noticeable only in local areas, where it could be easily coped with. Meanwhile, on regional and global scales the environment was maintained by the natural biota in a healthy and stable state. Ecology as a science was mainly concerned with interactions of species of natural ecological communities with their environment and was considered as a branch of biology.

When the rate of anthropogenic pollution began to increase together with the rate of human-induced degradation of natural biological communities, the natural biota began to lose its ability to stabilise the environment on regional and global scales. People realised that the current state of the environment strongly depends on the anthropogenic impact imposed on it. As a result, ecology turned into a much wider branch of knowledge than it used to be, and began to comprise not only biological, but also economic, political and ethical issues.

Ecological problems of civilisation manifested themselves clearly when the anthropogenic disturbance of natural biological communities closely approached a certain threshold level, beyond which biota loses its ability to stabilise the global environment. Evidently, the below-threshold stable existence of human civilisation that took place in the past is possible in the future as well. To ensure it, it is necessary to substantially restrict anthropogenic activities on the planet and, inevitably, impose constraints on the growth and sustainable number of the world human population. The concept of unlimited economic growth, inherent in modern thinking, is in apparent contradiction with the above-mentioned principles of the stable existence of civilisation. It is evident that unlimited growth will finally lead to over-exploitation of natural biota and complete loss of its stabilising properties. Then people would have to maintain a suitable-for-life environment themselves, using available scientific and technological knowledge. We have seen in the previous section that such a perspective appears to be absolutely improbable.

Functioning of natural species that assures biotic regulation of the environment is determined by the genetic information coded in genomes of these species. This information is hereditary and remains unchanged during the whole period of species existence. The genetic programme of any species is tightly coupled to the natural environmental conditions where the species exists, i.e. to its natural ecological niche. The work performed by species within a community remains meaningful and contributes to the environmental stability only if such work is performed in the natural ecological niche.

Homo sapiens, as well as all other species, originated in the process of biological evolution. Modern paleontological data provide an opportunity to trace the whole succession of evolutionary events that has led to the origin of *Homo sapiens*. Thus,

Homo sapiens, as all other species, originally belonged to a certain ecological community. Inside the community *Homo sapiens*, as well as all other species, did a certain amount of work aimed at stabilisation of the environment.

When a species is placed under distorted environmental conditions, actions of individuals that remain meaningful in the natural ecological niche no longer correspond to the task of environmental stabilisation. Moreover, under distorted conditions, the activity of living individuals may even enhance adverse environmental changes and cause complete degradation of the environment. Such a situation is often observed when a certain species invades an alien ecological community. In the framework of its own ecological community this species performs work aimed at the most effective stabilisation of the environment in concert with all other species of the community. But within an alien community the same behavioural programme of actions may come in conflict with programmes of the aboriginal species. This leads to disintegration of community, degradation of environment and, finally, death of the population of the alien species, together with the supporting ecological community. Examples of such on-going local ecological catastrophes include, amoung others, introduction of alien mammals (goats, rabbits) into ecosystems of small islands, invasion of alien seaweed into waters of countries with active foreign trade, etc. Also, it is a well-known fact that in many countries public opinion is seriously concerned with the destructive influence that is imposed on aboriginal natural ecological communities by alien species that penetrate into the country with purchased biological products.

Behavioural patterns of animals have a genetic basis, i.e. they are written into the species genome. Thus, when an animal is placed under the unnatural conditions of a foreign ecological niche, the strategy of its behaviour does not change significantly. In other words, animals proceed to work according to the old genetic programme, which under distorted conditions may cause degradation of the environment. The same effect is achieved when under normal environmental conditions the species genome is artificially modified. In such a case the genetic programme of a species no longer corresponds to environmental stabilisation and its carriers are forced out from normal ecological communities. For example, artificial breeds of domesticated animals and sorts of plants may exist only when directly supported by people. Left alone, they are forced out by their wild competitors (e.g. planted cultures are choked by weeds).

Within the natural ecological niche the genetic programme of *Homo sapiens* should ensure correct behaviour of people that would allow them to exist sustainably and contribute to the environmental stability of their own niche. Such a sustainable way of living can be still observed in indigenous societies. Behaviour of people is based on the genetically-encoded programme of positive and negative emotions. All things modern people tend to do, e.g. improve their living standard or care for children, are based on that genetic programme of emotions.

During the preindustrial era those actions of people that brought about positive emotions corresponded to environmental stability of the ecological niche of humans. Development of civilisation has not changed the genetic programme of humans, so that the same actions still bring about positive emotions. However, development of civilisation drastically changed the environment where people now live, as compared to the natural ecological niche of humans. Meanwhile human activities, including political and economical ones, continue to be governed by a genetically-programmed system of emotions. Those ideals and goals of people that had fitted the natural ecological niche and ensured its stability, prove to be destabilising in the modern world. Such a situation leads to further degradation of natural biota, regional and global environmental changes, continuing population growth, growing social tension, decay of genetic information of the *Homo sapiens* species (manifested in increasing frequency of all kinds of physical and mental disorders).

So far, civilisation has been developing spontaneously in the direction determined by the genetically-coded emotional programme of people. The human brain has been continuously solving the problem of bringing about positive emotions. It is this emotion-based spontaneous development of civilisation that has resulted in the modern unfavourable environmental situation.

However, during the process of cultural evolution people have discovered an important factor that allows them to work out emotion-independent strategies of behaviour. It is the scientific approach. This method of exploring the environment formed the basis for the scientific and technological revolution of the last two centuries that predestined current ecological problems of humankind. It is not unlikely that all major discoveries that could help people to effectively transform their natural environment have been already made by scientists (Ginzburg, 1985). The fundamental science is now tackling issues of doubtful potential importance to the modern problems of the humanity. This is presumably the reason for the observed decreasing enthusiasm of many governments to support fundamental scientific research.

However, fundamental scientific research is the only method that could help people to find a way out from the present-day ecological crisis. Using this method, people can find out what life style is optimal for *Homo sapiens* as a species. What is more favourable for people—living in an industrial landscape with all the modern conveniences of civilisation but deprived of free communication with nature due to high population density and high degree of industrialisation or living inside natural ecological communities having, possibly, lower levels of consumption. To answer this question, serious fundamental research is needed. At present, however, such questions are not even put forward. The predominating viewpoint in most countries is that priority should be by all means given to a high standard of living, which implies good food, clothes and housing, at the sacrifice of surrounding nature, global ecological safety and ethic and health requirements of people with respect to communication with nature. Yet only fundamental investigations of the above questions and working out an emotion-free but reason-based strategy of development of our civilisation can help people overcome the global ecological crisis.

1.15 DEMOGRAPHY

It is clear to everybody that the global population of people cannot increase infinitely, because the size of our planet is finite. Yet Earth is very large, and there is the impression that the limits on population growth imposed by the finite size of the planet will only become tangible in a distant future.

More rigid limits on the global human population are usually associated with the necessity to feed all people (Daily *et al.*, 1998). Today, more than 10 per cent of the human population is starving. Unfair distribution of food and other resources over different countries is usually considered as the main reason for such a situation. Solution to the food problem is sought in the social sphere. A more intensive exploitation of Earth's resources is also considered as a possible way out.

Very different limitations on the global population growth are imposed by the condition of conservation of the regulatory potential of natural biota on a global scale. These limitations are dictated by the task of ensuring long-term environmental stability for secure development of an infinite number of future generations of people. To quantify these limitations is one of the ultimate goals of this book (see Section 6.5).

Until now, the importance of the global demographic problem has been underestimated both by the general public in most countries of the world and by specialists in different fields of science and culture. So far, evaluation of scientific discoveries in different fields of knowledge has been based on the criterion of catering for positive emotions of people. Such use of scientific knowledge has led to further cultivation of the natural biota, population growth and degradation of the environment.

Overpopulation of Earth presents the most important global ecological problem of today. Until now, the on-going exponential population growth of humankind (not encountered in any other natural species of the biosphere) was regarded as a natural state of affairs (Figure 1.2). In the 1980s, all humankind rejoiced on the occasion of the birth of the five billionth human being. The birth of a six billionth human inhabitant of Earth in the end of 1999 was met with a more pronounced reservation.

However, only India and China have attempted to regulate population growth on a state level. In other countries, negative population growth may be envisaged as a catastrophe on a national scale. Calls for depopulation are from time to time compared to genocide. The most important thing is that such an attitude towards the overpopulation problem is shared by the overwhelming majority of general public in most countries. This becomes especially clear from the fact that very few political leaders touch upon that problem in their election campaigns.

What causes such an attitude towards one of the most important problems of today? Why are any attempts to analyse the problem on scientific grounds either ignored, hushed up or repulsed? Those few people who do come to think seriously about the problem and immediately realise its importance and the vital necessity of most urgent measures, may get an emotional impression that there exists a global-scale conscious plot of subversive actions aimed at undermining the stability of humankind's existence by means of well-conceived propaganda against limits to population growth.

However, as soon as the destabilising attitude towards population growth is basically the same in all countries and threatens the existence of the whole of humanity, it seems unreasonable to look for some mystical external forces that are consciously working in that direction. Rather, the reasons of such a situation are to

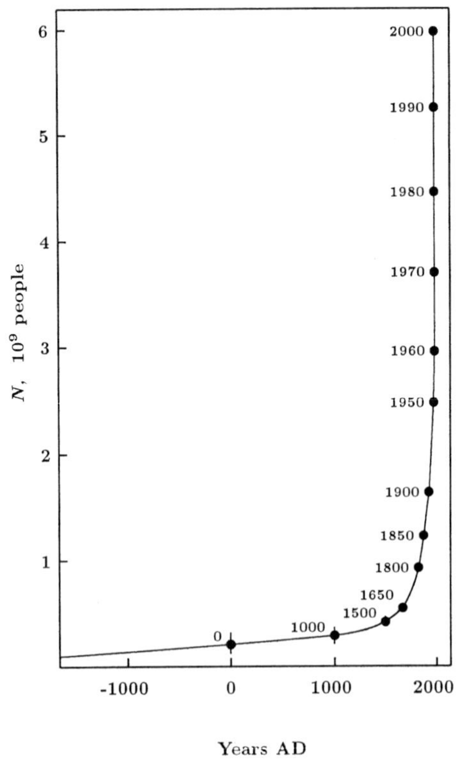

Figure 1.2. World human population growth.

be sought in the genetically-programmed system of emotions governing the behaviour of *Homo sapiens* individuals.

Humankind as a whole is composed of several races and a great variety of nations, that may be compared to subspecies and populations of other species of the biosphere. Different subspecies and populations of a single species may have morphological and behavioural differences. However, as shown in Section 1.6, these differences are caused by different localisation of random deviations from the normal genetic programme of the species in different populations and subspecies of that species (see also Chapters 9 and 10). Meanwhile, the meaningful genetic information is the same in all populations and subspecies of a single species. The total number of genetic deviations is limited by the stabilising selection and is also approximately the same in different populations of the same species. On this basis, all populations of a single species may be considered as genetically equivalent.

The same way of reasoning may be applied to races and nations of people. Therefore, racism, as well as nationalism, that assumes the existence of genetic differences in the levels of organisation between different races and nations, does not have any scientific grounds and has been rejected by the overwhelming majority of human society.

Unlike many other animals, humans are characterised by a very high degree of sociality genetically encoded into the species genome. People that inhabit the same territory tend to form associations at different levels of hierarchy. Within such associations (families, communities, nations, states) people are divided according to their professional specialisation. As a result of specialisation, all members of a single association become dependent upon each other. The internal correlation of states (association of the highest level) is maintained by competitive interaction of states in political, economical and other arenas.

Sociality of humans consists in correlated interactions of people. As well as all other properties of living objects, sociality of humans is genetically encoded into our species. Decay of the genetic information responsible for sociality is prevented by competitive interaction of different associations of people, for simplicity, let us call them countries. Sociality is maintained due to the fact that countries with high degree of sociality (i.e. high degree of correlation between its members) appear to be more competitive than countries with low degree of sociality. If competitive interaction of nations were forbidden, sociality as a genetic property of people would undergo rapid decay.

Sociality of humans is manifested in the fact that people enjoy social contacts. The great number of existing parties, communities, movements, etc. show that being within a correlated association of friends, co-citizens or co-thinkers brings about positive emotions in the majority of people. Naturally, most people try to increase the competitive capacity of the association to which they belong, be that a football team, a political party or a country.

Sociality forms the basis for such noble feelings as patriotism, which makes people give their lives for the happiness of their motherland. However, sociality is also responsible for such phenomena as wars, racism and nationalism. In the modern world, such manifestations of sociality threaten the existence of the whole of civilisation. Another genetically pre-programmed property of humans—their ability to critically analyse the observed phenomena and accordingly correct their behaviour—helped people to realise that threat and begin a world-wide campaign against military conflicts and any manifestations of racism and nationalism.

Sociality is ultimately responsible for the unwillingness of modern humankind to impose limitations on the population growth. People act to increase the competitive capacity of their country. The competitive capacity of a country is to a large degree determined by the total number of citizens. World history of the last several thousand years may be reduced to a succession of events when more rapidly-growing nations invaded territories of slowly-growing (sustainable) aboriginal nations and quickly forced the latter out from their territories. In such a manner territories of the Americas, Australia, Siberia and the Far East in Russia were invaded.

There exists, therefore, a contradiction between the emotional desire of people to maintain the high competitive capacity of their state or nation on the one hand, and on the other, the understanding that unlimited population growth threatens the existence of humankind and should be struggled with. A solution to the problem should be sought in finding different ways of competitive interaction of countries, which may be elaborated at the level of United Nations and assume the form of international laws. Such safe ways of competition may involve interactions of people in the spheres of sports, science and culture. However, such a way of civilisation development will remain a utopia until the majority of people begin to realise the danger of unlimited population growth and become capable of conscious, emotion-free actions in that regard. This stresses the need of ecological education of people all over the world.

On the above basis, the following conclusions can be made:

- Like social animals, people form large, internally-correlated associations (countries) and maintain their level of organisation via competitive interaction of countries.
- Competitive capacity of a country is tightly coupled to the population number of its citizens. As a result, the more rapidly-growing nations gradually acquire advantage in the world's political arena.
- Although human population growth presents the most serious threat to the global environmental stability, it is at present disadvantageous to most countries to impose limits on population growth of its citizens.
- While such a situation persists, humanity is rapidly moving towards a global ecological catastrophe.
- Competitive interaction of countries cannot be done away with, as far as sociality is a genetically-programmed property of humans.
- The way out is to be sought in uncoupling the competitive capacity of countries from the population number of its citizens by means of finding new forms of competitive interaction between countries and elaboration of new norms of international law. To decide how to do it should become the most urgent task of the modern world community.

2

What is Life?

In this chapter, the fundamental properties of living objects that form the basis for the biotic regulation concept are discussed. Life has invented a peculiar way of maintaining stability that is absent in inanimate nature. Any particular living being represents a super-organised object that inevitably decays with time. All living beings necessarily exist in the form of numerous sets but not in single numbers. Competitive interaction between homologous biological individuals within such a set (population) makes it possible to select out those that suffer decay. The vacancies that appear are filled by the progeny of remaining normal individuals. Such a scheme is absolutely unique to life and makes it possible to maintain the organisation of living objects at an arbitrarily high level.

2.1 DISTINCTIVE PROPERTIES OF LIFE

Living beings are drastically different from lifeless objects. The striking difference is due to the absence of any transient forms whatsoever between life and non-life. It seems that living and lifeless objects are separated by an insurmountable quantitative and qualitative gulf. On the other hand, detailed analysis of life properties performed by natural science during the last two centuries has shown that all processes taking place in living organisms conform to the same physical laws as processes in inanimate nature. There seem to be no characteristics that would be inherent solely to life and could not be encountered in lifeless objects. Thus, in spite of apparent, readily-discovered differences between non-life and life, it is difficult to provide a clear formal definition of the latter.

Indispensability of nutrition and *capability to reproduce* are often named as the distinctive features of living objects. However, nutrition is nothing else than transformation of one type of ordered energy into another, which is common to inanimate nature as well. One can say that rivers, for example, nourish themselves with water of rains. Hurricanes and tornadoes move by feeding on latent energy that is released in the course of atmospheric water condensation. Artificial mechanisms created by humans feed predominantly on chemical energy stored in organic matter of fossil fuels.

Capability of reproduction leading to an exponential growth of the number of offspring represents copying and multiplying existing structures. This property can be observed in inanimate nature as well. Any chain reaction can be envisaged as the reproduction of the reaction products. A chain reaction leads to exponential accumulation of its products, which is perceived as an explosion. For example, neutrons that normally do not exist in a free state due to rapid spontaneous decay, may accumulate in substantial quantities during a nuclear explosion. Autocatalytic reactions present another example of reproduction of non-living objects. In such reactions one of the reaction products is the catalyst itself, i.e. a substance that further accelerates the reaction. However, it is evident that all the above processes bear no relation to life.

Sometimes *self-preservation* or even *thinking* are named among the inherent properties of life. However, these are not common properties of all living objects. For example, trees and fungi are not capable of either self-preservation or thinking. However, there is no doubt that both trees and fungi are living objects.

Sharp distinctions between life and non-life pertain to structural rather than physical peculiarities of organisation. The most important and well-known distinctive features of living objects that can be readily observed in everyday life are as follows:

1. **An important, eye-catching property of living objects is their fantastic complexity.** In inanimate nature there is no physical object that could be compared to any living organism, as far as the degree of complexity is concerned. The most complex inanimate objects such as hurricanes, tornadoes, ball lightning and others arise spontaneously under the action of external flux of solar energy. No living object would ever arise spontaneously, although feeding and reproduction of living objects are ultimately based on the same solar energy. For the first time, this fact was described and understood by Pasteur and can be therefore called Pasteur's law. Nowadays, when much information about complex molecular structure of living objects has been accumulated, Pasteur's law looks self-evident or even trivial. However, this law grasps one of the most important differences between life and non-life: **no living object ever arises spontaneously in whatsoever flux of external energy**.
2. Another important and also well-known feature of living objects is their **existence in the form of organisms with strictly defined sizes of their bodies within which all structures are rigidly correlated, while there is no correlation among bodies of different organisms**. Different organisms are not correlated with each other. Death of one organism does not generally influence the well-being of others. On the contrary, all internal structures of one organism are strongly dependent upon each other. In a cell, such correlated structures are represented by molecules and organelles, and by internal organs in multicellular organisms. Degradation of any internal structure of the organism, may lead to various adverse changes in the organism, including its death.

This property of living objects differs greatly from the known properties of ordered processes in inanimate nature. Ordered processes like winds, oceanic waves, mixing of oceanic waters with different temperature or salinity, mixing of air masses with different temperature or humidity and others can stretch over territories of any size, even of the whole surface of the planet. On the other hand, these processes are characterised by a low degree of internal correlation. Disturbances of such process in a given local area do not have any considerable impact on the flow of the process in other local areas.

3. The third important life property lies in the fact that **all living objects exist in the form of populations**. There are no biological species that would consist of a single individual in the case of parthenogenetic species or a single pair of individuals in the case of sexually reproducing species. Population represents a set of interacting but non-correlated individuals, which means that death of one or a few individuals does not affect the population as a whole. Natural biota of Earth consists of discrete biological species. In the overwhelming majority of cases, individuals of a certain species can be unambiguously told apart from individuals of another species.

   This property of life stands in the most dramatic contrast with the observed features of highly-ordered physical objects. Such rare phenomena like, for example, ball lightning, arise spontaneously in the flux of external energy as single objects and decay long before an analogous object arises once again. Thus, unlike living objects, all highly-ordered inanimate objects necessarily exist in singular numbers.

4. Finally, there is another common property of life observed everywhere. This is **expansion, the inherent tendency of every biological species to spread over all suitable territories**, where fluxes of matter and energy necessary for life are present in sufficient quantities. Expansion is observed in absolutely all biological species, including humans with their discoveries of new continents in the past and attempts to conquer the outer space in the present. Due to the intrinsic capability of expansion, domestic sparrows became abundant in most cities of our planet. Existence of infectious diseases and epidemics are due to the expansive properties of certain bacteria and viruses.

   When all the territories suitable for existence of a species are colonised in the course of expansion, the species continues to exist sustainably on the newly-occupied territories. Expansion of living objects is therefore drastically different from explosion-like processes of exponential propagation of ordered objects and processes in inanimate nature. All such processes necessarily decay with time and the initial pre-explosion state is finally restored.

2.2 PHYSICAL AND BIOLOGICAL STABILITY

All processes and phenomena observed in nature are characterised by a certain degree of stability, otherwise they simply could not be observed. Any process changes the state of the system where they occur. In the absence of an external flux of energy, any system tends to a state of thermodynamic equilibrium. Thermodynamic equilibrium is characterised by the maximum degree of chaos possible in a given system. Thus, the state of thermodynamic equilibrium is stable, because there are no processes that could spontaneously arise out of complete chaos and change the state of the system.

Any processes in systems in the state of thermodynamic equilibrium can only be generated when an external flux energy is delivered to the system. Stability of the generated processes corresponds to their continuity in time or to a stable frequency of their occurrence and subsequent decay. Flow of water in rivers that is ultimately due to solar energy, tidal waves in oceans caused by the gravitational attraction of the Moon, functioning of geysers maintained by the geothermal power are examples of continuous processes due to external fluxes of energy. Decay of the ordered energy of these processes due to its transformation into heat is continuously compensated from the corresponding external source of ordered energy. The majority of other ordered processes in inanimate nature are characterised by stable average frequencies of their occurrence and decay. Among such processes one can mention winds, cyclones, thunderstorms, and hurricanes generated by solar energy and volcanic eruptions and earthquakes generated by geothermal power.

Some processes in inanimate nature, like ball-lightning or tornadoes occur very rarely. Nevertheless, the average low frequency of their occurrence is characterised by a certain stable value that is determined by peculiarities of the interaction between solar energy and the Earth's environment. Finally, there are sporadic processes with such a low frequency that their repeat occurrence is highly improbable. Such processes can be envisaged as unstable.

All ordered processes in inanimate nature are generated and maintained by external energy—solar energy, gravitational energy of planets and geothermal energy. **The frequency of occurrence of ordered physical processes is inversely proportional to their complexity**. Low-ordered processes and events are common, highly-ordered processes are rare. As mentioned above, rare processes always arise in singular numbers and decay well before another such process occurs. Thus, interaction between analogous highly-ordered inanimate objects is impossible. The frequency of occurrence of highly-ordered processes is determined by various characteristics of the environment where they occur and by the flux of external energy supporting them. The average frequency of occurrence of such processes remains stable unless the external flux of energy changes. When the latter is changed, frequency of highly-ordered processes assumes some other stable value determined by the new characteristics of the energy flux. Below we will refer to this type of stability as *physical stability*.

The nature of life stability is completely different. It can be understood on the basis of the four major properties of life outlined in Section 2.1. Any living individual represents a super-ordered object that cannot arise spontaneously in whatsoever fluxes of external energy. As a result of its high complexity, any living individual inevitably decays (dies) in spite of a continuous supply of ordered external energy (nutrition). During decay, the complex organisation of living objects loses its orderliness and tends to a more simple organisation of physical objects common in a given flux of external energy. It is important to note, however, that death of organisms occurs long before their complexity reduces to that of physical objects. Thus, living objects are not physically stable, unlike inanimate objects that, after decay, can spontaneously arise again in the external flux of energy.

The ability of living individuals to reproduce does not save the situation. Reproduction consists of copying the genetic material of the parents and the development of a new individual on this basis. Due to an unavoidable erosion of information before and during copying, the newly-born individuals will be always characterised by a lower degree of organisation than their parents, which will finally lead to loss of viability and extinction of life as a whole. The very ability to reproduce is also subject to decay, as well as all other highly-organised properties of life.

Stability of life organisation does not characterise a single individual or a succession of its offspring. Stability of organisation refers only to the population level. Formation of populations by individuals of absolutely all species suggests that the major function of population is to ensure a possibility of interaction of individuals with each other. On the other hand, the fact that the death of one or a few individuals does not normally affect the state of population, means that interaction of individuals within the population is competitive rather than correlated.

Competitive interaction essentially means comparison of levels of organisation of all individuals in the population. Due to processes of decay some individuals are characterised by a lower degree of organisation than others. Such individuals are less competitive than the rest of the population. In the course of competitive interaction they are forced out from the population. (That does not necessarily imply the death of the less competitive individuals. They can be, for example, prevented from reproducing.) Vacancies appearing after elimination of the non-competitive individuals are filled by the offspring of normal individuals remaining in the population (Figure 2.1). Such a mechanism maintains the population number as well as the level of organisation of individuals in a stable state. The complexity of the maintained organisation of living individuals can be as high as is wished. Unlike inanimate objects, the complexity of biological objects is absolutely independent of the degree of orderliness of the external flux of energy.

Detection of non-competitive individuals, their elimination and the filling up of the vacancies with offspring of normal individuals constitutes the essence of stabilising selection. Stabilising selection makes it possible to maintain stable organisation of living objects at an arbitrarily high level. Such a mechanism of maintaining stability is unique to living systems. Thus, we will refer to this type of stability as *biological stability*.

Stabilising selection operates given the following essential conditions are met. Individuals within the population must not be correlated with each other, so that elimination of any relatively small number of individuals does not affect the population's condition. This statement is very important for understanding the bases of life stability. Elimination of a non-competitive individual should occur quickly enough not to allow it to produce offspring which, naturally, will be also non-competitive. It means that competitive interaction should be intensive and

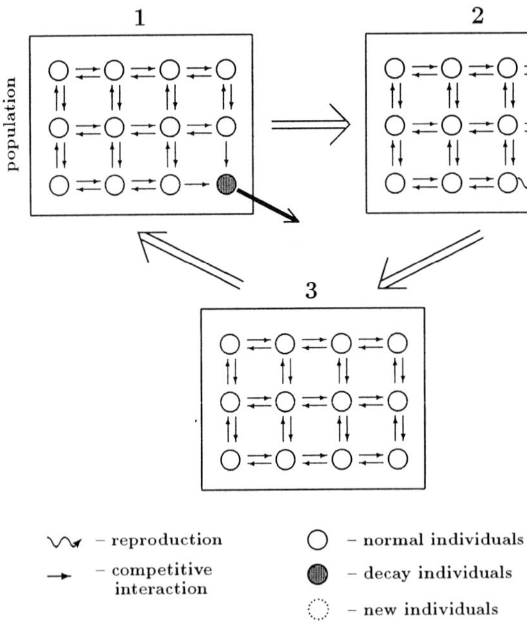

Figure 2.1. Competitive interaction of individuals in a population. Competitive interaction of normal (equally competitive) individuals is balanced, an action is equal to its counteraction. A decay individual cannot stand the competition and is forced out from the population (1). The vacancy is filled by offspring of one of the normal individuals (2). When adult, the new normal individual is able to compete with other normal individuals in the population (3). When another decay individual appears, the circle 1—2—3—1 is closed.

aggressive. Intensity is necessary for quick detection of all defects present in individuals. Aggressiveness of competitive interaction means that competitive interaction represents a continuous process independent of the availability or unavailability of food, territory, sexual partners or other resources.

It is sometimes stated that individuals within a population compete for some limited resource. Such consideration implies that if such a population is placed under conditions where this resource is abundant the competitive interaction would cease to operate. However, processes of decay of super-organised biological objects (living organisms) occur irrespective of whether resources are abundant or not (see also Section 5.7). It means that competitive interaction eliminating decay individuals is necessary always and everywhere. Thus, it cannot be dependent on the degree of availability of one or another resource and is in this sense simply aggressive, being a competition for its own sake.

Non-competitiveness may be caused either by somatic defects appearing during the individual's life span, for example, physical injuries, or by inherited genetic defects that are due to erosion of some part of genetic information. Any kind of deviation from the normal level of organisation is perceived by normal individuals as a malformation. However, the word 'malformation' typically applies to significant departures from the normal level of organisation. Thus, everywhere below we will refer to non-competitive individuals as *decay individuals*, bearing in mind that the word 'decay' stands for one or another degree of malformation.

To support a high level of organisation of every biological species, it is necessary to ensure that decay non-competitive individuals eliminated from the population are not given a chance to reproduce and accumulate in some other place and form populations built up entirely of decay individuals with a lowered level of organisation. This may be achieved only when there are no vacant territories suitable for the existence of a given species. Such a situation is realised as a result of the expansion of normal competitive individuals of the initial population, that is, rapid growth of population and simultaneous colonisation of all suitable territories.

Absence of vacant territories during the most time of population existence makes it possible to conserve the ability of individual organisms to form populations, which is one of the most essential properties of living objects. Decay of this ability would completely violate the mechanism of biological stability and lead to extinction of life as a whole.

Indeed, consider a situation where due to some decay changes, individuals in a population ceased to interact competitively with each other and went away from the population, each to its own territory. All their decay offspring will do the same and so forth, because in the considered situation the capability of expansion is not inherent to biological organisms and, consequently, there will be many vacant territories. As a result, the process of decay will go unimpeded in all further generations, because the only thing that could stop it is competitive interaction of individuals. Level of organisation of decay individuals will be continuously decreasing from generation to generation until it becomes incompatible with the viability of the organisms, causing complete extinction of the species. In contrast, when all the available territories are occupied by competitive conspecific individuals as the result of expansion, a situation when some individuals may escape competitive interaction is improbable.

2.3 SEXUAL DIMORPHISM

Continuous processes of decay result in the fact that in any population there is always a certain number of decay individuals that have to some degree lost the normal level of organisation characteristic of the species but still remain viable and fertile. The average level of organisation of the population is determined by the relative frequency of decay individuals. A high level of organisation corresponds to a low relative frequency of decay individuals.

Two strategies of maintaining a high level of organisation are possible. If the rate of decay processes is high, it is necessary to quickly eliminate decay individuals shortly after their birth. Such a strategy is realised in many plants. Among numerous seedlings, only very few grow to maturity. If the rate of decay is low, it is sufficient just to prevent decay individuals from reproduction. In such cases, elimination of decay individuals from the population is due to their natural death in the age of maturity.

Decay individuals can only be prevented from reproduction by normal individuals in the course of competitive interaction. This is not an easy task to perform, because decay individuals may try to escape competitive interaction, increasing the period of time between two successive acts of competitive interaction and producing offspring in the meantime. As a result, the relative number of decay individuals will grow up, while the level of organisation of the population will be lowered. The only way to avoid such a situation is to create a mechanism preventing all individuals from reproduction unless special permission is granted. Such permission, a sort of *carte blanche*, can only be granted to and by normal individuals that have been tested in the course of competitive interaction.

Such a mechanism of granting and receiving permission to reproduce was created by life in the course of evolution and is known as sexual dimorphism, that is, division of all individuals in the population into two distinctive classes, males and females.[1] Morphological differences between males and females are determined genetically. In some cases the genetic difference between males and females may even exceed the average genetic difference between closely-related species.

The major functional property of sexual dimorphism is that males are not capable of reproduction at all, while females are able to produce both females and males, but only with the permission of males. Intensive competitive interaction of males detects normal and decay male individuals. Sexual activity of decay males is further suppressed by normal males preventing the former from sexual contacts with all females in the population. Normal competitive males choose for their sexual contacts normal females only and ignore decay ones. As a result, all decay individuals in the population—both males and females—are excluded from the process of reproduction, although they are allowed to live in the population until their natural death occurs (Figure 2.2).

Among the two mentioned strategies of maintaining a high level of organisation of living beings, the strategy based on sexual dimorphism has a number of advantages compared to the strategy of physical elimination (killing) of the decay individuals shortly after their birth. Evaluation of competitiveness of any individual is performed by other individuals on the basis of its morphological and behavioural properties, that is, characteristics of its phenotype. Phenotypic characteristics generally reflect the genetic properties of the individual, that is, characteristics of its genotype. Correspondence between genotypic and phenotypic properties of the individual cannot be very informative in young individuals at various stages of uncompleted ontogenesis. Young that look like decay individuals when cubs or seedlings may prove to be quite normal as adults. Thus, it proves to be costly to eliminate the young that look like decay individuals shortly after their birth, because this leads to needless elimination of a certain percentage of normal individuals.

[1] Sexual dimorphism should not be confused or identified with exchange of genetic material between two individuals when producing offspring. Such an exchange pursues different goals and will be discussed in detail in the chapters that follow (Chapter 9). It can be observed in the absence of sexual dimorphism as well, e.g. in so-called hermaphroditic species, where any individual may act both as male and female.

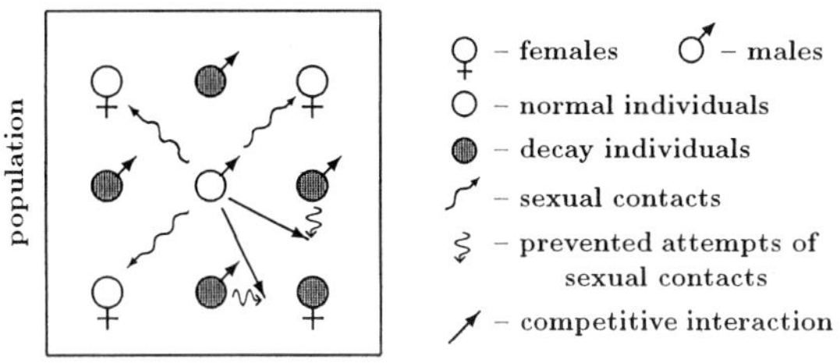

Figure 2.2. Competitive interaction of individuals in a population with sexual dimorphism. Competitive interaction between males detects normal and decay individuals. Competitive capacity of females is determined by normal males that choose normal females for sexual contacts and ignore decay ones. Competitive interaction between females can be completely switched off. Normal females ignore decay males. Normal males prevent decay males from sexual contact with decay females. All individuals, normal as well as decay, remain in the population. Reproduction of decay individuals is prohibited.

The second strategy, based on sexual dimorphism, does not possess this disadvantage. Evaluation of individual competitiveness occurs during the whole life span of the individual. Individuals that cannot demonstrate high competitiveness when young but achieve a normal level of competitiveness later in their life occupy a worthy place in the population and are allowed to produce offspring.

Another disadvantage of the first strategy is that normal individuals have to spend substantial amounts of their energy on physical elimination of decay individuals. In the second strategy, decay individuals are not exposed to any energetic impacts.

The apparent advantages of the second strategy, based on preventing reproduction of decay individuals, has resulted in the widespread occurrence of sexual dimorphism—division of the population into males and females—in most biological species. For example, among about 100 000 vertebrate species, asexual reproduction is encountered in 25 species only, including 3 fishes, 2 salamanders and 20 lizards (White, 1973).

2.4 COMPETITIVENESS AND ORGANISATION OF LIFE

Apart from providing for stability of organisation of the extant biological species, functioning of the stabilising selection mechanism opens a possibility of progressive biological evolution.

If, along with frequent decay changes of the normal genetic programme, there sometimes arises a new progressive change ensuring higher competitiveness of the corresponding individual, offspring of the latter will most probably force out from the population all the other, less competitive, individuals. As a result, the whole population changes genetically. The genome containing the progressive change will become a new standard for the population. Former normal individuals will be regarded as decay ones and eliminated from the population along with true decay offspring of the individual that initiated the process. This process represents a possible mechanism of the origin of new biological species that could explain the observed succession of species' origins and extinctions known from paleodata. This possible mechanism of speciation may be called *progressive natural selection* in contrast to stabilising selection discussed in the previous sections.

It should be noted that, while stabilising selection that prevents accumulation of decay changes in a population can be readily illustrated with a great number of empirical data accumulated during the last century of active development of genetics, there is not a single contemporary example that could unambiguously document the existence of progressive natural selection leading to formation of a new species. Such a situation is due to an extraordinarily low frequency of appearance of progressive genetic changes and will be discussed in detail in Chapter 11.

Stabilising and progressive natural selection ensure the stability of a high level of organisation of biological objects provided that the following important condition is met. An increased level of organisation should be paralleled by a corresponding increase in competitiveness, while decreased level of organisation should bring about lowered competitiveness. Such correlation of the level of organisation and competitiveness is not self-evident and is not always observed. One can envision a variety of situations where high competitiveness is compatible with a low level of organisation or even a low degree of viability of organisms and vice versa. Consider, for example, a situation when competitiveness of animals determined by their physical power increases at the expense of their fertility. In reality, such a situation can be observed in many animal hybrids that are often extraordinarily strong, being at the same time partially or completely sterile. The strongest individuals would then force all other individuals out from the population and become extinct, being unable to maintain a stable population number. Similarly, in technology it is often the case that increased complexity of a mechanism (i.e. its level of organisation) adversely affects the reliability of the mechanism and, consequently, its competitiveness.

It is important to note that all spontaneous genetic changes are decay in nature, that is, they arise in a random and unpredictable manner in the course of spontaneous decay of the high level of organisation of biological objects. Those decay changes that ensure increased competitiveness of the individual carriers are by definition progressive and bring about transition to a new stable genetic state of the population. In the absence of mechanisms forbidding progressive evolutionary changes that are accompanied by a lowered level of organisation, a series of such changes would finally result in loss of viability of individuals and extinction of the species. Stable existence of life during the last four billion years points unambiguously to the fact that progressive evolutionary changes of this type are strictly limited. This limitation is ecological in nature and pertains to the inability of biological objects with lowered level of organisation to participate in the maintenance of suitable-for-life environment.

To sum up, physical stability of inanimate objects is dictated by external fluxes of energy. Level of organisation of inanimate objects is uniquely determined by the degree of orderliness of the supporting external energy and cannot increase with time. Random, highly-ordered physical processes occur sporadically, undergo decay and, in most cases, never arise again. Mechanism of biological stability is based on stabilising selection. It does not only ensure maintenance of any level of organisation, however high it may be, but also contains a potential for progressive evolution accompanied by increasing levels of organisation.

Life supported by the mechanism of biological stability could not have been existing eternally in the ever-expanding universe. Essentials for the existence of life conditions could not be met at early stages of the universe's development. Thus, life and the mechanism of biological stability appeared spontaneously as a result of a certain physical process that already contained in itself all the necessary conditions of biological stability, that is, capability of physical objects to form populations, to compete and to reproduce. These properties made it possible for stabilising selection to operate which prevented the newly arisen life from rapid decay and extinction. They also opened a possibility for evolution. Since that time, life has been existing and evolving on the basis of the mechanism of biological stability during nearly four billion years. In the whole universe this is the only example of what might be called *sustainable development*.

Note that biological stability at any *relatively* low level of organisation of biological objects requires an extraordinarily high *absolute* level of organisation (orderliness) to make it possible for the complex processes of stabilising selection to occur. Artificially created robots that are often considered as cybernetic analogues of life are not able to ensure such processes and cannot therefore maintain their organisation at a stable level. All possible concerns that a stable, artificial civilisation of robots may arise and force out humans are groundless.

One of the most complex properties of life is aggressive competitive interaction, that is, the inherent desire or tendency to compete, which is essential to all living individuals. In the absence of this property, the mechanism of maintenance of biological stability of living objects will be undermined, which will lead to rapid degradation and extinction of life as a whole.

2.5 ALTRUISTIC INTERACTION OF INDIVIDUALS

As was mentioned above, absence of correlation between individuals in a population is a necessary condition for stabilising selection to operate. Elimination of one individual should not influence the wellbeing of others. This statement seems to be apparently contradictory to altruistic interactions of two or more individuals that are often observed in populations of living organisms. There cannot be aggressive competitive interaction between a male and a female during the breeding season, between parents and offspring during period of parental care, between the queen and working ants in an anthill. Interaction between these individuals is correlated, which means that these individuals are mutually dependent on each other and their interaction is performed along strictly specified behavioural paths.

Stabilising selection provides for the maintenance of a given level of organisation or orderliness (which is essentially the same) of individuals in a population. Orderliness of an individual is determined by the degree of correlation between various parts of the individual. Stabilising selection prevents this correlation from decay. In the absence of stabilising selection, correlation of any kind will be inevitably lost.

Internal correlation of macromolecules and organelles inside a living cell is maintained by stabilising selection in a population of unicellular organisms, only if the latter are not correlated with each other.

Correlation of cells with different specialisations in a multicellular organism can be considered as their altruistic interaction ensuring the wellbeing of the whole organism. Such altruistic interaction cannot be maintained by a stabilising selection of cells inside the organism, because high competitiveness of individual cells does not automatically ensure high competitiveness of the whole organism. Moreover, selection of the most competitive cells inside the organism necessarily disrupts the existing correlation of different cells vitally important for the organism as a whole. To avoid it, genomes of cells of multicellular organisms must contain information forbidding infinite reproduction (division) of individual cells and preventing the most productive cells from forcing out less productive ones, i.e. those cells that divide more slowly. Decay of this preventive genetic information may result in cancer, which is one of the most serious diseases encountered in multicellular organisms.

Internal correlation of a multicellular organism (i.e. correlation between internal organs, correlation between cells of different types inside organs and correlation of organelles and macromolecules inside individual cells) can be only maintained by stabilising selection in a population of non-correlated multicellular organisms. It is highly improbable that a multicellular organism as a whole would remain competitive if, for example, in the result of decay processes organelles in a substantial part of its cells ceased to function in a coordinated manner. Thus, the hierarchical structure of multicellular organisms automatically guarantees maintenance of correlation at all levels of organisation (from cells and tissues to organs) provided that stabilising selection operates in the population of multicellular individuals.

As well as unicellular organisms, multicellular organisms also maintain certain internal milieux within the organism. Objects like beehives, anthills and others all represent internally correlated spatial associations that consist of individual multicellular organisms with different specialisation (queens and drones in bees, foragers and soldiers in ants, etc.) and can be also characterised by a specific internal milieu. Although individual members of such associations are capable of independent travelling outside that internal milieu, they cannot sustainably exist outside the framework of strictly-specified correlated interactions with other members of the corresponding eusocial structure. Competitive interaction between individuals with different specialisations is impossible for very much the same reasons as there cannot be competitive interaction between different organs of one and the same organism. Thus, in spite of the absence of a continuous envelope that separates various parts of the objects like a beehive from the external environment, biological objects like anthills and beehives feature the same major properties as true multicellular organisms.

Decay of the correlation between individual members of anthills cannot be prevented by any means inside one isolated anthill. This can be only achieved in the course of stabilising selection that would detect decay and normal anthills in a

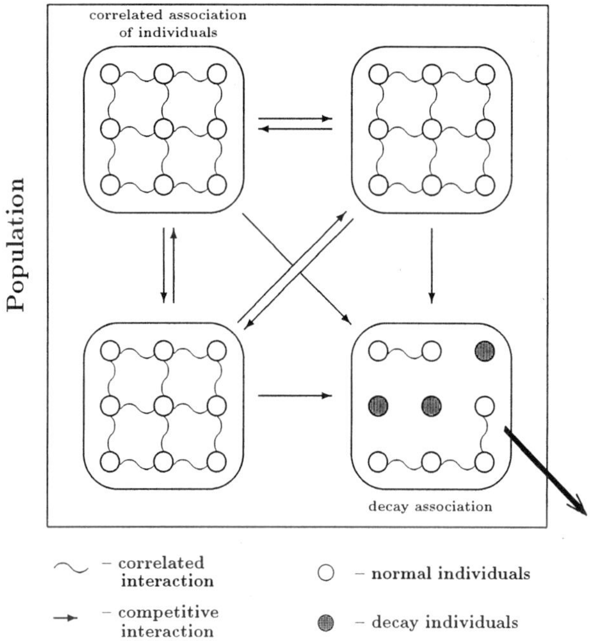

Figure 2.3. Competitive interaction of correlated associations of biological objects. Correlated interaction between individuals of one and the same association is prevented from decay due to the formation of a population of such associations and switching on competitive interaction between them. Decay of internal correlation of any association manifested in disruption of correlated links between its constituents lowers competitive capacity of that association. As a result, the decay association is forced out from the population by normal associations. Any type of correlated interaction between biological objects, both strong (cells in a multicellular organism, species in an ecological community) and weak (animals in a herd) may be maintained only in the course of competitive interaction of respective associations within a population.

population of non-correlated anthills, which is completely analogous to stabilising selection of multicellular organisms.

Using the same way of reasoning one comes to the conclusion that the only way to prevent decay of correlation between objects of any level of organisation is to form a population of internally correlated associations of these objects and allow stabilising selection to operate in this population eliminating defective associations with impaired internal correlation of their constituents. Here the word *associations* may stand for cells, organisms, eusocial structures like anthills, hierarchical herds of mammals and, as we will see below, ecological communities of species.

The main property of associations in the population should be the absence of correlated links *between* different associations, so that aggressive competitive interaction of associations could be switched on. In this sense such associations represent generalised individuals (Figure 2.3).

Subdivision of populations into internally-correlated groups of individuals with different specialisations is observed not only in eusocial insects. Sexual dimorphism—division of population into females and males—is common to the majority of the extant species. During the breeding season, females and males are correlated in pairs and cannot aggressively compete with each other. Correlation of male and female is as subject to decay as any other type of correlation. It may be only conserved if correlated associations—male–female pairs in this example—compete with each other, forcing out from the population decay individuals that are not able to perform normal sexual interactions.

Finally, one of the most important types of correlation encountered in the biosphere is correlated interaction of individuals from different species. This type of correlation is fundamental for formation of ecological communities that are the only objects with a level of organisation high enough to enable them to maintain a suitable-for-life environment on local and global scales. The most important characteristics of the organisation of ecological communities are considered in the chapters that follow. Here we stress that ecological communities represent correlated associations of species with different specialisations, the most crude division being into species-autotrophs that are primary producers of organic matter and species-heterotrophs that destroy it. As with all correlated associations, there cannot be aggressive competitive interaction between individuals of different species, i.e. under normal ecological conditions interspecific competition is completely absent from ecological communities. Interspecific competition may only take place in disturbed communities during the period of their recovery.

An essential prerequisite for stabilising selection is the limited size of competing associations. Cells and organisms have well-pronounced envelopes that define the spatial extent of the internal correlation of these objects. In the majority of cases ecological communities, as well as eusocial structures of insects and pairs of sexual partners, do not have a visible boundary. For associations such as an ecological community, the condition of limited size means that correlation of individuals from different species inside the community should be characterised by a certain finite radius, i.e. correlation should become weaker with distance and die out at a certain critical value of it.

In multicellular organisms, correlated cells are characterised by one and the same genome, though its active parts are different in cells with different specialisations. In eusocial structures of insects, many individuals feature similar genomes, being offspring of one and the same queen. Meanwhile, in ecological communities correlated individuals from different species naturally possess different genomes. Different genomes of correlated objects may be also observed in unicellular organisms, where different organelles like, for example, chloroplasts and mitochondria, function in a correlated fashion on the basis of different genetic material in concert with the main nuclear genome of the cell. Different genomes characterise males and females in correlated pairs of sexual partners.

Thus, highly-ordered internal correlation of an association of biological objects does not necessarily imply either visible boundaries of the association or identical genomes of the correlated biological objects. Absence of these properties should not be an obstacle to noticing the general biological principle of maintenance of any kind of correlation, i.e. stabilising selection of independent competing associations, whatever their nature may be.

2.6 NOTORIOUS GROUP SELECTION

Stabilising selection operating in a population of several correlated associations of biological objects is sometimes called *group selection*. Most often this term refers to stabilising selection of anthills, beehives, termitaria and other structures of eusocial insects. Stabilising selection of communities may be also classified as group selection (see, e.g. Wynne-Edwards, 1986). The term emerges every time researchers encounter a phenomenon of altruistic interaction of different individuals. Many investigators deny the existence of group selection, trying to explain the peculiarities of altruistic behaviour by individual or kin selection of organisms taking part in altruistic interactions (Hamilton, 1964; Maynard Smith, 1964). The term *group selection* is never applied to selection of internally correlated and, consequently, altruistic associations of cells and organs of multicellular organisms. On the other hand, sometimes one speaks about group selection of population or even species (Wilson, 1980; 1997).

To our mind, the reason for such indefinite status of this term lies in the absence of its clear definition. First of all, contradictions arise because of vague and ambiguous interpretations of the term *population*.

Population is often used to denote a set of similar, weakly-correlated components belonging to one and the same individual (in a general sense), be the latter an organism or a social structure. One may speak about a population of erythrocytes or leukocytes in blood, a population of working ants in an anthill or working bees in a beehive, a population of animals in a herd or leaves in a tree. There may be elements of competitive interaction between such components belonging to the same individual, but the essential thing is that they are always partially correlated. All blood cells belong to one and the same circulatory system. All working ants belong to one and the same anthill and contain genetic information about it. There is no competitive interaction between erythrocytes or working ants from the same anthill. Hierarchical correlation of animals within a herd may, on the contrary, make competitive interaction between individuals more intensive. All leaves of a tree are connected through branches with one and the same trunk. One can say that there is a non-aggressive competition for light between different leaves and branches.

A low degree of correlation between such numerous components within the individual makes it possible to substantially decrease the frequency of random fluctuations in major characteristics of the organism. This effect is known as the law of great numbers. Each erythrocyte plays a small role in supplying oxygen to the various organs. When there is a large absolute number of non-correlated erythrocytes, relative fluctuations of oxygen supply in the organism become very low, making possible reliable functioning of the whole organism. Similarly, a large number of working ants ensures a reliable supply of food to the anthill. A large number of leaves guarantees the process of photosynthesis against sharp fluctuations (see Section 3.5).

Sometimes the word *population* simply refers to a group of sexually-reproducing organisms. As we have seen, sexual reproduction represents a special type of correlation between individuals. The fact that the term *population* is used to denote a set of objects that may compete but are at the same time correlated, is a source of confusion. To avoid it, we will everywhere below use the word *population* to denote a set of aggressively competing, mutually independent (non-correlated) objects of any degree of complexity.

Thus, in the above definition the possibility of correlated interaction between individuals is ruled out. On the other hand, unlike traditional definitions of the term (Raven and Johnson, 1988), the above definition can be also applied to sets of such objects as asexual organisms, social groups and ecological communities. Individual organisms of the same or different species may form internally-correlated associations in the same manner as cells and organs constitute the structure of multicellular organisms. Such internally-correlated associations can be considered as individuals in a general sense of the word (generalised individuals). Stability of the internal organisation of associations can be maintained by stabilising selection in a population of such associations.

We emphasise that correlation of any type is characteristic of a (generalised) individual but not of the population. No properties that would be absent in an individual emerge when we consider a population of such individuals. For example, the ability of ecological communities to ensure close biogeochemical cycles of biogens and maintain a favourable environment is inherent to every individual community instead of emerging at the level of a large population of communities or biosphere as a whole. Similarly, flight is a property of an individual bird or insect, rather than a property emerging at the population level. Hierarchical organisation of herds of animals is not an example of *emergent properties*. Due to substantial correlation of individuals within a herd, the latter cannot be considered as a population. Any single herd represents a generalised individual. The internal correlation of herds (i.e. their hierarchical organisation) can be only maintained by competitive interaction of different herds.

Note, however, that one and the same object may be sometimes envisaged both as a population of non-correlated components or as an internally-correlated structure. It depends on what properties of the object are under consideration. For example, competitive interaction between drones belonging to the same beehive may serve to prevent some important properties of drones (e.g. their ability to fertilise the queen) from decay. In this sense, one may speak about a population of drones within a beehive and about stabilising selection operating in this population and supporting the property in question. In other words, internal correlation of drones themselves can be supported (at least to some degree) by competitive interaction between drones, which, in this sense, represent a population. On the other hand, competitive interaction between drones or other bees within a single beehive cannot prevent the decay of the property of bees to combine into a correlated structure like a beehive. Internal correlation of beehives can be supported only in the course of competitive interaction of beehives so that, with respect to this property, a beehive cannot be envisaged as a population of bees. The same refers to herds of animals.

Stabilising selection provides for stability of population in spite of the continuous decay of its individual members. Population stability is a fundamentally new phenomenon that has no analogues in any physical or chemical systems.

Earlier, it was called *biological stability*. Note that biological stability refers only to populations, but not to individuals. All individual biological objects—be they organisms, social structures or communities—are unstable and inevitably decay with time. Stabilising selection acts by definition on individuals, not on populations.

Competitive interaction is performed by non-correlated individuals in a population. By the above definition, different populations cannot compete with each other. Competition between different populations would mean that individuals from the same population treat alien individuals in a different way than they treat their population mates. Such a situation implies a certain correlation between individuals of the same population, which is impossible by the above definition. Interaction of two or more populations can therefore be reduced to competitive interaction between individuals that is performed regardless of what populations the interacting individuals belong to. Interaction of two or more populations is simply equivalent to an increase in the number of competing individuals. Thus, stabilising selection or group selection of populations is impossible.

Group selection of populations makes sense only if the word *population* refers to an internally-correlated object like, for example, a beehive. However, as we do not apply the word *population* to internally-correlated objects of any level of organisation, we will not use the term *group selection* anywhere below.

2.7 THE BASIC PRINCIPLE OF BIOLOGY

Modern physics is based on a few well-established, fundamental empirical laws. One such law is the Second Law of Thermodynamics. According to this law, entropy of a closed system never decreases. In other words, the degree of disorder (entropy) of any closed system may only increase. No type of correlation (order) between different components of a closed system may arise spontaneously. The Second Law of Thermodynamics does not apply to open systems, i.e. systems existing in an external flux of energy. The frequency of spontaneous appearance of ordered structures in an open system is inversely proportional to the degree of orderliness of these structures. In all biological objects, the degree of orderliness is so high that no biological objects can arise spontaneously in any external fluxes of energy, which constitutes the essence of Pasteur's law (see Section 2.1). In this sense, the degree of orderliness of the external energy does not affect the orderliness of biological objects. Thus, though biological objects are in essence open systems, their orderliness is not exposed to any external influence, as well as in close systems of inanimate objects. There should be, therefore, a biological analogue of the Second Law of Thermodynamics, which we discuss below.

Let us now formulate the basic principle of biology, which makes biological objects fundamentally different from objects of inanimate nature.

All types of biological correlation are characterised by super-orderliness. They cannot be supported by any fluxes of external energy, are subject to inevitable decay and never spontaneously arise again. Stability of biological objects is maintained due to detection of defective objects with a lowered degree of organisation and their substitution for copies of normal objects retaining the initially high level of organisation. Normal objects must have time to produce their copies before they decay. The process of detection and elimination of decay objects requires energy and matter expenditures that are consumed by living beings from the environment. Thus, life can only exist on the basis of continuous metabolic processes of energy and matter exchange that take place within living objects. Detection of decay of the level of organisation of one biological object is ensured by competitive interaction of living objects. All these processes constitute the essence of stabilising selection. Any level of internal correlation of individuals in a population can be maintained over indefinitely long periods of time by stabilising selection.

Individuals in a population are not correlated with each other. Competitive interaction between them is aggressive and occurs irrespective of abundance or shortage of resources. Each individual is characterised by a certain probability of losing the initial level of organisation (decay probability). A stationary population number is maintained due to reproduction of normal individuals retaining their competitiveness at the maximum level. In the absence of population and competitive interaction, any type of internal correlation of individuals decays and never arises again spontaneously. All the aforesaid refers equally to all types of biological correlation from molecular level up to ecological communities.

It is impossible to maintain any correlated (altruistic) links between individuals on the basis of any type of individual selection, e.g. kin selection (Hamilton, 1964; Maynard Smith, 1964). Individuals by themselves cannot possess information about what sort of correlation between them could increase competitiveness of their correlated association. This can be found out only in the course of competitive interaction between different associations. In an isolated correlated association any type of internal correlation of its components inevitably decays. The last statement represents an extension of the Second Law of Thermodynamics on super-ordered biological systems existing in external fluxes of energy.

Internal correlation of associations of living individuals from the same or different species can be maintained only by stabilising selection in a population of independent associations. Defective associations with impaired or totally degraded internal correlation become less competitive and are forced out from the population by normal associations (Figure 2.3). As soon as associations in the population are independent, elimination of one or a few defective associations does not affect the wellbeing of the others.

2.8 IMPOSSIBILITY OF GLOBALLY-CORRELATED LIVING OBJECTS

Let us now consider a speculative possibility of stabilisation of a super-ordered macroscopic system without competitive interaction with similar systems. Such a system could in principle be created on the basis of centralised control that would include recovery and regeneration of those parts of the system which have suffered decay. Decay of the control function itself can be prevented if the control organs are present in the system in the form of multiple copies.

Such a totalitarian system capable of regeneration of its decay parts would not need to replicate. Unitary as it is, it would not need competitive interaction for maintenance of its stability. Using external fluxes of ordered energy, such a system could control the environment, compensating adverse changes on the basis of the principle of homeostasis programmed into it. Neither would it need to expand. Under conditions of energy and material abundance such a system could exists as a single individual for an unlimited period of time. *The Black Cloud* by Fred Hoyle, and The Ocean in Stanislav Lem's *Solaris* are fictitious examples of such systems. Another example of such a hypothetical globally-correlated biosystem is Gaia (Lovelock, 1988).

However, such a system would not be able to evolve. Due to the absence of competitive interaction that could test properties of the system as a whole, there would be no possibility for the system to discern between progressive and regressive changes. Every possible change in the state of such a system would be equally probable. Since the overwhelming majority of spontaneous changes are of decay nature and naturally result in a decreased level of organisation, change in such a system would be a synonym of degradation. Thus, stability of such system can be attained only in the state of complete conservation where all changes are forbidden. However, due to its high complexity, such a system could not have arisen spontaneously as it is. It could only appear as a product of evolution, which is necessarily based on competitive interaction of independent individuals in a population. Thus, a globally-correlated biosystem is impossible.

In all organisms a mechanism of regeneration is present in one form or another. In the course of progressive natural selection, the stability of DNA molecules in some organisms can be increased due to the evolutionary appearance of a genetic programme of DNA repair (Lewin, 1987) capable of restoration of impaired parts of DNA molecules. Multiple back-up of genetic material achieved by polyploidisation of chromosomes and gene duplication makes it possible, in principle, to produce individuals with a hereditary system absolutely immune to decay. A genetic programme of regeneration of biochemically-defective parts of the organism together with polyploidisation of the programme itself makes it possible, in principle, to achieve an unlimited lifespan. However, a system that is immune to decay changes is also immune to any kind of progressive changes including evolutionary ones. Thus transition to a single, immortal, centrally-controlled super-organism immune to decay would deprive this organism of the possibility of progressive development and render it competitively inefficient, in contrast to an evolving set of competitively interacting independent organisms.

Hence, nature knows no immortal individuals and there are no biological species where all individuals were correlated with each other.

2.9 NORM AND DEFECT

The degree of orderliness of organisation of a given species is determined by the amount of genetic information of this species. The amount of genetic information is roughly proportional to the length of the species' genome. The character of the genetic information stored in the genome is determined by the particular structure of local parts of the genome. Most local parts of genome are the same in all individuals in a population. Random differences appearing in the course of decay processes are observed in a small percentage of local parts of the genome (Chapter 9). These genetic differences are responsible for the observed morphological and behavioural diversity of normal individuals in a population. As they are not numerous, they do not affect competitive capacity of individuals. Decrease of competitiveness in decay individuals is due to a critically large number of such differences, i.e. a large number of genetic defects (Chapter 9).

Decay individuals with various defects in their organisation may remain perfectly viable and capable of reproduction outside their natural population, while inside the population their activities are suppressed by normal competitive individuals. Many domestic animals, e.g. cows or pigs, appear completely uncompetitive when deprived of human care. They thus cannot compete with their wild relatives and are not capable of maintaining stable populations in the wild.

This brings up the following important question: What is the defect (malformation, ugliness) and what is the norm? Perhaps this opposition is purely conventional, and what seems to be ugly under certain conditions becomes the norm in some others, and vice versa.

A possible way to get an insight into this problem is to analyse the state of the matter in the human population. In humans a variety of defects of individual organisation are known. Most of them are called diseases and pertain to mental and morphological properties of the individuals. There is no doubt that diseases lower the competitive capacity of individuals under all conditions, and thus are absolute defects. This is very clearly manifested in the fact that no person would ever prefer to be affected by any known disease. There are also behavioural defects observed in murderers, thieves, terrorists and other criminals. The wellbeing of any society can be monitored by its capacity to struggle effectively with such defective individuals. No society will be prosperous and self-sustainable if the majority of its population consists of murderers or thieves. Thus, these behavioural defects are also of an absolute nature.

On the other hand, people classify as *normal* all individuals capable of performing a certain amount of work aimed at maintaining the stability and wellbeing of society. It is due to work of normal individuals that the artificial environment created by modern civilisation does not degrade and demonstrates a tendency for further progress. Work performed by normal individuals in any society is not chaotic, but can be characterised by specific direction and quantity. Any deviation from the generally prescribed fashion may undermine the existing stability of society. Thus, for the human population, the notions *norm* and *defect* (or decay) can be defined with sufficient clarity, i.e. as those pertaining to stability and degradation, respectively. As we will now show, an analysis of populations of biological species living in the wild yields essentially the same result.

All organisms are characterised by a specific internal milieu. Healthy internal milieu of the organism determines correct interaction of the organism with the external environment that comprises, among other things, individuals of the same as well as of different species. The internal milieu of any organism is kept in a healthy state thanks to the correlated work of macromolecules and organelles in unicellular organisms and of internal organs in multicellular organisms. This correlated work is made possible as a result of the consumption of ordered energy supplied from the external environment.

Ecological communities represent correlated associations of individuals belonging to different species. Correlated functioning of such individuals (community members) keeps the internal milieu of the community (e.g. such parameters as temperature, light regime, humidity, availability of biogens, see Section 6.7.1) in a stable state suitable for life for all its members. However, ecological communities, unlike multicellular organisms, cannot be characterised by a pronounced closed boundary delimiting the internal milieu of the community from its external environment. Internal milieu of ecological communities represents an external environment for all types of correlated associations of lower levels of organisation, e.g. multicellular organisms. Thus, by maintaining their own internal milieu, ecological communities stabilise the external environment for all organisms of the biosphere. Correlated interaction of individuals in ecological communities is, therefore, the most complex work performed by living objects in the biosphere.

Viability of cells is maintained by a strictly specified set of macromolecules and organelles. Viability of a multicellular organism is maintained by a strictly specified set of internal organs. Elimination of any of them or their substitution for alien structures adversely affects the state of the organism and may even cause its death. In very much the same manner, normal functioning of an ecological community can be maintained only by a strictly specified set of species characterised by strictly specified population densities of their individuals. Elimination of an aboriginal species or introduction of an alien one disturbs the normal functioning of the community and, consequently, impairs its internal milieu. This leads to a decrease in competitiveness of this particular community. As a result, the latter is forced out from the population of communities by a normal community which contains all the necessary aboriginal species and does not include any alien ones.

Consequently, not all the species that are able to adapt to an external environment are able to survive in the biosphere. Only those species that are able to perform certain specific work on the stabilisation of the environment in the framework of some ecological community have a chance to persist. It is the correlation of an individual of a given species with other individuals in a community that determines the meaning of the notion *norm* for this particular species. Any substantial disturbance of such correlation is a *defect*. Thus, as well as in the human population and biosphere as a whole, the notions *norm* and *defect* refer to the stabilisation or destabilisation of a given environment, be this environment of social (human society) or biogeochemical nature.

2.10 THE QUANTUM NATURE OF LIFE

An indispensable condition of population stability is the ability of individuals to make copies of themselves, i.e. to reproduce. As mentioned above, this ability is not unique to living objects (see Section 2.1).

All living objects are super-organised structures and, inevitably, are subject to rapid decay. On the other hand, life has been existing during several billions of years. At first sight, there is a contradiction between these two statements. It seems puzzling how copying of continuously-decaying, super-organised objects might help to support the existing level of organisation. If we, for example, tried to build new technological devices by copying devices already in operation, the obtained copies would necessarily feature a lower degree of organisation than the original device due to continuous ageing and wear of the latter. Making copies of absolutely new devices does not rescue the situation, because the process of ageing (decay) is of a continuous nature. Therefore, every device already contains certain elements of decay almost immediately upon completion of its construction. The same statement is true for all living organisms. Objects changing noticeably over time periods shorter than the possible time of their copying are traditionally known as *classical* objects.

Conservation of the existing level of organisation by stabilising selection and even its increase in the course of evolution, are only possible due to the quantum nature of molecular processes within the cell. Decay of excited states of molecules, i.e. transition from excited to ground states, occurs by quantum jumps. The probability of the decay of an excited molecule per unit time is a fundamental constant independent of time or prehistory of the current state of the molecule. Time-independence of the decay constant is a well-established empirical fact. The decay constant is proportional to the so-called half-life period of molecules, i.e. to the time needed for half the initially existing excited molecules to decay. Quantum character of decay means that a molecule not yet decayed does not change in any way. It neither ages nor wears away. A copy of such a molecule, made at any arbitrary moment of its life, is absolutely identical to the original. In that sense a molecule does not have any age. That non-trivial empirical fact forms the basis for the quantum-mechanical theory (Schrödinger, 1945).

Identity is a characteristic that can be tested experimentally. The identity of two objects separated by space or time means that no increase in experimental accuracy may reveal any difference between them. If there were no such phenomenon as identity of atoms and molecules, chemical elements and substances would be incapable of conserving their properties. For example, gold would not be the same as it had been several hundred or a thousand years ago. Two objects not identical to each other may be experimentally resolved. For example, two molecules of hydrogen, one of them containing two identical protons and another a proton and a deuteron (a nucleus consisting of a proton and a neutron) differ in their electronic energy levels and hence in their radiation spectra.

It is the quantum nature of DNA molecules—major carriers of hereditary information in the organism—that makes it possible to arrange copying of biological objects not accompanied by loss of level of organisation.

Functioning of unicellular organisms is governed by information written in their DNA molecules. Activity of DNA molecules is regulated by catalysts, the synthesis of the latter being also coded in the DNA. In the course of their activities, DNA molecules may suffer quantifiable decay changes that determine their ageing and wear. (Decay, wear and ageing all mean essentially the same in this context.) Ageing of DNA and ageing of living cells are of a different nature. Working elements of the cell—molecular catalysts and macroscopic organelles—are synthesised in the cell according to the information written in the DNA. They wear out during functioning and no longer correspond to the DNA information. Molecular catalysts and organelles wear out much faster than DNA. The lifetime of catalysts is generally much shorter than that of the cell. It is therefore possible to decrease the relative number of decay catalysts in the cell by increasing the rate of their replacement with new copies made on the basis of the DNA. Meanwhile, the lifetime of cell organelles and other macroscopic structures like cell membrane is of the order of lifetime of the cell itself. The only way to reduce the relative number of decay organelles is by cell division. In the daughter cells, about half of their macroscopic structures are newly synthesised on the basis of the relatively stable information of their DNA. Thus, regulation of the rate of cell division makes it possible to keep the relative number of decay organelles at a constant low level. Note that such a process does not require formation of a population of cells. In the absence of divisions, the cell rapidly ages owing to decay of its organelles and finally dies.

Absence of ageing of organelles in continuously-dividing unicellular organisms is often mistakenly interpreted as immortality. In reality, however, division of cells is accompanied by the inevitable decay of their DNA molecules. The rate of DNA decay is slow and does not catch the eye, provoking erroneous conclusions about immortality. However, in the absence of stabilising selection of organisms, decay of DNA molecules, slow as it is, eventually results in loss of viability and death of cells. Decay of DNA can be only prevented by competitive interaction of individuals in a population. This possibility of preventing DNA decay is due wholly to the quantum, saltatorial nature of change of DNA molecules. In a sufficiently large population of unicellular organisms that are offspring of a single cell, there are always some cells containing DNA molecules that are absolutely identical to the DNA of the initial normal cell. Competitive interaction of cells results in elimination of cells with decay DNA. As a result, the level of organisation determined by the normal DNA remains unchanged for an arbitrarily long period of time.

An important point is that if the nature of DNA decay were not quantum, any part of the DNA of absolutely all offspring of a single individual would necessarily contain some elements of decay. There would be no offspring with any part of the DNA identical to that of the parent cell. No competitive interaction would then help to maintain the initial level of organisation (Eigen, 1971; Orgel, 1992).

Competitive interaction of cells may assume various forms. In particular, it is possible that a cell generates an 'escort' of rigidly-correlated subordinate cells that create and maintain conditions favourable for the main cell that contains genetic information necessary for reproduction of the whole system of cells. Such a structure is characteristic for multicellular organisms, where the most part of the work necessary to maintain everyday existence of the organism is performed by somatic cells that do not take part in the reproduction of the organism. The 'main' cells, i.e. cells transmitting genetic information to the offspring (e.g. ovicells in females of mammals), are maintained under favourable mild conditions, where the rate of DNA decay is minimised.

Several different strategies of existence of a multicellular organism are observed in nature. In some organisms such as birds or mammals, the growth of the body stops by a certain age. In most organs of such organisms, division of somatic cells is strongly suppressed or completely stopped. In such organisms, ageing is due to accumulation of macroscopic decay structures in somatic cells. Such a process occurs more rapidly than decay of the cell DNA. Lifespan of these organisms is determined by the lifespan of their somatic cells.

Bodies of individuals of other biological species (e.g. most fishes), grow throughout their entire lifespan. In their organs, the division of somatic cells is not suppressed. Decay organelles of old somatic cells are replaced by normal, newly-synthesised organelles of new somatic cells. The lifespan of such organisms is determined not by the rate of decay of organelles as in the first group of organisms, but by the rate of decay of cellular DNA. As a result, the lifespan of such organisms can be substantially longer than that of birds and mammals characterised by the same average metabolic rate. However, decay of the DNA itself cannot be prevented in multicellular animals, because it is impossible to arrange a competitive interaction of somatic cells within an animal organism without impairing its normal functioning which is based on rigidly correlated work of somatic cells. When such competitive interaction does appear in cancer tumours, it often leads to death of the organism.

Finally, weakly-correlated organisms like plants can afford to arrange competitive interaction of some parts of the organism (leaves, branches, roots). Somatic cells with decay DNA can be eliminated from the organism without causing major changes in it. That is why tumours in plants seldom lead to the death of the organism. The same type of organisation is characteristic for correlated associations of multicellular organisms, e.g., social structures of insects, herds of mammals and, finally, ecological communities. The lifespan of such weakly-correlated biological objects is generally longer than in the first two groups of organisms. It is determined by the time of normal functioning of the unifying part of the object (e.g. trunk or stem in vascular plants or queen in a beehive).

Let us now consider major principles of information storage and transmission in biological and other systems. Clearly, DNA molecules can only be used by life due to the fact that they decay more slowly than the time between two successive acts of DNA copying. Were DNA molecules to represent classical objects that decay faster than they can be copied, transmission of genetic information from parents to offspring could not be performed on the DNA basis. Similarly, it is fundamentally impossible to build life on the basis of copying purely classical, continuously-ageing multicellular individuals, without recourse to the quantum nature of a DNA molecule of a single cell. Thus, development of any multicellular organism necessarily starts from a single cell and inevitably goes through stages of infancy, adolescence, adulthood and death.

In very much the same manner, any type of non-genetic information written in memory cells of classical nature cannot be prevented from decay by stabilising selection. Information of individual memory of animals and humans that is accumulated during the lifespan of individuals is an example of such information. The information of individual memory cannot be copied without loss.

In modern computers, loss of information is prevented owing to the macroscopic size of memory cells. A reliable storage of information in a computer memory cell is achieved due to collective interactions of a huge number of molecules of the condensed matter that constitutes the cell. For example, storage of one bit of information may correspond to a state when all molecules of the computer memory cell are characterised by the same magnetic moment imparting certain non-zero value of magnetisation to the memory cell as a whole. One or a few molecules may, of course, change their magnetic moment either spontaneously or being induced to do so by computer operations in which the memory cell is involved. However, due to a huge number of molecules in the macroscopic memory cells, changes pertaining to a small number of molecules cannot affect the state of the cell as a whole and will not lead to loss of information. In that sense, the macroscopic character of information-carriers is essentially equal to multiple back-up of stored information.[2] Thus, macroscopic memory cells make it possible to ensure reliable storage of information. However, the macroscopic size of memory cells rigidly restricts both memory capacity and the information-processing capacity of modern computers, as compared to molecular memory cells used by biological objects. Transition to molecular memory cells in computer technology (i.e. abandonment of the principle of multiple back-up of stored information) will have to face the problem of rapid erosion of information. This can be only helped by creation of mechanisms specialised in recovery of the partially-lost information, similar to the catalytic system of DNA reparation used by life in higher organisms.

2.11 THE ECOLOGICAL COMMUNITY AS THE HIGHEST LEVEL OF BIOLOGICAL ORGANISATION

As noted in Section 2.4, high competitiveness and high level of orderliness of a biological object do not automatically follow from each other. Competitiveness essentially means the ability of the object to dominate over similar objects, suppressing their activities in one way or another. Orderliness means the inherent complexity of biological objects that enables them to persist in the environment. It includes their ability to form populations, to feed, to reproduce, to escape danger, etc.

It follows that competitiveness represents a more simple property of living beings than their orderliness. One can say that competitiveness is as simple as the process of destruction (elimination, forcing out, suppressing, killing, etc. of rivals), while

[2] Human language and other types of cultural heritage in humans represent another example of multiple back-up of information aimed at prevention of its loss. Most members of any society possess a certain basic knowledge of language and culture. Misuse of language by small group of individuals cannot lead the degradation of the language system as a whole.

orderliness is as complex as the process of creation. For example, in a hypothetical highly-ordered society consisting of honest people only, a liar suddenly appearing will become the most competitive individual, because all the people, used to believing each other, are vulnerable to lies. However, such a liar will not contribute to the stability of the society and will only lower its level of organisation. In this sense, the highly-competitive liar contributes to degradation of the society.

Thus, the important question that arises is, how could the high organisation of living beings persist and even increase through time, if among all possible ways of increasing competitiveness of a biological object, only few correspond to an increase in its orderliness and higher level of organisation? What made it possible to couple competitiveness and the high level of organisation of living beings?

Competitiveness of a living individual is a function of two variables, $C = f(P, E)$. Firstly, it is a function of the properties of the individual itself, i.e. its phenotype P. To be competitive, the individual must be healthy, strong, fertile, etc. Secondly, competitiveness of the individual is a function of the state of the environment E where it exists (see Section 9.8). Individuals capable of swimming are competitive in an aquatic environment, while those capable of walking are competitive on land, but not vice versa. The maximum competitiveness corresponds to a combination of an optimal phenotype with an optimal environment, $C_{max} = f(P_{opt}, E_{opt})$.

As noted in Section 2.7., life is impossible without metabolic processes of matter and energy exchange taking place within the body of living beings. Thus, any living being necessarily imposes a certain impact on its environment and changes it. Being highly competitive in the initial moment, individuals may lose their competitiveness and become extinct when the environment (i.e. the second factor determining their competitiveness) considerably changes, even though the optimal phenotype P_{opt} remains the same. For example, the presently most-competitive branches of industry based on fossil fuel burning will inevitably become extinct as soon as the fossil fuel is exhausted. In this sense, they undermine their competitiveness, themselves changing the optimal environment E_{opt} where they may remain competitive.

In such a situation it is clear that only such life has a chance to persist that is represented by individuals able to support both P_{opt} and E_{opt}, i.e. able to control both factors that determine high competitiveness of living beings. Thus, an inherent property of life must be the ability of living beings to control their optimal environment. Individuals achieve the highest competitiveness C_{max} possible under the given environment choosing the most favourable phenotype P_{opt} and further maintain their optimal environment E_{opt} in a stable state. Only then, their maximum competitiveness is guaranteed a long-term stability.

As soon as regulation of the environment cannot be performed by individuals of one species only (that function on the basis of open matter cycles), we come to the conclusion that living beings should be organised in ecological communities that are able to ensure complex reactions to random perturbations of the optimal environment returning it to the initial optimal state. To maintain the environment in a stable optimal state, a very high level of organisation is needed. Thus, the ecological community represents the highest level of biological organisation tightly coupled with high competitiveness (due to control of E_{opt}). The fact that competitiveness is a function of environmental conditions makes it possible to couple competitiveness and high level of organisation necessary for existence of life.

The major principles of organisation of ecological communities are addressed in the following chapters.

3

Ecology of Organisms with Different Body Sizes

Why is the natural biota composed of such differently-sized organisms, from microscopic bacteria to huge whales? What role does this diversity of sizes play in the biotic regulation of the environment? How are the biotic energy fluxes distributed over differently-sized organisms? What is the size of ecological community? These and related questions are addressed in this chapter.

3.1 METABOLIC POWER OF INDIVIDUALS

Life of all individuals is maintained as a result of the continuous consumption of an external flux of ordered energy. When external energy is supplied to the organism in the form of energy-rich matter, this supporting flux of energy is called *nutrition*. Plants are able to consume solar energy which is linked to massless photons. Within living organisms the external flux of energy is used to generate ordered processes that assume the form of biochemical reactions. The ordered, non-chaotic nature of such reactions is dictated by molecules of catalysts synthesised by the organism. A catalyst is a chemical substance capable of accelerating a particular reaction without being spent in the course of the reaction itself.

Reaction rate is proportional to the relative number of molecules having energy E above a certain potential barrier B, $E > B$, which characterises each particular reaction under given conditions. At a given temperature T the number of molecules having energy E follows Boltzmann distribution and is proportional to $e^{-E/k_B T}$, where k_B is the Boltzmann constant and $k_B T$ characterises the average thermal energy of molecules. Expression B/k_B can be interpreted as the effective 'barrier' temperature T_B of the reaction, $T_B \equiv B/k_B$. For most biochemical reactions, this temperature is close to the temperature of the Sun, $T_B \sim 6000$ K, see below (3.1.4). Body temperature is of the order of $T \approx 300$ K, so that $T_B/T \approx 20$. Thus, the relative number of molecules that could spontaneously enter the reaction is of the order of $e^{-T_B/T} \approx 10^{-10}$, which makes the reaction absolutely impossible.

A catalyst lowers the potential barrier B down to the average thermal energy of molecules $k_B T$. This enables practically all molecules to enter the reaction, so that the reaction rate is increased by $e^{T_B/T} \approx 10^{10}$ times. In the absence of a catalyst, spontaneous biochemical reactions would never take place. Thus, synthesising necessary catalysts, the organism controls all the biochemical reactions within it, and strictly routes the decay of consumed ordered energy into prescribed channels. Uncontrollable reactions that run spontaneously in the absence of catalysts cannot be used by life.

All biochemical processes in the biosphere may be reduced to synthesis and destruction of organic matter. During the synthesis of organic substances from inorganic ones, energy is absorbed from the environment. Conversely, energy is released in the process of the destruction of organic matter. The energy content of organic substances (the calorific coefficient) naturally differs from substance to substance (hydrocarbons, fats, proteins, etc.). However, on average, the energy content per unit mass of organic carbon K is approximately constant for most organisms in the biosphere and constitutes (Odum, 1983):

$$K = 42\ \text{kJ}\ (\text{gC})^{-1} \tag{3.1.1}$$

On average, the body mass of living organisms exceeds by a factor of ten the mass of organic carbon, if we account for free water content of live matter. This relationship is used in all the computations below. Thus we may assume an energy content of a unit live biomass K_{lb}:

$$K_{lb} = 4.2\ \text{kJ}\ \text{g}^{-1} \tag{3.1.2}$$

Deviations from this average may, in the extreme, reach a factor of two (Whittaker, 1975; Winberg 1979; Odum 1983).

The energy content coefficient K_{lb} may be given an interesting interpretation. Its dimension $\text{kJ}\,\text{g}^{-1}$ coincides with that of squared velocity. We thus may formally write:

$$K_{lb} \equiv \frac{\omega^2}{2} = 4.2 \cdot 10^6\ \text{m}^2\ \text{s}^{-2}, \qquad \omega = 3 \cdot 10^3\ \text{m}\ \text{s}^{-1}$$

If all the energy content of the living body of an individual were transformed into kinetic energy of its movement, the individual would move at a speed of $\omega = 3 \cdot 10^3\ \text{m}\,\text{s}^{-1}$, which is 10 times faster than the speed of sound. Another interpretation is that the obtained value of ω is equal to the average speed of molecules after sudden combustion of all the organic matter of the living body. Recalling that water accounts for about three-quarters of the living matter, we may relate the obtained velocity ω to temperature, following standard relationships of the kinetic theory of gases according to which the average kinetic energy of molecules is proportional to their temperature:

$$\tfrac{3}{2} R T^{+} = M \frac{\omega^2}{2}, \tag{3.1.3}$$

where $R = 8.3\ \text{J}\ \text{K}^{-1}\ \text{mole}^{-1}$ is the gas constant, $M = 18\ \text{g}\,\text{mole}^{-1}$ is the molar mass of H_2O. The maximum possible efficiency of energy use η is given by the well-known Carnot formula and is equal to the relative difference between the effective temperatures of the initial highly-ordered and the final low-ordered states, T^+ and T_E, respectively. Taking T_E equal to the average temperature of the Earth's surface, $T_E \approx 288$ K $\approx 15°$C, and using (3.1.3) and the Carnot formula we obtain for the maximum efficiency of energy use that is characteristic for organic matter:

$$T^+ = \frac{M\omega^2}{3R} \approx 6000\ \text{K}, \qquad \eta = \frac{T^+ - T_E}{T^+} \approx 0.95 \tag{3.1.4}$$

The temperature T^+ that characterises orderliness of the chemical energy of organic matter, is equal to the temperature of the Sun. Thus, the maximum possible efficiency at which such energy can be used coincides with that for solar energy, see Sections 7.2 and 8.7.

To support biochemical reactions within the living body, each individual uses energy released during decomposition of organic matter. This leads to an exchange of substances between the individual and its environment. Rate of food consumption by an individual is usually measured as the rate of consumption of organic carbon, Q, kg C year^{-1}which is called *metabolic rate* (Brody, 1945; King, 1974). The rate of energy consumption by an individual (respiration) may be called metabolic power q, W. Variables q and Q are interrelated as follows:

$$q = QK \tag{3.1.5}$$

Using the calorific coefficient K (3.1.1) we find that the metabolic rate of 1 kg C year^{-1} or of about 10 kg of food per year corresponds to the metabolic power of 1.3 W.

In spite of the fact that catalysts decrease the potential barrier of all catalytic biochemical reactions very substantially, a certain low energetic barrier still persists. Thus, rates of all biochemical reactions in the organism increase exponentially with rising temperature in accordance with a Boltzmann distribution. This effect is usually described using the empirical coefficient Q_{10} which describes the relative increase in metabolic power $q(T)$ per 10°C of temperature increase:

$$Q_{10} \equiv q(T_0 + 10°\text{C})/q(T_0)$$

For most species the value of Q_{10} varies between two and three (Winberg, 1976).

The individual's metabolic power may vary over quite a wide range. However, it cannot drop below a certain minimal value q_0 known as the basal metabolic power (Brody, 1945; Winberg, 1976; McNab, 1983).[1] Basal metabolic power is observed in resting individuals some time after food consumption. The maximum metabolic power achieved by an individual per unit of its body mass is determined by the maximum rate of biochemical reactions in the living cell and should be universal for all kinds of living beings. The average metabolic power of existence for most species

[1] Note that in literature both Q and q are sometimes called metabolic rate, in spite of the fact that these two variables have different dimensions. The basal metabolic power is often called basal metabolic rate. However, we consider it more correct to discriminate between metabolic *power* which refers to energy consumption and metabolic *rate* which refers to consumption of nutrients.

of the biosphere approximates $2q_0$ (Calder, 1974; King, 1974; Winberg, 1976; Panteleyev, 1983; Gavrilov, 1994). The maximum possible metabolic power maintained by continuous oxygen consumption is limited by the maximum possible rate of food consumption by the individual. In most cases the maximum metabolic power reaches $4q_0$, which is twice as large as the average metabolic power of existence (Gavrilov, 1994).

An average-sized man is characterised by basal metabolic power of about $q_0 = 80$ W and average metabolic power of existence of about 140 W. In the regime of oxygen debt a sprinting man may put out power in excess of 10 kW (Brody, 1945).

3.2 BODY SIZE LIMITS

Metabolic processes must support functioning of all the living tissues of an individual. Of principal importance is the value of metabolic power per unit *mass* or *volume* of an individual (the mass-specific rate) (Brain and McNab, 1980). Body mass m is equal to ρV, where $\rho \approx 1\,\mathrm{g\,cm^{-1}}$ is the body density, which may differ from water density by no more than several per cent. Thus, in the common units of measurement, body mass is numerically equal to body volume. Since all the external energy enters the individual's body through its surface, it is also important to consider metabolic power per unit *area* of the average living body projection upon the Earth's surface. (Area of such projection differs from the total surface area of the body by a constant multiplier.)

We therefore introduce *effective vertical size* of an individual, l, and its *projection area at the Earth's surface*, s, as follows:

$$l \equiv V/s, \tag{3.2.1}$$

where V stands for the volume of metabolically active parts of the individual, i.e. the volume where biochemical reactions actually take place. For example, wood of a tree trunk, where biochemical reactions are practically absent, should not be included into the metabolically active volume V of the tree. For animals, the metabolically active volume approximately coincides with the total body volume, because metabolically inactive tissues (fur, antlers, claws) constitute a minor part of the animal organism. For locomotive animals their projection area s, averaged over a period of movement, is of the order of l^2. In contrast, for green plants and fungi their effective vertical (l) and horizontal ($\sqrt{s}$) sizes differ by several orders of magnitude. The volume of space occupied by a green plant consists overwhelmingly of air and metabolically inactive tissues forming the frame of the plant, e.g. wood in trees. The average effective vertical size of terrestrial vegetation l appears to be of the order of several millimetres (!), that is much lower than the visible height of green plants H, which is usually of the order of $\sqrt{s}$.

Metabolic powers per unit volume (λ, volume-specific or mass-specific metabolic rate) and projection area (j, surface-specific metabolic rate) of an individual are given by:

$$\lambda \equiv q_0/V, \qquad j \equiv q_0/s \equiv \lambda l \tag{3.2.2}$$

Note that $\lambda \equiv j/l$ by definition. If the fluxes of energy and matter consumed by an individual were dictated solely by the external environmental conditions (e.g. if nutrients were delivered to the individual by physical diffusion), the value of metabolic power per unit projection area j would be common for individuals of any body size. Hence, the metabolic power per unit volume λ would then decrease with growing body size l as $1/l$.

Meanwhile, it is naturally appealing to the universal biochemical organisation of life that the power of biochemical reactions in a unit volume of the living body, i.e. λ, should be approximately constant for all living beings. Thus, with environmental conditions dictating the value of j, and the value of λ remaining biochemically fixed and universal, we may then have organisms of only one strictly determined size

$$l = \lambda/j. \qquad (3.2.3)$$

In fact, the metabolic power per unit body volume varies within quite wide margins depending on the species. However, by all appearances, these margins are universal for all the living world, so that the body size l may only vary within strictly determined limits for a given value of j, see Figure 3.1. Second, the majority of species in the biosphere tend to have a value of λ close to an optimal value that seems to be universal for life as a whole. Therefore, at fixed j one may speak about an optimal body size for living beings (Figure 3.1).

Heterotrophic individuals, i.e. those feeding on organic matter synthesised by other individuals, may change the value of surface-specific metabolic power j by radically altering the organisation of their bodies. They may transit from passive feeding based on physical diffusion of nutrients (as is the case for bacteria and fungi), to active mechanical consumption of food with creation of digestive and distributing systems of internal organs (as in locomotive animals).

In the process of evolution, several fundamental principles of increasing the value of j and, consequently, of the optimal body size, were discovered. As a result, the following five taxonomic groups of organisms have principally different organisations of metabolic processes: bacteria, fungi and plants; invertebrate animals; cold-blooded vertebrate animals; warm-blooded vertebrate animals excluding passerine birds; and passerine birds (*Aves passerines*) (Figure 3.1). Each of these groups is characterised by its own value of surface-specific metabolic power j, and by an optimal body size at fixed optimal value of volume specific metabolic power λ_{opt} (Figure 3.1). In each of these groups, the body size may vary within the limits defined by the minimum and maximum observed values of λ, λ_{min} and λ_{max}, see (3.2.3). The available empirical data for the presently- and previously-existing animals of a given taxonomic group give good grounds to assert that the observed values λ_{min} and λ_{max} are universal for life as a whole.

The observed minimum values λ_{min} are obviously determined by the basal metabolic power q_0. These values limit the maximum possible individual body size in each taxonomic group (Figure 3.1). Extending the obtained basal metabolic rate lines at their known slopes to the maximum known body size ever encountered in organisms of a given taxonomic group, we find that the values of λ_{min} for three out of the four taxonomic groups coincide (see lines 1, 2, 3 in Figure 3.1). This fact

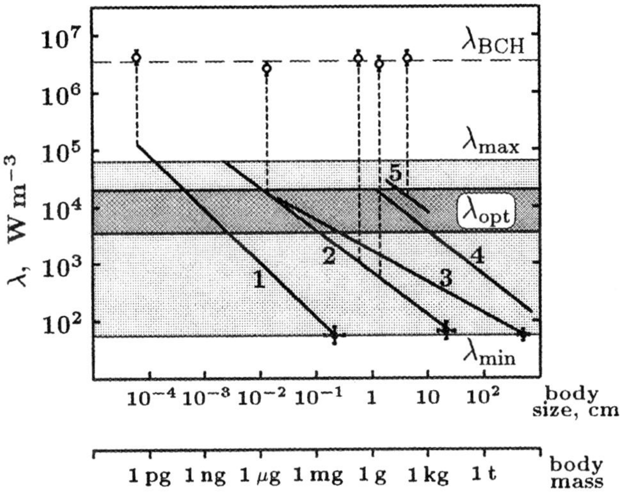

Figure 3.1. The dependence of the volume-specific basal metabolic power, $\lambda \equiv q_0/V$, on individual effective vertical body size l. 1 plants, bacteria, fungi, protozoa (without infusoria); 2 infusoria and multicellular invertebrates; 3 cold-blooded vertebrates and cephalopods; 4 warm-blooded animals (without passerine birds); 5 passerine birds (Brody, 1945; Hemmingsen, 1960; Kleiber, 1961; Winberg, 1976; McNab, 1983; Gorshkov, 1981; 1983a; 1984a; Kanwisher and Ridgway, 1983). For plants the basal metabolic power was taken equal to that of dark respiration (Govindjee, 1982). The dark shaded band λ_{opt} includes more than 95% of species in each taxonomic group (Chislenko, 1981). The minimum value of volume-specific metabolic power λ_{min} was obtained by extrapolating lines 1, 2 and 3 to the maximum effective vertical body sizes (filled circles) observed within each taxonomic group. Open circles show the maximum possible volume-specific metabolic power (the biochemical limit) λ_{BCH} ever achieved by living beings. It corresponds to metabolic power of bacterial division and the power achieved per unit volume of the push muscle during the highest possible jumps of different animals (Gorshkov, 1983a).

suggests that the obtained value of λ_{min} is indeed the sought-for universal limit of volume-specific metabolic power. It is interesting to compare the obtained universal value of λ_{min} (Figure 3.1) with volume-specific 'metabolism' of the Sun, which constitutes 0.27 $\mathrm{W m^{-3}}$ (Allen, 1955), i.e. only 5% of λ_{min}.

In some situations, individuals have to put out their maximum possible metabolic power. This power cannot exceed a certain biochemical limit λ_{BCH}, which must, by all appearances, be universal for life in general being determined by basic properties of the living cell. The maximum power exerted by locomotive animals can be measured during their longest or highest jumps. In bacteria the maximum possible power corresponds to the highest rate of cell division,[2] which is about one division per fifteen minutes (Lewin, 1987). The available experimental data show that the maximum volume-specific metabolic power exerted by bacteria in the process of dividing, and by animals of different body size during jumps, e.g. flea, grasshopper,

[2] In order to increase the mass of the body twice, an organism must consume food in quantities not less than its own mass. Thus, the mass (or volume) specific metabolic power λ and time of cell division τ are related as $\lambda = mK_{lb}/\tau$, see (3.1.2).

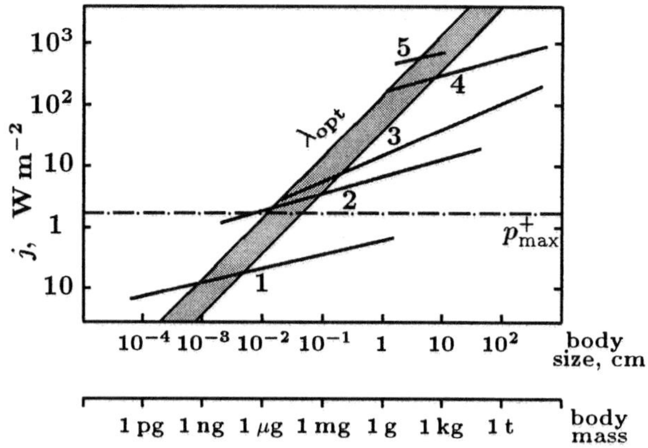

Figure 3.2. The dependence of the surface-specific basal metabolic power, $j \equiv \lambda l$, on individual effective vertical body size l. Solid lines correspond to the same taxonomic groups as in Figure 3.1. The shaded band corresponds to optimal volume specific metabolic rate λ_{opt}, see Figure 3.1. The dash-dot line corresponds to the maximum net primary productivity encountered in the biosphere. Note that organisms that exist above the dash-dot line (i.e. organisms of all the taxonomic groups except for the first one) cannot in principle be immobile, because their metabolic rate of consumption exceeds the maximum possible productivity of immobile plants. These organisms have to move and destroy the biomass accumulated over longer periods of time (see also Section 3.6).

locust, kangaroo rat and bushbaby, all coincide (Gorshkov, 1983a). This fact makes it possible to draw the biochemical limit line, λ_{BCH}, in Figure 3.1.

The maximum metabolic power of existence, λ_{max}, observed in each taxonomic group, should be substantially lower than the biochemical limit λ_{BCH}. It cannot differ greatly from the optimal metabolic power λ_{opt}. The value of λ_{max} in Figure 3.1 is assumed to be equal to the maximum metabolic power of existence observed for eucaryotes, i.e. to the metabolic power of existence observed in infusoria of the smallest size (Gorshkov, 1981).

If the surface-specific metabolic power j did not change at all within each group, the volume-specific metabolic power λ would have decreased as $1/l$ with increasing body size, see note to Eq. (3.2.2). In reality, however, a slight growth of j is observed in each taxonomic group (Figure 3.2), which slows down the decline of λ with increasing body size. As a result, individuals of larger body size than that determined by λ_{max} (3.2.2) may appear within a given taxonomic group. It has been empirically found that all the dependencies of j (and, consequently, of λ) upon body size l are governed by allometric relationships, that are represented as straight lines in bilogarithmic scales (Figures 3.1 and 3.2). Slopes of these lines depend on particular morphological properties of multicellular individuals within each taxonomic group.

Figure 3.2 makes it explicit that surface-specific power of existence j, i.e. metabolic power per unit area of body's projection at the Earth's surface, exceeds the maximum productivity of plants (the dash-dot line p^+_{max} in Figure 3.2) for all taxonomic groups except the first one. This means that species belonging to the four groups must be locomotive and feed at the expense of the destruction of the community's *biomass*. Only species of the first group (bacteria, fungi and plants themselves) can afford to be immobile and feed on primary *production* (see also Section 3.6).

3.3 ENERGETICS AND BODY SIZE OF PHOTOSYNTHESISING PLANTS

The metabolic power of photosynthesising plants is due to the absorbed solar radiation. The absorbed light is spent on synthesis of organic substances from inorganic ones in the course of photoreactions. Thus the amount of solar power consumed by a green plant should be proportional to the import of inorganic substances needed for such synthesis. The known calorific coefficients obtained from matter exchange measurements make it possible to relate import of inorganic substances to the gross primary production of organic matter. Thus, one may retrieve the value of the metabolic power of green plants from the known value of the gross primary production of organic carbon. Its density per unit Earth's surface (*gross primary productivity*) may be denoted as P_g^+.

If mass of a plant remains constant, all the synthesised organic matter is spent on respiration within the plant itself. In such a case, import and export of nutrients from and to the environment can be completely compensated. However, in contemporary plant species no such compensation is observed and, if metabolic processes do take place, the mass of a plant continuously increases. The plant mass increment is its net primary production (this is the part that can be consumed by heterotrophs), and we shall denote its density per unit Earth's surface (*net primary productivity*) by P^+ and the respective power density by $p^+ = KP^+$ see (3.1.1). Here '+' refers to synthesis.

Plant respiration R^- supporting existence of the plant itself is limited from above by the amount of incoming solar energy. Plants are immobile and their way of feeding is passive. This means that they may only consume the needed nutrients from the environment via physical diffusion, similar to fungi and bacteria. The available empirical data suggest that for photosynthesising plants the dependence of their metabolism (respiration) per unit projection area ($j \equiv KR^-$) and unit body volume (λ) on the effective vertical size l, coincides with the respective dependence for bacteria and fungi (Gorshkov, 1981, and see Figures 3.1 and 3.2). In contrast to animals, no advancements in morphology can enable plants to increase their value of j, i.e. the amount of energy consumed per unit projection area. Due to this fact, the effective vertical size of plants l cannot be extended beyond the limits of the first taxonomic group. The effective vertical size of plants cannot thus exceed 1 cm (Figure 3.1). The average effective thickness of metabolically active vegetation layer in the biosphere is of the order of 3 mm. In that sense, plants are dwarfs as compared to higher animals. The visible giant (as compared to l) vertical size of most plant species (e.g. trees), determined by their height, H, may be reached only because the most part of the volume, occupied by a plant, consists of metabolically-inactive tissues.

Solar energy may be transformed into the energy of organic matter with an efficiency close to unity, see (3.1.4) and Section 8.7. However, the observed efficiency

Table 3.1. Net primary production and live biomass of the biosphere.

Ecosystem	S 10^9 ha	P $\mathrm{t\,C\,ha^{-1}\,yr^{-1}}$	$\boldsymbol{P}$ $\mathrm{Gt\,C\,yr^{-1}}$	B $\mathrm{t\,C\,ha^{-1}}$	$\boldsymbol{B}$ Gt C	τ yr	d
Forests, bogs, swamps, marshes	3.5 (6.4)	10	34	140	500	15	8
Grasslands, shrubs	4.7 (3.5)	4	20	15	70	4	4
Arable lands	1.4 (0.1)	4	5	4	5	1	4
Lakes, rivers	0.2	3	0.5	0.4	0.1	0.2	2
Deserts, tundras	5.2 (4.8)	0.2	1	0.4	2	2	0.7
Continents, total	15	4	60	40	600	10	4
Open ocean (pelageal)	33	1	30	0.01	0.5	0.01	0.6
Coastal waters (shelf)	3	3	9	0.3	1	0.2	2
Ocean, total	36	1	40	0.04	2	0.05	0.8
Earth, total	51	2	100	12	600	6	2

Notes to Table 3.1:
- 1 ha = $10^4\,\mathrm{m}^2$; 1 Gt = 10^9 t = 10^{12} kg
- S is the area occupied by the ecosystem
- P is the net primary productivity of the ecosystem (Whittaker and Likens, 1975; Ajtay *et al.*, 1979);
- $\boldsymbol{P}$ is net primary production of the ecosystem ($\boldsymbol{P} \equiv PS$) (Whittaker and Likens, 1975; Ajtay *et al.*, 1979; Degens *et al.*, 1984; Platt *et al.*, 1989; Townshend *et al.*, 1991; Townshend, 1992)
- B is the live biomass per unit surface area in the ecosystem
- $\boldsymbol{B}$ is the total live biomass in the ecosystem ($\boldsymbol{B} \equiv BS$)
- τ is the residence time of carbon in live biomass ($\tau = \boldsymbol{B}/\boldsymbol{P} = B/P$)
- d is the leaf area index (the average number of times the tree's leaves overlap each other along a vertical axis, i.e. how thick (in terms of leaves) would be the tree's foliage when all the leaves are put on the projection area of the tree)
- Bogs, marshes and swamps cover an area of about 10% of the area occupied by forests. In the table they are joined with forests because these ecosystems feature similar values of productivity and biomass (Whittaker and Likens, 1975). Tundra occupying an area of about 10% of that of deserts is joined with the latter for the same reason
- Preindustrial values for land ecosystems are given in parentheses. At present nonperturbed forests occupy about $1.3 \cdot 10^{13}\,\mathrm{m}^2$, i.e. they account for about 37% of the total area occupied by forests and bogs (Bryant *et al.*, 1997)

of photosynthesis $\eta_{ph} \equiv KP^+/I$ never exceeds 7%, while the observed annual average efficiency of photosynthesis for the whole biosphere constitutes about 0.3% (Govindjee, 1982). The available estimates of net primary production and of plant biomass for various Earth ecosystems are given in Table 3.1.

All the nutrients, except gaseous CO_2 and O_2, can be consumed and transported along vessels of terrestrial plants only in the form of water solutions. The amount of water consumed by plants exceeds their net biochemical needs by many orders of magnitude. Therefore, pure water must be constantly removed from plants. This need is satisfied through transpiration—evaporation of water by leaves. The amount of nutrients consumed by land plants and, consequently, their net primary production, thus varies in proportion to transpiration. Due to the high leaf area index of plants, i.e. their high effective evaporating surface, plant transpiration may significantly exceed the level of regular physical evaporation from open water surface. Additional enhancement of transpiration in comparison to physical evaporation is due to the increase of the eddy diffusion in air with height. Higher coefficient of eddy diffusion at higher altitude enables trees to evaporate significantly more water from the surface of their leaves than, for example, do herbs.

The ratio of the mean annual rate of transpired moisture to the annual net primary productivity of plants (measured in $\mathrm{kg\,C\,m^{-2}\,yr^{-1}}$) is known as transpiration coefficient k_T. It is measured empirically and is of the order of 1000 kg H_2O per kgC (Larcher, 1980). Transpiration is proportional to total evaporation from a given surface, which also includes physical evaporation from soil. The transpiration coefficient may be expressed as follows:

$$k_T = \alpha_T \rho E / P^+, \tag{3.3.1}$$

where α_T is the share of transpiration in total evaporation ρE ($\mathrm{kg\,m^{-2}}$) from a unit surface area; E is the observed evaporation rate ($\mathrm{m\,year^{-1}}$); ρ is the water density ($\mathrm{kg\,m^{-3}}$); P^+ is the net primary productivity in units of live mass ($\mathrm{kg\,C\,m^{-2}yr^{-1}}$).

The share of the incident solar radiation power spent on transpiration, η_T, is equal to:

$$\eta_T \equiv \alpha_T \rho E \frac{L_w}{I} = \alpha_T \frac{E}{E_{\max}}, \qquad E_{\max} = I/(\rho L_w) \approx 2000\ \mathrm{mm\ year^{-1}} \tag{3.3.2}$$

Here $L_w = 2.5\ \mathrm{kJ(g\,H_2O)^{-1}}$ is the latent heat of water evaporation (the amount of energy needed to evaporate 1 g of water); I is the flux of solar radiation reaching the Earth's surface; $E_{\max}$ represents the maximum possible rate of evaporation from unit surface area. Owing to (3.3.1) the efficiency of photosynthesis η_{ph} appears to be unambiguously related to η_T:

$$\eta_{ph} \equiv KP^+/I = \frac{K\eta_T}{L_w k_T} \equiv \frac{K}{L_w k_T} \alpha_T \frac{E}{E_{\max}} \tag{3.3.3}$$

where KP^+ is the power of net primary productivity ($\mathrm{Wm^{-2}}$); K ($\mathrm{kJ\,(g\,C)^{-1}}$) is the energy content of living cells (3.1.1). (Note that energy content of wood differs from that of metabolically active plant tissues. However, production of wood constitutes no more than 10% of the net plant production. Account of wood production would not therefore change the results above.)

The values of L_w, K and k_T are fundamental physical and biophysical constants which cannot be changed by the natural biota. Therefore the efficiency of photosynthesis of terrestrial plants is completely determined by the rate of evaporation, E, which depends on the state of continental biota.

In humid areas such as tropical rain forests, plant transpiration accounts for an overwhelming part of total evaporation, and more than 90% of the available solar radiation power is spent on transpiration (Larcher, 1980; Govindjee, 1982). As a result, in these areas the photosynthesis efficiency approaches its observed maximum (Larcher, 1980). Conversely, the efficiency of photosynthesis is sharply reduced in the areas of insufficient evaporation like deserts.

The value of the total land net primary production may be estimated from the known values of the average land evaporation rate $\overline{E_b} = 500\ \mathrm{mm\ year^{-1}}$ (see Figure

6.7), the land surface area (see Table 3.1), and the average annual value of the share of transpiration in total evaporation $\overline{\alpha_T} = 0.6$ (Gorshkov, 1980). Using these values in Eq. (3.3.3) we obtain the average annual photosynthesis efficiency $\overline{\eta_{ph}} = 0.3\%$, the average power of net primary productivity $p^+ = KP^+ \approx 0.5\ \mathrm{W\ m^{-2}}$ and the average land net primary productivity in carbon units $\overline{P^+} = 4\ \mathrm{t\,C\,ha^{-1}year^{-1}}$ in accordance with the independently obtained data of Table 3.1.

As follows from (3.3.2), the maximum possible evaporation rate $E_{\max}$ is four times as large as the observed mean annual rate of evaporation on land $\overline{E_b}$ (see Figure 6.7). The maximum possible rate of evaporation is realised in a situation when all the solar energy incident upon the Earth's surface is spent on evaporation. The maximum values of η_{ph}, p^+ and P^+ corresponding to $E_{\max}$ are 1.2%, $2\ \mathrm{W\,m^{-2}}$ and $16\ \mathrm{t\,C\,ha^{-1}year^{-1}}$, respectively. The observed rate of evaporation from the ocean surface ($E_s \approx 1.3\ \mathrm{m\,year^{-1}}$) is more than twice as large as the contemporary evaporation rate from the total land surface (see Figure 6.7). Artificial restoration of vegetation cover, especially in deserts, could significantly change the global evaporation regime. This could lead to unpredictable changes in the overall temperature regime of the planet (Henderson-Sellers and Cogley, 1982).

3.4 SENSITIVITY OF THE BIOTA

In this small section we address a conceptual issue that is of critical importance to the biotic regulation concept. Biotic regulation is only possible if the biota reacts to changes in the environmental characteristics. It is evident that very small relative changes in these do not alter the functioning of the biota. In simple words, the biota does not notice very small changes. A biotic response to environmental perturbations may only follow if the relative change of environmental characteristics at such perturbations exceeds the sensitivity of the biota. We can define the sensitivity of the biota ε_b as the minimum relative change of a characteristic X (concentration of a biogen, temperature, pressure, etc.) after which the mechanism of biotic regulation is switched on: $\varepsilon_b = \Delta \mathrm{X}_{\min}/\mathrm{X}$. The sensitivity ε_b is a most important characteristic of individual organisms as well as of ecological communities as a whole.

In spite of possible differences in biotic sensitivity with respect to different components of the environment, the value of sensitivity (or, as it sometimes called, resolution) of the biota is on average of the order of 10^{-4}. For example, the body temperature of warm-blooded animals is fixed to the accuracy of a tenth of degree Kelvin. It means that organisms with their average body temperature significantly deviating from the optimum lose competitive capacity. A person with body temperature of 36.6°C is healthy, while a person with body temperature close to 37.0°C already is not. At average body temperature of 36.6°C (324.6 K), reaction of the organism to a change of 0.1°C (0.1 K) corresponds to sensitivity of $\varepsilon_b \approx 3 \cdot 10^{-4}$. Similar changes of environmental temperature are likely to be noticed by the majority of organisms.

During the vegetation season, especially at the beginning of it, we are able to trace daily changes in plant biomass. With total forest biomass of about $100\ \mathrm{t\,C\,ha^{-1}}$ and productivity of about 10 t C $ha^{-1}yr^{-1}$ see Table 3.1, the relative daily change of plant biomass is of the order of $\varepsilon_b \sim$ (10 t C $ha^{-1}yr^{-1}$)/(1 yr/365 days)/(100 t C ha^{-1})$\sim 3 \cdot 10^{-4}$, i.e. the same value as for the temperature sensitivity.

In the next section, we return to the problem of independent estimates of ε_b and confirm that it is indeed of the order of 10^{-4}. Thus we use the value

$$\varepsilon_b \equiv \frac{\Delta X_{min}}{X} \approx 3 \cdot 10^{-4} \qquad (3.4.1)$$

in all subsequent calculations.

Changes in concentrations of inorganic and organic substances in the environment are due to changes in the rates of production or destruction of organic matter, which is accompanied by transition of the major biogens used by life—carbon (C), nitrogen (N), phosphorous (P) and others from organic state to inorganic, and vice versa. To ensure environmental stability and suppress fluctuations of environmental characteristics, relative fluctuations of productivity P^+ and destructivity P^- in the ecological community should be low. Quantitatively we may write it as follows:

$$\overline{P^+ - P^-}\tau \leq \varepsilon_b M \qquad (3.4.2)$$

Here $\overline{P^+ - P^-}$ is the difference between mean values of rates of production and destruction of a certain biogen averaged over time τ. $\overline{P^+ - P^-}$ describes the average rate of deviation of environmental characteristics from their optimum (if the value of P^+precisely coincides with that of P^-, nothing changes in the environment). M is the store of the respective biogen in the environment. Expression (3.4.2) means that during time τ the store of the biogen should not deviate from the optimal value M by more than $\varepsilon_b M$. The characteristic time τ is actually equal to the turnover time of the considered biogen in the environment, $\tau = M/P^+$, which makes it possible to rewrite (3.4.2) as

$$\frac{\overline{P^+ - P^-}}{P^+} \leq \varepsilon_b \qquad (3.4.3)$$

When the deviation from the optimum reaches the critical value of $\varepsilon_b M$, the biota notices this unfavourable change and begins to compensate, returning the corresponding environmental characteristic back to the optimum. If ε_b were equal to unity, i.e. the biota were only able to notice the environmental change when the store of the biogen would completely extinguish (or double) $\Delta M = M$, then the environmental store M would fluctuate between 0 and M_{max} each τ years. Accordingly, when ε_b is small, the value of M fluctuates only slightly.

3.5 FLUCTUATIONS OF SYNTHESIS AND DESTRUCTION OF ORGANIC MATTER

To ensure a high degree of correlation between rates of synthesis (production) and destruction of biogens, the ecological community should be characterised by a high degree of correlation between functioning of different organisms. As soon as conspecific individuals are mostly involved in the process of competitive interaction (Chapter 2), correlated functioning within the community is inherent to individuals from different species. Such is, first of all, a strict correlation between the species synthesising and destroying the organic matter (i.e. between plants on the one hand, and bacteria, fungi and animals, on the other), which is superimposed by correlation between carnivores and herbivores (their prey) in the animal kingdom.

In the interest of stabilising selection, the interaction of individuals within a species must be competitive. Since individuals of the same species are uncorrelated, there remains only one way to diminish the possible fluctuations of matter and energy fluxes. According to the well-known statistical law of large numbers, relative fluctuation of a measurable variable in a non-correlated system is of the order of $1/\sqrt{N}$, where N is the number of mutually uncorrelated parts of the system. Therefore, to satisfy the condition (3.4.3) within a local ecosystem, organic matter should be synthesised and destroyed by $N \sim 1/\varepsilon_b^2$ uncorrelated elements. Note that it is useless for the ecological community to keep N significantly larger than $1/\varepsilon_b^2$, because the corresponding increase in accuracy of control of P^+ and P^- will not be noticed by the biota anyway and will not give the community any competitive advantage.

Different patterns of organisation of synthesising and reducing components of ecological community are employed in nature to satisfy condition (3.4.3). Plants (synthesisers) are immobile and form a continuous cover. The strongest fluxes of synthesis usually flow through the largest plants in the community. Let us denote the average correlation radius for plants (i.e. the average distance between various parts of a plant functioning in a correlated mode) by r^+. We thus obtain for the number N of uncorrelated parts of plants within an ecological community of size L_C:

$$N = \frac{dL_C^2}{(r^+)^2} \sim \frac{1}{\varepsilon_b^2}, \qquad r^+ \sim \varepsilon_b L_C \sqrt{d}, \tag{3.5.1}$$

where d is the leaf area index (Table 3.1), i.e. the effective vertical size of plants in terms of r^+. Fluctuations in plant productivity are kept at a low level due to the peculiar morphological organisation of large plants that function as weakly internally-correlated modular individuals (Begon *et al.*, 1986) in a mode similar to that of an equivalent colony of completely uncorrelated single cell algae.

The second relationship in (3.5.1) makes it possible to obtain an upper estimate of the biotic sensitivity ε_b. The local ecosystem size L_C cannot be smaller than that occupied by a single tree species that accounts for the major part of the net primary production in the ecosystem. (Different local ecosystems are uncorrelated by definition, while within a tree all biochemical fluxes are to a significant degree correlated thanks to biochemical transport in roots, trunk and branches.) In boreal forests one such dominant species is spruce. Its surface projection area is of the order of $S \sim 10\,\text{m}^2$. The average projection area of a single spruce needle is of the order of $(r^+)^2 \sim 5 \cdot 10^{-6}\,\text{m}^2$. Leaf area index for spruce is of the order of 10. Thus, from the condition that $L_C \geq \sqrt{S}$ and (3.5.1) we obtain

$$\varepsilon_b \leq [5 \cdot 10^{-6}\,\text{m}^2/(10\,\text{m}^2 \cdot 10)]^{1/2} \approx 2 \cdot 10^{-4},$$

i.e. the same order of magnitude as in Section 3.4. In other words, we assumed that separate needles represent almost independent synthesising parts of a plant, and estimated ε_b as the relative fluctuation of the number of needles N (3.5.1) from tree to tree of the same surface projection area.

Deciduous trees are characterised by larger projection areas r^+ of their synthesising parts (leaves). However, separate parts of a leaf function in an uncorrelated manner. This is manifested by the fact that even significantly damaged leaves still retain the ability to synthesise organic matter. The correlation radius r^+ of a leaf is presumably of the order of its thickness, which approximately coincides with that of a spruce needle. Thus, for broad-leaved trees as well we obtain the same estimate of ε_b.

The biotic sensitivity ε_b can be estimated also for epilithic lichens that represent the simplest ecological communities of the biosphere, consisting of individuals of algae (synthesisers) and fungi (reducers). The average diameter of algae cells and fungi hyphae (independently functioning units) is of the order of $r^+ \sim 10^{-5}$ m, while the surface projection of a single correlated lichen community is of the order of $(L_C)^2 \sim 1\ \text{cm}^2$. Vertical size of epilithic lichens l is of the order of 0.1 mm, which corresponds to the leaf area index of $d \sim l/r^+ \sim 10$ (see Section 3.8 below). Hence, for epilithic lichens we obtain

$$\varepsilon_b \sim r^+(L_C\sqrt{d}) \sim 3 \cdot 10^{-4}$$

This value is a representative rather than upper estimate, because for lichens the average community size L_C is known, while for trees we used the lower estimate of community size, $\sqrt{S}$, where S is the area occupied by a single tree. Note that all the above estimates give the same order of magnitude for ε_b coinciding with that obtained in Section 3.4, $\varepsilon_b \sim 3 \cdot 10^{-4}$.

If there are n^+ plant species in a given ecological community, that give approximately equal contributions to the net primary production, occupy different niches and, consequently, function in uncorrelated mode, then the number N of uncorrelated patches increases by the factor of n^+ and the expression for the average correlation radius r^+ transforms into:

$$r^+ \sim \varepsilon_b L\sqrt{n^+ d} \qquad (3.5.2)$$

We can see that diversification of species in a community makes it possible to increase the average correlation radius of plants without unbalancing the matter cycles within the local ecosystem. The radius of a local ecosystem as a whole (i.e. environment occupied by a single ecological community), L_C, must be one and the same for all species inhabiting the local ecosystem, both plants (synthesisers), and bacteria, fungi and animals (reducers). Therefore, the average correlation radius for plants, Eq. (3.5.2), should also be treated as a universal characteristic, which does not depend on the size of separate plants in the ecosystem. Universality of r^+ is manifested in the fact that, irrespective of plants' total size, their different weakly-correlated parts (leaves, branches, roots, etc.) feature similar spatial distribution. In other words, low trees do not necessarily have smaller leaves or thinner roots than large trees. Roughly, all plants can be envisaged as differently sized assemblages consisting of one and the same basic element characterised by linear size r^+.

Relationship (3.5.1) allows one to assess the size of a local ecosystem, L_C, from the average correlation radius of plants r^+. Assuming that for forest ecosystems the value of r^+ is of the order of thickness of a green leaf (i.e. $r^+ \approx 0.1$–1 mm) we obtain from (3.5.1) using $\varepsilon_b \sim 10^4$ and $d \sim 8$ (Table 3.1) the following rough estimate:

$$L_C \approx 1 \div 10\,\text{m} \tag{3.5.3}$$

This means that ecological communities occupying local ecosystems within which local biochemical cycles take place independently, are rather small. Thus, a forest consists of a huge number of local ecosystems that are presumably grouped around separate trees that ensure the major part of biochemical transport. Note that the finite size of ecological communities is absolutely indispensable for ensuring functioning of stabilising selection and competitive interaction of communities. A globally-correlated ecological community (were such to exist) would lose its internal correlation in particular due to uncorrelated evolution of its separate components. The internal correlation of ecological communities can be sustained only due to stabilising selection of communities in a population of mutually independent communities, the latter therefore necessarily having limited spatial size (Section 2.8). The estimate (3.5.3) provides empirical evidence for the statement that Earth cannot represent a globally-correlated ecological system, as it is sometimes stated. This is not an obstacle, however, for biotic regulation on a global scale (see Section 5.6).

Immobile fungi and bacteria incapable of active locomotion are organised in a manner similar to that of plants to diminish fluctuations in the process of destruction of organic matter. Like higher plants, fungi have a small correlation radius, which is of the order of thickness of their hyphae, while bacteria exist as a completely non-correlated set of separate single-cell organisms. Natural fluctuations of destruction of organic matter by fungi and bacteria are very low and cannot unbalance functioning of the community.

Thus, ecological problems arise only after locomotive animals have evolved and entered the ecosystem. As we have seen in Figure 3.2, locomotive animals inevitably cause sharp fluctuations of the biomass in the community.

3.6 IMMOBILE AND LOCOMOTIVE ORGANISMS

Similarly to other macroscopic physically-ordered processes, life is sustained due to consumption of ordered energy in the form of solar radiation. The maximum flux of solar radiation reaching the Earth's surface does not exceed $1\,\text{kWm}^{-2}$ (see Table 8.1). Hence, no macroscopic process directly driven by the radiative solar energy may exceed this energy flux threshold.

In contrast, release of energy flux during thunderstorms, hurricanes and tornadoes exceeds that value by a factor of several tens or even hundreds.[3] Such high energy flux may be due either to concentration of the solar power incident upon a larger area and transformed into mechanical energy, into a smaller surface area, or to a quick release of potential energy previously accumulated over a long time period.

One way of achieving high physical power density, i.e. concentration of solar energy incident upon large territories, into small ones, is observed in processes associated with precipitation of water. Residual precipitation, which fails to reevaporate, accumulates into rivers, and the resultant hydropower density significantly exceeds the solar energy flux. But the global hydropower only constitutes a minor part of the total sunpower incident upon Earth (see Table 7.1).

Another way of achieving high energy flux (i.e. quick release of energy previously accumulated over a long time period) owes itself to the possibility to accumulate the absorbed solar energy in the form of evaporation/melting latent heat. Water vapour that evaporates slowly over a long period of time stores large amounts of energy. That energy may then be released in a pulse by a tornado-like formation. Having spent within a second the energy that has been accumulating in a given volume for several hours or even days, the tornado has to move further in space in order to consume energy from a newly-occupied volume, and so forth. The higher the tornado power (i.e. its rate of energy release per unit volume), the more quickly it has to move. Without movement, the tornado would quickly dissipate all its power and fade away.

The fundamental peculiarity in using the potential energy accumulated in unit volume of the environment over a certain period of time is that the object using that energy, whatever its nature may be, must move. If it were possible to concentrate potential energy from a large surface, local immobile tornadoes or hurricanes could appear which (like hydropower plants) would remain spatially stable (Jones, 1990), but this is not actually observed.

Fires, rock falls and avalanches are phenomena of similar nature. During a forest fire the biochemical energy of organic matter synthesised for many previous years is spent. Fire may only sustain itself if it spreads quickly. Rock falls and avalanches release gravitational energy accumulated during prolonged processes of snowfall and/or mountain formation. In all such cases, high density power fluxes only become possible due to the previous accumulation of potential energy in various forms.

Energy of sunlight cannot be immediately stored in the form of photons, since photons are massless particles possessing kinetic energy only. Since plants live off the absorbed flux of solar radiation, they cannot achieve power densities higher than

[3] Flux of wind energy is equal to the product of its velocity, u, and the density of wind energy $\rho u^2/2$, where ρ is air density. Dissipation flux of wind energy (its destructive power per unit area) is equal to $c\rho u^3/2$, where $c \sim 1$ is the coefficient of dissipation, practically independent of the velocity. The average global wind speed and the wind energy flux are equal, respectively, to $7\,\mathrm{m\,s^{-1}}$ and $3\,\mathrm{W\,m^{-2}}$ (Gustavson, 1979). In the vicinity of a typhoon centre the wind speed u may be as high as $200\,\mathrm{m\,s^{-1}}$, exceeding the global mean value almost by a factor of 30. Wind energy flux increases in proportion to u^3, that is by the factor of $3 \cdot 10^4$, reaching $100\,\mathrm{kW\,m^{-2}}$. In the centre of a tornado the energy flux is presumably much higher.

that of incident light. Active locomotion would have not helped plants absorb more light. That is why plants are immobile and may form a continuous vegetation cover of the Earth's surface. This fundamental property of plants may be called the *law of plants' immobility*.

Plants store the absorbed solar energy in the form of potential biochemical energy of the organic matter that they synthesise. The observed maximum power flux of photosynthesis is about 2.0 W m^{-2} (Section 3.3), which corresponds to about 160 t of live organic matter or about 16 t of organic carbon per ha per year. The average energy flux of net primary productivity is about 0.5 W m^{-2} and 0.25 W m^{-2} for land and Earth as a whole, respectively (Table 3.1).

Animals feed off the organic matter accumulated by plants. Energy flux of organic matter consumed by an animal considerably exceeds the plant productivity (Figure 3.2). For mammals and birds, the consumption fluxes per unit average projection area exceed 1 k Wm^{-2} (see the end of Section 3.1, Figure 3.2 and Section 3.2), i.e. are above plant productivity by a factor of several thousand. Thus, life of animals may be sustained only through consumption of organic nutrients which are synthesised by plants on a territory exceeding the animal's individual average projection area by the same factor. Passive consumption of organic matter synthesised at a large area surrounding the animal's body is only possible in a liquid environment. Such feeding mode in combination with immobile existence is employed by several marine and fresh water animals (actinias, ascidias, some molluscs, and balanuses). In the absence of water-transported fluxes of organic matter, such an animal is capable of eating within an hour the organic matter synthesised in a whole year by all the plants around it. After that the animal has to move to a new area to be able to feed off the organic matter available there, and so it goes on. Return of the animal to the starting area must not occur until the consumed biomass has been regenerated there. This obliges the animal to protect its feeding area, preventing invasions of other individuals of the same species.

Life of animals is based on consumption of previously accumulated biomass. This cannot be ensured without locomotion. Thus, locomotion is an indispensable prerequisite of animal life. This fundamental property of animals may be called the *law of animals' locomotion*.

The average correlation radius for locomotive animals r^- is close to their body size: $r^- \approx l \equiv (m/\rho)^{1/3}$, where m is the animal's body mass, and ρ is its body density, see Section 3.2. The area of correlated food consumption by an animal is of the same order of magnitude as its body projection area, $s = l^2$. Correlated consumption of food on a given area means that food is mostly consumed on this area and neglected on the neighbouring territories. For example, an elephant may consume food within an area where it may reach with its trunk and nowhere else.

The relative fluctuation of food consumption by an animal on a territory similar to its body projection area is close to unity. If a large animal of body size $l \approx L_C$ completely consumes production from a local ecosystem of size L_C, then the relative fluctuation of production and, consequently, the fluctuation of biomass in the local ecosystem, would also be of the order of unity, which is incompatible with stable existence of the ecosystem (see Section 3.5).

The only way for large animals to exist without threatening the integrity of the ecological community and its environment is to decrease their quota of consumption. An animal of body size $l \approx L_C$ should be allowed to deplete no more than $\varepsilon_b \approx 10^{-4}$ of the net primary production from a local ecosystem, Figure 3.5.

The more diverse the animal species in a given range of body sizes, the more weakly their consumption of plant production is correlated (because they have different food habits), and, hence, the lower are the overall fluctuations of that consumption. If a ceiling is prescribed for possible fluctuations (which is determined by the biotic sensitivity ε_b), diversification of animal species makes it possible to increase the total mass of animals in the given body size range.

Let us introduce spectral density per given body size of the consumption quota, β, and of the number of species, n^-. Then, similar to Eqs. (3.5.1) and (3.5.2), we obtain for a given body size the number of uncorrelated animal units in a local ecosystem, $N = (L_C^2/l^2)n^-$, and the relative fluctuation of consumption (destruction) introduced by animals of that body size, $\beta/\sqrt{N}$. The latter should not exceed the biotic sensitivity ε_b. We thus have:

$$\beta \sim \frac{\sqrt{n^-}}{l} \varepsilon_b L_C \tag{3.6.1}$$

In plants, the analogous condition (3.5.2) of the low fluctuations of primary production dictates the internally-uncorrelated structure of plant bodies limiting the linear size r^+ of a relatively independently functioning part of a plant. In animals, condition (3.6.1) imposes restrictions on the quota of biomass destruction and prescribes its specific distribution over animals with different body sizes. The strong, internally-correlated structure of animals' bodies is conditioned by the necessity of locomotion and cannot be modified. In other words, unlike plants, animals cannot function as weakly-correlated modules.

The quota of consumption allocated to large animals with body size in excess of $l = 10$ cm can be estimated from (3.6.1): at $\varepsilon_b \sim 10^{-4}, L_C < 10$ m, $n^- \approx 1$ we retrieve $\beta \sim 0.01$, which means that large animals altogether cannot consume more than 1% of the net primary production of the ecosystem. This estimate agrees well with the empirical data available for different natural ecosystems unperturbed by anthropogenic activities (see Figure 3.3 and Section 3.7).

It is worth noting that according to Eqs. (3.5.2) and (3.6.1) an increase in species diversity does not enhance stability of local ecosystems, but rather ensures a possibility of existence of a larger total number of individuals of different plant and animal species on a given area without threatening the integrity of the local ecosystem. The fact that the most stable tropical ecosystems (Wilson, 1988) are characterised by the highest species diversity encountered in the biosphere can be explained as follows. Higher species diversity per unit surface area means lower population numbers of each particular species. Low population number makes a species vulnerable to drastic fluctuations of the environment. Thus, low population numbers and, consequently, higher species diversity, can be sustained only in a mild environment that can be maintained by the biota in a stable state with only small fluctuations, e.g. the tropics. In contrast, the highly fluctuating severe environment of temperate and polar zones favours high population numbers and, consequently, low species diversity. Therefore it is not the stability of tropical ecosystems that owes itself to high species diversity, but vice versa.

To find an exact dependence of β on l one has to know how the number of animal species n^- in Eq. (3.6.1) depends on l. At small body sizes $l < 1$ cm n^- linearly decreases with l (Chislenko, 1981) and β should decrease proportionally to $l^{-1/2}$, which is in good agreement with the available empirical data (see Section 3.7).

Relationships (3.5.2) and (3.6.1) describe the general principles of the organisation of local ecosystems. They demonstrate that, on the whole, the biosphere may be envisaged as a non-correlated set of living internally-correlated biosystems devoid of centralised control and averaging about 0.1–1 mm in effective size. Examples of such biosystems are bacteria, fungi and uncorrelated parts of plants (e.g. leaves, needles). Such a structure guarantees that the fluxes of organic matter synthesis and destruction do not fluctuate too strongly and remain equal to each other with a relative accuracy of 10^{-4}. Large animals may exist within a stable stationary biosphere only on condition that their quota of consumption of organic matter produced in the biosphere does not exceed 1% (see Section 3.7).

With respect to large animals the existing biosphere may be considered just as an energy producing engine, working to provide them with food and to stabilise their environment at an optimal level. However, this engine functions at an efficiency level of no more than 1%. The other 99% of energy fluxes must be consumed by the other species in the community, and that way they should be considered as indispensable 'overhead expenses'.

3.7 DISTRIBUTION OF CONSUMPTION BY HETEROTROPHS WITH RESPECT TO THEIR BODY SIZE

The share β of primary production consumed by all animals of a given body size is equal to the ratio of consumption by those animals to the net primary production of the ecosystem. It is a convenient characteristic for comparing various ecosystems with different values of net primary production against each other. Summing up consumption by all individuals within the body size interval from l to $l + \Delta l$, one may plot a distribution of the share of consumption β with body size l.

However, the density of consumption per an absolute unit body size interval, Δl, has no sense within a wide range covering several orders of magnitude. For example, consumption by bacteria having body size of about 1 μm changes noticeably while the body size is changed by $\Delta l \approx 1$ μm. Consumption by large mammals of the body size of $l \approx 1$ m noticeably changes with the body size changing by $\Delta l \approx 1$ m, but does not change at all while that size changes by $\Delta l \approx 1$ μm.

Therefore, a distribution function retrieved for a unit relative interval $\Delta z \sim \Delta l / l$, $\beta(z)$, is the only sensible function. Such type distribution is known as *size spectrum* (Hutchinson and MacArthur, 1959). We define $\beta(z)$ as follows:

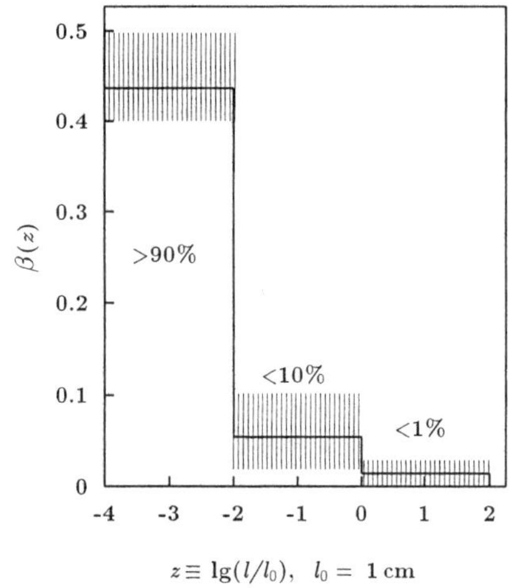

Figure 3.3. Distribution of the relative share of consumption $\beta(z)$ of the net primary production of terrestrial ecosystems over heterotrophs of different body sizes. Numbers in per cent represent inputs from various parts of the histogram. Deviations from the mean distribution observed in natural ecosystems are hatched. See (3.7.1) and text.

$$\beta(z) \equiv \frac{1}{\Delta z}\sum_{i}^{z+\Delta z} \beta_i \equiv 23\, l\left(\frac{1}{\Delta l}\sum_{i}^{l+\Delta l} \beta_i\right), \qquad \int_{z_{\min}}^{z_{\max}} \beta(z)\,\mathrm{d}z = 1$$

$$z \equiv \lg\left(\frac{l}{l_0}\right), \qquad l_0 = 1\ \mathrm{cm}, \qquad \Delta z \equiv 0.43\frac{\Delta l}{l} \tag{3.7.1}$$

Summation here is conducted over all the species entering the body size range from l to $l + \Delta l$. β_i is the i-th species' share of consumption of the net primary production P^+ of the ecosystem, $\beta_i \equiv P_i^-/P^+$. Various age groups of individuals of one and the same species, strongly differing in their body size (e.g. adult fishes and their fry), refer to different relative body size groups. The factor 0.43 is due to using decimal logarithms instead of natural ones. Function $\beta(z)$ is the relative body size spectrum, i.e. the density of the consumption share of the net primary production by individuals of a given body size, entering a unit range of relative size $\Delta z = 1$.

Figure 3.3 plots the distribution $\beta(z)$ based on published field studies from a large number of natural non-perturbed ecosystems. The solid line presents the average distribution over all the studied ecosystems (Gorshkov, 1981). Scatter of the measured values in various ecosystems is shown by hatched areas.

Histogram intervals are chosen to correspond to the taxonomic groups 1–5 presented in Figures 3.1 and 3.2. The first interval, $10^{-6}\ \mathrm{m} \le l \le 10^{-4}\ \mathrm{m}$, is mostly occupied by bacteria, protozoa, saprophytophagous fungi (those destroying dead plants) and mycorrhiza (fungi being in symbiosis with plant roots). Body size of fungi is assumed to be equal to that of hyphae (strings of mycelium composing the body of each fungus), which may be assumed to be only weakly correlated with each other. In the second interval, $10^{-4}\ \mathrm{m} \le l \le 10^{-2}\ \mathrm{m}$, consumption of the net primary production is mostly determined by invertebrates. The largest bacteria and fungi entering the second interval do not play any significant role due to their low volume-specific metabolic rate of consumption λ (see Figure 3.1), even though their cumulative biomass may exceed that of invertebrates. The third interval, $10^{-2}\,\text{m} \leq l \leq 1\,\text{m}$, is completely dominated by vertebrates, mainly mammals. Input from consumption by the largest invertebrates to the third interval is insignificant for the same reason as above.

As follows from Figure 3.3, in natural non-perturbed ecosystems, bacteria and fungi consume more than 90% of all the net primary production, invertebrates less than 10%, and vertebrates less than 1% of that production. A distribution similar to that presented in Figure 3.3 is also observed in shallow closed water ecosystems, where the principal part of primary production is controlled by large multicellular plants having large biomass (Kamenir and Khailov, 1987; Kamenir, 1991). Note that in flowing waters, such as littoral, estuaries and rivers, the distribution of consumption may be different from that in Figure 3.3 due to the high influx of nutrients into local ecosystems (see also Section 4.6).

In the open ocean and other water surface areas where the total production is controlled by microscopic phytoplankton (its individual biomass is three to four orders of magnitude less than that of higher plants), there are no large herbivores, and all the net primary production is consumed by bacteria and zooplankton, which occupy the first part of the histogram in Figure 3.7.1. A similar situation is observed for lichen ecosystems. In other words, where there is no biomass, there are no large animals, and the histogram (Figure 3.3) reduces to one single bar at the smallest sizes.

Thus, with the account of plant respiration, more than 95% of the gross (total) primary production consumed by individuals in the community flows through immobile individuals. All the locomotive animals together represent only a 5% addition to the energy consumption structure of the community. However, being large animals with strongly slowed metabolism (see Figure 3.1), they may comprise a considerable (though metabolically inactive) part of the total community biomass.

The fact that practically all the energy flux goes through immobile individuals, and that the principal part of plant biomass is consumed after their natural death, together provide for stability of matter cycles without any need for horizontal transport of biogens, and also for highly effective absorption of solar energy in the community. The main condition the community's immobile heterotrophs must observe is that their continuous layer should not shade the community's producers from the Sun. Immobile heterotrophs should find a place below the photosynthesising parts of vegetation (either bacteria and fungi in the soil horizon, or bacteria and zooplankton in oceanic depths).

We stress once again the important property of higher plants which makes possible the existence of large herbivores that significantly destroy the particular plants they feed on. While all the single-cell individuals and all the locomotive animals function on the basis of strictly-specified correlated interactions between different parts of their bodies, and execute centralised control of that body as a whole, immobile multicellular individuals (higher plants and fungi) are deprived of such control. They do not possess a head and consist of practically independent, weakly-correlated parts of small sizes (e.g. leaves of a plant, hyphae of a fungus). Moreover, such separate parts may compete between themselves for solar light and nutrients. When one part of a plant perishes (e.g. when eaten up by a herbivore), that does not entail death of the other parts and of the whole plant, but may even stimulate development of the other parts. In that sense, higher plants and fungi are more in the nature of a set of independent individuals (leaves, hyphae), which are partly correlated via a common woody trunk and the vessel system supplying nutrients, water and food from soil. Although weak, it is this correlation that provides for a variety of important properties of plants, e.g. vertical orientation of plant bodies.

Locomotive herbivores are capable of eating only the metabolically-active parts of plants which contain nutrients in proportions required for their metabolism. Higher plants are noted for the maximum thickness of their metabolically-active biomass layer and for lack of correlation between separate parts of each individual. These features are particularly favourable for animals to feed off certain specific parts of plants, which may constitute a considerable part of a particular plant. Meanwhile, consumption of any small part of the metabolically-active mass of a rigidly correlated individual (particularly of a single-cell one) is impossible, since it would lead to that individual's death or significant impairment of its state.

Were all the large multicellular plants as rigidly internally correlated as the animals, large animals would produce extremely high fluctuations in the natural distribution of vegetation, even with the share of consumption by those animals remaining quite small. (That, for example, is what happens when the vessel network in the cambium is violated along the whole perimeter of the tree trunk.) That is why the appearance of locomotive herbivores and of carnivores feeding off the former could only take place in the course of evolution after higher plants had appeared.

The distribution $\beta(z)$ (Figure 3.3) may be used to estimate empirically the correlation radius of the community. Denoting the share of consumption of plant production by a heterotrophic species i as β_i^- and using the normalisation $\Sigma_i \beta_i^- = 1$ (summing is extended to cover all the heterotrophic species in the community) one may estimate the correlation radius for the community from the linear distance at which the sum is saturated reaching some value close enough to unity, taking, for example, the distance at which $\Sigma_i \beta_i^- = 0.95$. It follows from the above analysis that the principal input to that sum is produced by a small number of bacteria and saprophytophagous fungi species. The same may be done for autotrophic plant species, understanding β_i^+ as the share of the solar energy or the share of inorganic nutrients consumed by the i-th plant species in the community ($\Sigma_i \beta_i^+ = 1$). Such an approach indicates a significantly smaller size of the community (Gorshkov and Gorshkov, 1992), than the saturation length for the whole number of species, independent of their consumption share (Giller, 1984) and appears (in the case of undisturbed ecosystems) to be close to the estimate of the correlation radius of the community obtained in Section 3.5.

3.8 DISTRIBUTION OF BIOMASS OF HETEROTROPHS WITH RESPECT TO THEIR BODY SIZE

Let us introduce the volume-specific and surface-specific metabolic rates $\Lambda \equiv K\lambda$, $J \equiv Kj$ that describe the power of metabolic processes in carbon mass instead of energy units (Sections 3.1 and 3.2). Note that $\Lambda = J/l$ according to (3.2.2) (Sections 3.1 and 3.2).

The average flux of consumption by all individuals of a given body size l of all species in the community is equal to βP^+, where β is the share of consumption by these individuals and P^+ is the net primary productivity.

On the other hand, this flux of consumption is equal to the product of the volume-specific metabolic rate of a heterotrophic individual, $\Lambda = J/l$, with its biomass per unit surface area B, i.e. $BJ/l = \beta P^+$. Such a presentation makes it possible to determine the biomass B from the equation:

$$B = \beta \frac{P^+}{\Lambda} = \beta \frac{P^+}{J} l \tag{3.8.1}$$

Distribution $\beta(z)$ (Figure 3.3) corresponds to an order of magnitude's decline of the share of consumption per two orders of magnitude's decline of body size l, that is, $\beta \sim 1/\sqrt{l}$. When all the heterotrophs are considered irrespective of the five taxonomic groups (Figures 3.1 and 3.2) they belong to, the value of volume-specific metabolic rate $\Lambda_{\text{opt}} = K\lambda_{\text{opt}}$ for the majority of species does not change with increasing body size l (see the dark shaded band in Figure 3.1), whereas the surface specific metabolic rate J grows proportionally to the body size l (see the shaded band in Figure 3.2). As a result, the biomass of heterotrophs of a given body size decreases as $1/\sqrt{l}$ in accordance with (3.8.1).

On the other hand, the surface-specific metabolic rate J can be assumed to be approximately constant within each of the five taxonomic groups. In this case, in accordance with (3.8.1), the cumulative biomass B of all the community species of a given size increases like $B \sim \sqrt{l}$ with increasing individual body size l within each taxonomic group (assuming that $\beta \sim 1/\sqrt{l}$).

Quite a different situation is observed for the *species-specific* share of consumption β_l (i.e. the share of consumption allocated to all individuals of a particular species of size l), which is related to β as:

$$\beta_l \equiv \beta/n \tag{3.8.2}$$

Here n is the number of species in a given interval of body sizes.

We obtain for β_l from (3.6.1):

$$\beta_l \sim \frac{1}{l\sqrt{n}} \varepsilon_b L_C \tag{3.8.3}$$

The biomass of a single species B_l is then determined by (3.8.1) substituting β for β_l.

The available empirical data indicate that in various terrestrial ecosystems the distributions of the number of heterotrophic species n over body size are similar (Chislenko, 1981). Thus it is natural to consider the task of finding the distributions

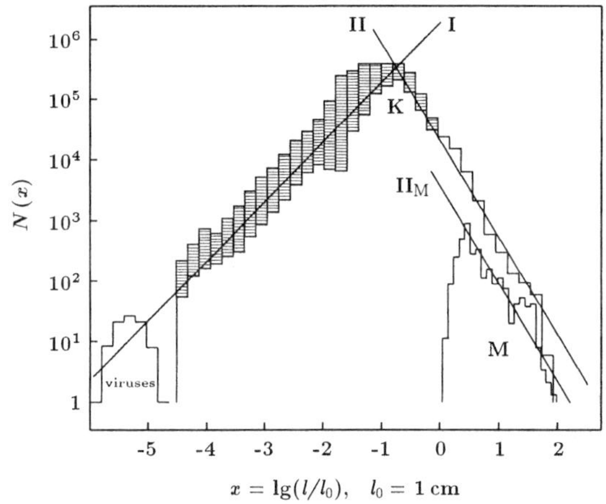

Figure 3.4. Size spectrum of the world's herbivore species. $N(x)$ is the density of the number of species per unit of the relative body size interval, $\Delta x; x = \lg(l/l_0); l \equiv (m/\rho)^{1/3}$, where l, m and ρ are the size, mass and density of the body, respectively. The histogram **K** is the distribution of terrestrial herbivore species of the world including reducers (Grant, 1977; Chislenko, 1981; Gorshkov, 1985a; 1985b; Wilson, 1988; May, 1990; Thomas, 1990). The histogram **M** is the distribution of mammals (Chislenko, 1981; Eisenberg, 1981). I, II, II_M are the allometric approximations. The slopes of the straight lines are 0.96 for I, −1.6 for II and II_M.

of β_l and n for a globally averaged local terrestrial ecosystem. The distribution n averaged over all ecosystems may be found from the distribution N similar to it and available for all the continental herbivore species in the global biota (Figure 3.4; Gorshkov, 1984a; 1985b). New data identifying the number of tropical insects more precisely (Thomas, 1990) in practice do not change the form of the distribution N (Figure 3.4).

As shown in Figure 3.4, the number n of mammalian species has a maximum in the vicinity of the smallest body size l_{min} and then drops approximately proportionally to l^{-2}. It follows from formula (3.8.1)–(3.8.3) that in the range $l > l_{min}$ the share of consumption of all mammals of a given body size β and their biomass B change in accordance with the approximation of a constant surface-specific metabolic rate J, see Figure 3.2, as $\beta \sim l^{-2}$ and $B \sim l^{-1}$ (Figure 3.5a). In the meantime, the species-specific share of consumption β_l and species-specific biomass B_l change as $\beta_l \sim l^0 = \text{const}$ and $B_l \sim l$ (Figure 3.5b). Note that this consideration makes it possible to estimate only slopes of the respective curves, which are in good concert with the available empirical data (Damuth, 1981a). Absolute values of the considered variables need to be determined independently, see notes to Figure 3.5.

Thus, the share of consumption of a single mammalian species does not change with body size, while the biomass increases proportionally to the body size. As a result, biomass of the largest animal species appears to be the most abundant in the community. This property of large animals readily catches the eye and makes an illusion of the dominating role of large animals in the ecosystem which, as we have seen, is far from being the case. In reality, the huge biomass of large animals is

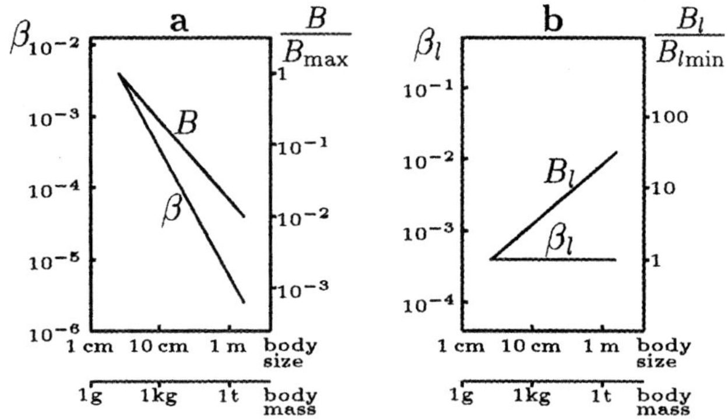

Figure 3.5. Share of consumption and biomass of natural species of herbivorous mammals of different body size (with the slight dependence of surface-specific metabolic rate J on body size l, Figure 3.2, taken into account). a: β and B are the share of consumption and biomass, respectively, of individuals of all mammalian species of a given body size. The absolute value $\beta_{\max} = 0.5 \cdot 10^{-2}$ corresponds to the average value observed in natural ecosystems (Gorshkov, 1981). b: β_l and B_l are share of consumption and biomass, respectively, per single mammalian species of a given body size. The absolute constant value of $\beta_l = 5 \cdot 10^{-4}$ corresponds to the average value observed in the biosphere (Damuth, 1981a). The biomass B and share of consumption β (as well as B_l and β_l) are related as in (3.8.1).

metabolically inactive compared to the tiny amount of highly-active biomass of smaller organisms (see Figures 3.1 and 3.2).

Let us now estimate population numbers of heterotrophic species belonging to different taxonomic groups. For heterotrophs, we may introduce the *projection area index* $d \equiv L/l \equiv s/S$, where L is the average thickness of the biomass layer of heterotrophs (i.e. thickness of the live biomass layer that would be formed were the whole biomass of heterotrophs evenly distributed (spread) over their feeding territories), l is the body size, s is the average projection surface area of one individual, S is the effective consumption area (home range, feeding territory) per single individual. Note that $B = L\rho$, where ρ is the live body density, $\rho \approx 1\,\mathrm{t\,m^{-3}}$. The projection area index d is equivalent to leaf area index for plants (Table 3.1).

Using (3.8.1) the projection area index for species of bacteria and fungi (taxonomic group 1, Figures 3.1 and 3.2) can be estimated as $d = \beta p^+/j \le p^+/j \le 50$, where p^+ is net primary production in energy units ($p^+ \le 2\,\mathrm{W\,m^{-2}}$, $j \ge 0.04\,\mathrm{W\,m^{-2}}$, see Figures 3.1 and 3.2). (This result is obtained assuming that for bacteria and fungi $\beta \sim 1$.) Thus, it is reasonable to assume that the average projection area index of a single species of bacteria and fungi is of the order of $\bar{d} = 5 \div 10$. The layer of biomass of bacteria and fungi in soil is not thick: the bacteria size is $l \approx 10^{-6}$ m, the effective size of a fungus (its hyphae thickness) is $l \approx 10^{-5}$ m, so that we obtain for bacteria $L = ld \le 5 \cdot 10^{-5}$ m ($B < 50\,\mathrm{g\,m^{-2}}$), for fungi $L_h \ge 5 \cdot 10^{-4}$ m ($B < 500\,\mathrm{gm^{-2}}$).

Using the obtained values we may now estimate the total number of bacteria inhabiting the biosphere. Under the assumption that bacteria form a continuous cover both on land and in the sea (Sieburth, 1976) of thickness $L = 5 \cdot 10^{-5}$m we obtain for the total live mass of all bacteria (spread across the Earth's surface of

$S_E = 5 \cdot 10^{14}\,\mathrm{m}^2$) a value of $M = \rho L S_E = 3 \cdot 10^9$ t. The average mass of a single bacteria is estimated at 10^{-18} t and their total number on Earth is given a value of $3 \cdot 10^{27}$. That value may simultaneously serve as an estimate of the total number of living beings in the biosphere, since the number of individuals of all the other species together is several orders of magnitude lower. The number of species among the bacteria being estimated as $3 \cdot 10^3$ (Chislenko, 1981; Raven and Johnson, 1988), the number of bacteria within a single species estimates as 10^{24}.

All the unicellular eukaryotes (apart from infusoria) belong to the same taxonomic group 1 (see Figures 3.1 and 3.2), as bacteria and fungi. The mean eukaryotic cell size l is about $50 \cdot 10^{-6}$ m, i.e. 50 times larger than an average bacterial cell. Hence, the global mass of unicellular eukaryotes in the biosphere should be about an order of magnitude ($B \sim \sqrt{l}$; $\sqrt{50} \approx 7$) larger than that of bacteria and approximately equal to that of fungi, i.e. $3 \cdot 10^{10}$ t. Estimating the mass of eukaryotic cells as 10^{-13} t, we obtain for the world population of unicellular eucaryotes a value of 10^{23} individuals. Given the total number of species of unicellular eukaryotes equal to 10^5 (Chislenko, 1981), the average population number of a species is about 10^{18} individuals.

Small invertebrates belong to the second taxonomic group that differs from the taxonomic group of bacteria, fungi and unicellular eukaryotes. According to Figures 3.3 and 3.4, the share of consumption β decreases as $\beta \sim 1/\sqrt{l}$ with growing body size l irrespective of the taxonomic group of the organisms until the body size l reaches $\sim 10^{-2}$ m. The cumulative biomass B of community species decreases then as $B \sim 1/\sqrt{l}$ (at constant Λ and $I \sim l$). Body size of the smallest invertebrates is about $l \sim 10^{-4}$ m, which corresponds to body mass of about $3 \cdot 10^{-12}$ t (Chislenko, 1981; Gorshkov, 1984a). The global biomass of the smallest invertebratres in the biosphere is about $\sqrt{10} \approx 3$ times less than the global biomass of fungi and constitutes about 10^{10} t. Hence, we obtain that the world population of invertebrates consists of about $3 \cdot 10^{21}$ individuals. The total number of species in the biosphere is of the order of 10^7 (Thomas, 1990; May, 1990), among which invertebrates are dominating. Thus, the average population number of an invertebrate species is about 10^{15} individuals. All the above estimates will be used below when comparing rates of evolutionary changes in different taxonomic groups (see Chapter 11).

We want to stress that all the quantitative results of Sections 3.5–3.8 confirmed by the available empirical data are ultimately based on the key relationship (3.4.3), which imposes limits on fluctuations of fluxes of production and consumption of organic matter in the community, which only makes sense within the biotic regulation concept. The ability of species to interact in such a manner that fluctuations of consumption and production are kept low may only be supported in the course of competitive interaction of ecological communities. Only stabilising selection of communities may impose such strong restrictions on populations of large animals. Communities where large animals begin to dominate, lose competitiveness, as soon as their environment becomes unstable. Thus, the results of this chapter (and Figure 3.3 in particular) present empirical support to the biotic regulation concept.

4

Ecology of Locomotive Animals

Locomotive animals feed on live biomass of plants and other animals. Small animals exist under conditions of energy abundance. Their consumption of live biomass does not go beyond natural fluctuations of the biomass itself. Large animals face energy shortage. Consumption of live biomass by large animals leads to significant degradation of local ecosystems. To keep the biota and environment stable, population numbers of large animals should be strictly limited.

4.1 DAILY AVERAGE TRAVELLING DISTANCE

Let us now consider the movement of land surface animals. The total metabolic power q of an animal moving at a speed u is usually estimated from the rate of oxygen consumption by the animal. It is convenient to relate the metabolic power q attained by an animal to the basal metabolic power q_0 (see Section 3.1) using a dimensionless total activity A:

$$q(u) \equiv [A(u) + 1]q_0 \tag{4.1.1}$$

At the average metabolic power of existence $\bar{q} = 2q_0$ (Section 3.1) the average total activity $\bar{A}$ is equal to unity.

Numerous experiments have demonstrated that the total activity $A(u)$ of an animal grows linearly with increasing u up to the maximum speeds developed by that animal (Schmidt-Nielsen, 1972, 1984). This relationship holds for all species studied. Another observed fact is that when the animal's speed of movement approaches zero, the value of $A(u)$ approaches a certain non-zero limit $A(0) > 0$, i.e. the total power of movement q does not grade smoothly into the basal power q_0 (Schmidt-Nielsen, 1972, 1984). It is natural to label this limiting value of activity $b \equiv A(0)$ as 'readiness' (for movement) (Gorshkov, 1983a, 1984a). The average value of b is close to unity (see Figure 4.1a). The available empirical data can be represented in the form

$$A(u) = a + b, \qquad a \equiv u/u_0, \tag{4.1.2}$$

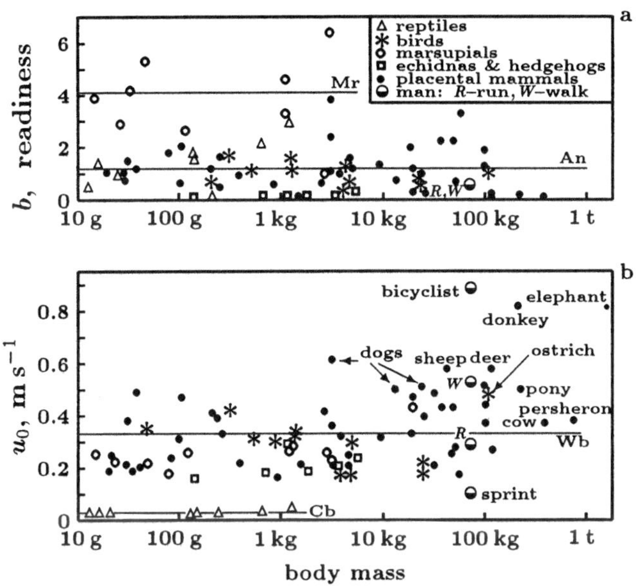

Figure 4.1. Readiness for movement, b, and available speed, u_0, for different taxonomic groups of animals (Schmidt-Nielsen, 1972, 1984; Gorshkov, 1983a). Readiness b is equal to the saltatory increment in metabolic power (in units of the basal metabolic power q_0) which corresponds to transition of an animal from the state of rest to movement at almost zero speeds. The available speed u_0 is the average daily speed that can be supported by the animal's metabolism, see also (4.1.2). Lines in the figure denote: **An**—average readiness for the majority of animals (reptiles, birds, mammals except for marsupials) (b = 1.0); **Mr**—average readiness for marsupials (b = 4.2); **Wb**—average available speed for warm-blooded (endothermic) animals ($u_0 = 0.3\ \mathrm{ms}^{-1}$); **Cb**—same for cold-blooded (ectothermic) animals ($u_0 = 0.003\ \mathrm{ms}^{-1}$). Man is a mediocre long-range runner, the worst sprinter among the warm-blooded animals, but one of the best walkers in the animal world.

where the value of a is the net movement activity, and the fundamental dimensional constant u_0 has the meaning of speed developed by the animal when its net activity a is equal to unity.

Speed u_0 determines the slope of the line presenting the dependence of the measured metabolic power, q, on the speed of movement u, Eqs. (4.1.1) and (4.1.2). Experimental data plotted in Figure 4.1b show that u_0 is a universal characteristic of movement, independent of the animal body size within a given taxonomic group. The average value is $u_0 = 0.3\ \mathrm{m\,s}^{-1} = 26\ \mathrm{km\,day}^{-1}$ for all warm-blooded animals from mouse to elephant. The record belongs to the donkey and elephant, which feature $u_0 = 0.8\ \mathrm{m\,s}^{-1}$ (Gorshkov, 1983a; Langman *et al.*, 1995).

Let us now analyse the meaning of speed u_0, which acquires considerable significance being a universal fundamental constant. The average daily activity $\bar{A}$ is limited by the average metabolic power of existence. According to the available empirical data, for most animals the average daily activity is close to $\bar{A} \approx 1$. We denote the relative duration of active state for an animal as $x_a \equiv t_a/T$, where t_a is duration of the time interval when the animal remains active during a day, and

$T = 24\,\text{h}$. The average daily activity can be thus written as

$$\bar{A} = Ax_a = \left(\frac{u}{u_0}b\right)x_a$$

from which we obtain:

$$x_a = \frac{\bar{A}}{\dfrac{u}{u_0} + b} \tag{4.1.3}$$

If $\bar{A} = 1$, $u = u_0$, and $b = 1$, we have $x_a = 1/2$ and $t_a = 12$ (a 12-hour working day). More information that can be retrieved from (4.1.3) for $\bar{A} = 1$ and $b = 1$ is that animals cannot move all day long at a speed different from zero (if $x_a \to 1, u \to 0$). The distance L_T travelled by an animal in a day is equal to $L_T = ut_a$, i.e. $x_a = L_T/uT$. We then find from Eq. (4.1.3) that:

$$L_T = L_{T_{\max}}\frac{u}{u + bu_0}, \qquad L_{T_{\max}} = \bar{A}u_0T. \tag{4.1.4}$$

It follows from Eq. (4.1.4) that the maximum daily travelling distance $L_{T_{\max}}$ is reached at speeds $u \gg u_0$, because the readiness b differs from zero. It means that animals should move quickly but in short bursts. Note that $b \approx 1$ for untrained people, for athletes $b < 1$ (see Figure 4.1). According to Eq. (4.1.4) the maximum distance walked in a day at the daily average activity of $\bar{A} = 1$ is about 26 km (for a donkey it amounts to $70\,\text{km}\,\text{day}^{-1}$). Finally, using Eq. (4.1.3), the average daily speed, $\mathbf{u_a} \equiv L_T/T = ux_a$, that the animal's metabolism may support (we may call it the available speed u_a), can be expressed as

$$u_a = \bar{A}u_0\left(1 - \frac{bx_a}{\bar{A}}\right)$$

At $\bar{A} = 1$ and $x_a \ll 1$, the available speed u_a is equal to the speed u_0. Everywhere below we assume that $\bar{A} = 1, x_a \ll 1$ and

$$u_a = u_0. \tag{4.1.5}$$

4.2 THE MAXIMUM SPEED OF MOVEMENT FOR ANIMALS

The speed u_0, however, cannot be totally independent of animal body size. We shall demonstrate from the law of energy conservation that, starting from some critical body size, the speed u_0 must start to decrease for larger body sizes l.

Metabolic power is transformed into mechanical power at a certain efficiency level α, which cannot be larger than the observed maximum efficiency of muscles. Within a living individual the latter does not exceed 25% (Hill, 1960; Cavagna and Kaneko, 1977; Heglund *et al.*, 1979). One may therefore assume that $\alpha \leq 0.25$. The mechanical power put out at a constant speed u is spent to compensate for energy dissipated to ground and air friction. The law of energy conservation is then expressed as equality between the mechanical power and the dissipative energy losses.

Energy dissipation during movement on the ground is proportional to the product of body weight, mg ($g = 9.8\ \mathrm{m\,s^{-1}}$ is the free fall acceleration), and the speed of movement u. It may be written as γmgu, where γ is the ground dissipation coefficient analogous to the coefficient of friction. The empirical data available on the maximum velocities of different animals show that the value of γ depends neither on the speed of movement, nor on the body size (i.e. mass) of the animal, and is, on average, equal to 0.04 for most animals (Gorshkov, 1983a).

Dissipation of energy due to air friction is equal to $cl^2\rho_c u^3/2$, where ρ_c is the air density ($\rho_c = 1.2 \cdot 10^{-3}\rho, \rho = 1\ \mathrm{t\,m^{-3}}$); $\rho_c u^3/2$ is the energy flux through a unit surface, equal to the product of energy density $\rho_c u^2/2$ and velocity u; cl^2 is the effective streamlined body surface. This value may be treated as the product of the resistance force $cl^2\rho_c u^2/2$ and the velocity u. The air resistance coefficient c may be measured (see references in Gorshkov, 1983a, 1984a). It is equal to 0.4 for the majority of land surface animals.

The body size l of locomotive animals is related to their body mass by the relationship:

$$l = (m/\rho)^{1/3} \tag{4.2.1}$$

The mechanical power of movement is equal to $\alpha a q_0$. Equating it to cumulative dissipative energy losses we have:

$$\alpha a q_0 = \gamma_{\text{tot}} mgu, \tag{4.2.2}$$

$$\gamma_{\text{tot}} \equiv \gamma + \gamma_c \equiv \gamma\left(1 + \frac{Fr}{k^2}\right), \tag{4.2.3}$$

$$\frac{\gamma_c}{\gamma} \equiv \frac{Fr}{k^2}, \qquad Fr \equiv \frac{u^2}{gl}, \qquad k^2 \equiv \frac{2\rho\gamma}{\rho_c c} \tag{4.2.4}$$

Here γ_{tot} is the total dissipation coefficient, equal, by definition, to the ratio of resistance force to individual body weight mg; Fr is known as Froude's number, γ_c is the air dissipation coefficient, which is relatively small when $Fr/k^2 \ll 1$.

Using the above relationships one may rewrite the expression for net activity a, Eq. (4.1.2), in the following form (for $Fr/k^2 \ll 1$):

$$u_0 = \frac{\alpha}{\gamma}\lambda_0, \qquad \lambda_0 \equiv \frac{q_0}{mg}. \tag{4.2.5}$$

Here λ_0 is the metabolic power per unit body weight mg. It has the dimension of velocity and differs from the volume-specific metabolic power used in Chapter 3 (Figure 3.2) by the constant factor $(\rho g)^{-1}$. We have therefore retained the same symbol for denoting it.

Since λ_0 drops for higher body sizes (see Figure 3.2), the observed constancy of u_0 for varying body size l means that either the efficiency α or the dissipation coefficient γ must change with body size l. Using calculations based on the available empirical data we shall demonstrate that the value of γ does not change with body size. Consequently, it is the value of efficiency α that must change.

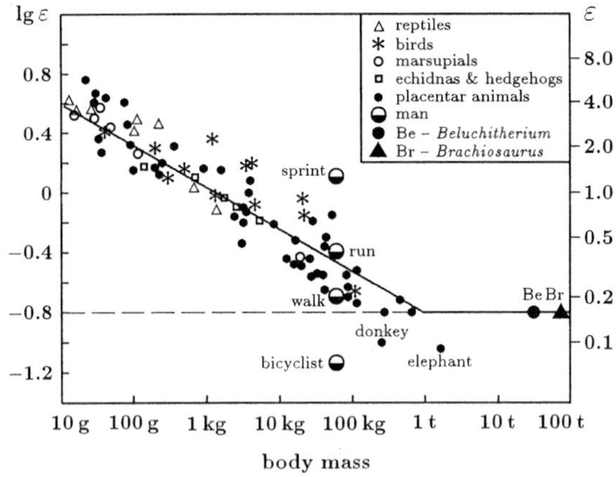

Figure 4.2. Energy cost of movement vs. the animal body size (mass) (Schmidt-Nielsen, 1972, 1984; Gorshkov, 1983a). The dimensionless energy cost of movement ε is equal to net expenditure of metabolic energy (the difference between the total metabolic energy and the metabolic energy put out at the zero speed limit) per unit body weight per unit distance:

$$\varepsilon = [q(u) - q(0)]/(mgu)$$

$q(u)$ is the metabolic movement power at the speed of u; mg is the animal body weight. The solid line is the average value of ε. It does not depend on the absolute animal metabolic power and is identical for mammals, birds and reptiles of equal body size. Due to higher movement efficiency α the value of $\varepsilon = \gamma/\alpha$ decreases at higher body size until the maximum possible value of α is reached. After achieving it, further decrease in ε for larger body sizes should stop. The solid line is numerically extended in the horizontal using values of the observed maximum efficiency α and the average dissipation coefficient γ for the mechanical movement energy (see Figure 4.3). Be—the maximum body mass for an extinct land mammal—*Beluchitherium* (30 t); Br—the maximum body mass for a terrestrial animal, dinosaur *Brachiosaurus* (80 t) (Schmidt-Nielsen, 1984).

The ratio

$$\varepsilon \equiv \frac{\lambda_0}{u_0} = \frac{\gamma}{\alpha} \tag{4.2.6}$$

represents the dimensionless net energy cost of moving a unit weight along a unit distance (it is similar to car mileage): $\lambda_0/u_0 = E/mgL = aq_0/mgu$, where $E = aq_0t$ is the net energy expenditure per travelled distance $L = ut$. The dependence of ε on body size l for animals is presented in Figure 4.2. Due to constancy of u_0 the energy cost ε linearly falls off for larger body sizes proportionally to λ_0 and α^{-1}. This drop goes on until the efficiency α reaches its maximum possible for muscles value. Since then the value of ε must remain constant and independent of body size (Gorshkov, 1983a).

According to observations λ_0, u_0 and hence the ratio α/γ do not depend on the speed of movement. This means that animals move slowly enough, so that one may neglect air resistance, i.e. the second u-dependent term in Eq. (4.2.3). Speeds at which the resistance of air becomes equal to ground friction should be the top limit, because further on energy expenditures (i.e. net activity a) would be mostly due to air

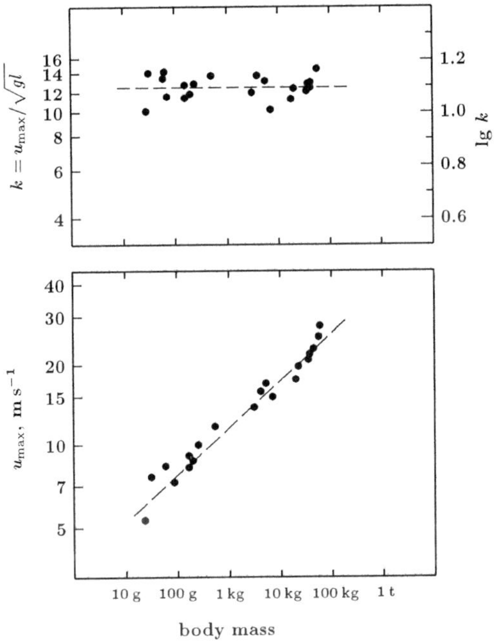

Figure 4.3. Land animals' maximum speeds of movement vs. their body size (mass). The maximum speeds of animal movement correspond to a constant value of Froude's number $k^2 = u_{max}^2/gl$. Equating energy expenditures for ground and air resistance, Eq. (1.2.7), and accounting for the definition of ground resistance, Eq. (4.2.4), the ground resistance, γ, may be retrieved from the known air resistance coefficient $c = 0.4$. It averages $\gamma = 0.04$ (Gorshkov, 1983a, 1984a). At a movement efficiency of $\alpha = 0.25$ we obtain the maximum energy cost of movement $\varepsilon = \gamma/\alpha \approx 0.16$, see Eq. (4.2.6) and Fig. 4.2.

resistance which grows very rapidly as the third power of u making movement at higher speeds extremely inefficient. Therefore the condition

$$\frac{1}{k^2}\frac{u_{max}^2}{gl} = 1, \qquad \text{or} \qquad u_{max} = k\sqrt{gl} \tag{4.2.7}$$

must correspond to maximum speeds recorded for animals of body size l. Empirical data on these speeds (see Figure 4.3) demonstrate that the maximum value of Froude's number u_{max}^2/gl remains constant for all body size intervals, i.e. depends neither on the speed u, nor on the size l. On these grounds one may state that the dissipation coefficient γ entering the definition of k, see (4.2.4), is also constant; as soon as the constancy of the air resistance coefficient c had been repeatedly tested in independent experiments (see references in Gorshkov, 1983a, 1984a).

The above leaves efficiency α as the only parameter to change with size l in the relationship (4.2.5). Its changes should be such that the product $\alpha\lambda_0$ remains constant due to the observed constancy of u_0 (Gorshkov, 1983a). As soon as λ_0 decreases with increasing body size l, the efficiency α must accordingly increase with the body size l growing. However, the efficiency α is limited from above by its maximum possible value $\alpha_{max} = 0.25$. After this value is reached, the energetic cost of movement ε assumes its minimum possible value and does not diminish further with increasing l, while the velocity u_0 begins to decrease with increasing body size l proportionally to λ_0, see (4.2.5). The data available on u_0, γ and λ_0 indicate that this limit is reached for the animal mass of $m \approx 100\text{–}300$ kg (see Figure 4.1). (The observed deviations from the average value of $u_0 = 0.3\,\mathrm{m\,s^{-1}}$ observed for the elephant and donkey do not go beyond the natural variability of u_0 in warm-blooded animals). Direct measurements made for horse and man showed that their efficiencies are actually close to the top limit (Brody, 1945; Atkins and Nicholson, 1963).

The decrease in efficiency of locomotion α for lower body sizes is only observed when locomotion occurs in the regime of oxygen balance, and energy expenditure is continuously compensated by oxygen consumed from the environment (Gorshkov, 1983a). All animals are capable of short bursts of locomotion in the regime of oxygen debt at maximum efficiency α, independent of their body size. That conclusion follows from the analysis of record jumps by animals of different body sizes (Gorshkov, 1983a, see Figure 3.1).

The observed independence of the mechanical power per unit body weight, $\alpha\lambda_0$, of body size ensures equal conditions for the existence of both small and large animals. Constancy of $\alpha\lambda_0$ means that smaller animals are characterised by a lower muscle efficiency α, because their weight-specific metabolic power λ_0 is greater than that of large animals (see Section 3.2). However, there are no physical reasons that could limit the efficiency of locomotion α of small animals. A possible way of decreasing α at smaller body size could be via reducing the ratio of muscle mass to body mass at still the same maximum efficiency of muscles themselves. However, such a morphological pattern is not actually observed (Gorshkov, 1983a). It means that small animals only use their muscles at the maximum possible efficiency in extreme situations of oxygen debt, e.g. during the highest or longest jumps. With the overwhelming part of their life occurring in conditions of oxygen balance, the efficiency of muscles α apparently drops in animals of smaller body sizes.

In the absence of physical reasons that could bring about such an effect, we thus conclude that the observed constancy of the mechanical power per unit body mass is apparently ecologically caused. It is only due to that feature that the energetic competitiveness of large animals, controlled by their mechanical power, is not inferior to the energetic competitiveness of a congregation of small animals of an equal mass (e.g. an elephant and a hundred thousand mice). Internal correlation of bodies of large animals and their ability to create and maintain a favourable internal milieu optimal for functioning of all cells and organs of the body, contributes additionally to the competitiveness of a large animal as compared to a non-correlated congregation of small animals. As a result, there appear ecological niches where large animals may exist alongside with smaller ones.

However, with body size increasing, the constancy of $\alpha\lambda_0$ is only supported up to a certain critical size, at which the efficiency of locomotion reaches its maximally admissible value $\alpha_{\max}$. Very large animals, with their body size exceeding the critical one, lose energetically to smaller animals and are forced to seek for some additional exotic means of increasing their competitiveness and of gaining an ecological niche fit for their existence. One such exotic means was the appearance of cultural knowledge in the *Homo sapiens* species.

4.3 MAXIMUM PERMISSIBLE SHARE OF BIOMASS CONSUMPTION BY LOCOMOTIVE ANIMALS

Let us denote here the metabolically-active, short-lived plant biomass per unit land surface as B. We further assume, that while moving across its feeding territory (home range), the animal eats up a part of the plant biomass equal to $B_L \equiv \beta_L B$, where β_L is the consumed share of plant biomass. The effective width of land band across which the animal eats up vegetation is close to that animal's body size l (4.2.1). The vegetation mass consumed in unit time with the animal moving at an average daily speed of u is equal to $B_L lu = \beta_L Blu$. The energy content of the eaten biomass is equal to $K\beta_L Blu$, where K is the energy content of live biomass, see (3.1.2) (we omit the low index '*lb*' in this chapter).

Coefficient of food assimilation (digestibility) for an animal may, for simplicity, be assumed equal to unity: for most animals it is actually equal to 0.8 (Gessaman, 1973; Kendeigh, 1974). Energy contained in the food, which is consumed during movement, must be equal to the average metabolic power of existence, i.e. $q = (\bar{A} + 1)q_0 = 2q_0$, where q_0 is the basal metabolic power (see (4.1.1) and Section 3.1):

$$K\beta_L Blu = (\bar{A} + 1)q_0 = 2q_0, \tag{4.3.1}$$

The relationship in Eq. (4.3.1) determines the speed of movement needed to support the animal's existence. It may be called the ecologically necessary speed $u_{\rm n}$:

$$u_{\rm n} = \frac{2q_0}{K\beta_L Bl} \tag{4.3.2}$$

The animal may only survive if its available speed $u_a = q_0/(mg\varepsilon)$, see Eqs. (4.1.5), (4.2.5) and (4.2.6), is larger than or equal to $u_{\rm n}$, $u_{\rm a} \geq u_{\rm n}$. In other words, **the energy provided to the animal by food should not be less than the energy spent by the animal during moving when searching for food and consuming it**.

Using Eqs. (4.1.2), (4.1.5), (4.2.5) and (4.2.6) this inequality may be rewritten as a limitation upon the consumed share of vegetation biomass:

$$\beta_L \geq \beta_{L_{\min}} = \beta_{L0}\varepsilon\left(\frac{l}{l_0}\right)^2, \qquad \varepsilon \equiv \frac{\lambda_0}{u_0} = \frac{\gamma}{\alpha}, \qquad \beta_{L0} \equiv \frac{2g\rho l_0^2}{KB} \sim 10^{-5} \tag{4.3.3}$$

or as a limitation upon the consumed biomass:

$$B_L \geq B_{L_{\min}} = B_{L0}\varepsilon\left(\frac{l}{l_0}\right)^2, \qquad \beta_{L0} \equiv B_{L0}B, \tag{4.3.4}$$

The estimate of the share of consumption β_{L0} of herbivorous animals was made taking the global average metabolically-active (edible) biomass equal to

$B = 4\,\text{kg m}^{-2}$ (Gorshkov, 1995). The relationship between scaling values of body size $l_0 = 10$ cm, body mass $m_0 \equiv \rho l^3 = 1$ kg and $\varepsilon = 1$ is retrieved from Figure 4.2. All the variables entering the right-hand part of Eqs. (4.3.3) and (4.3.4) are well-known. The available estimates of metabolically-active, short-lived plant biomass B for different ecosystems are presented in Gorshkov (1995).

Note that the basal metabolic rate q_0 has cancelled out from the relationships (4.3.3) and (4.3.4). The energetic cost of locomotion ε is determined by the dissipation coefficient, γ, and the muscle efficiency, α, and hence it cannot depend on the taxonomic group the animal belongs to, being only dependent on body size l. As a result, the limitations (4.3.3) and (4.3.4) are equally justified for all the land surface animals: insects, amphibians, reptiles, mammals and birds.

4.4 SETTLED AND NOMADIC LIFESTYLE OF LOCOMOTIVE ANIMALS

There is another way of interpreting of the necessary speed of movement u_n determined by Eq.(4.3.1). The metabolically-active biomass of vegetation per unit surface area $B = \rho L$ (L is the thickness of the layer of metabolically-active biomass when evenly spread over the considered territory, $\rho = 1\,\text{t m}^{-2}$, $B \sim 4\,\text{kg m}^{-2}$, $L \sim 4$ mm) may be expressed as $B = P^+\tau$, where τ is the turnover time for the metabolically-active (short-lived) biomass; P^+ is the net primary productivity in units of live biomass $\text{kg m}^{-2}\text{year}^{-1}$. The area of feeding territory S of a single animal, approximately equal to its 'home range' (Harestad and Bunnell, 1979; Damuth, 1981a, 1981b), may be expressed as $S \equiv N^{-1}$, where N is the population density of animals of a particular species per unit surface area.

The distance travelled by an animal across its feeding territory is of the order of S/l (l is body size). Denote the time in which the animal makes a round of the whole feeding territory as τ_S. (The band of width l must scan the whole territory of area S in time τ_S.) Then the necessary speed of movement u_n may be expressed as $u_n = S/(l\tau_S)$.

The rate of food consumption $Q \equiv (\bar{A} + 1)q_0/K$ by animals of a given species and a given body size l may be written using the value of consumed species-specific share β_l of plant production P^+ over the total feeding territory S, $Q = \beta_l P^+ S$. Substituting these values into Eq. (4.3.1) and cancelling identical terms in both parts of the equation we obtain the relationship between the species consumption share of vegetation biomass β_L and the species consumption share of the net primary production β_l:

$$\beta_L = \beta_l \tau_S/\tau \tag{4.4.1}$$

The share of consumption of net primary production $\beta_l = Q/(P^+S)$ may be expressed using the value of cumulative biomass B_l of animals of a given species of body size l, $B_l = \rho L_l$ (L_l is the layer thickness for animal biomass). Feeding territory S is related to biomass B_l by relationship $SB_l = m$, ($SL_l = l^3, m = \rho l^3$), where m is the average body mass of a single animal. We thus obtain for a species

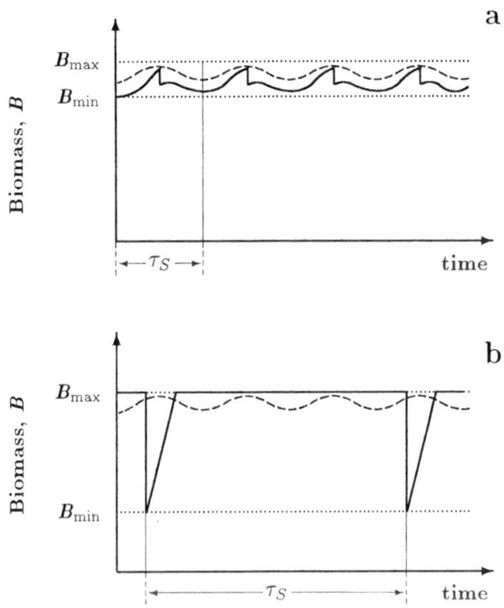

Figure 4.4. Oscillations of the natural community biomass caused by small settled (a) and large nomadic (b) animals. The dashed line indicates the range of natural fluctuations of plant biomass independent of the presence or absence of animals. τ_S is the time in which the animal makes a round of the whole of its territory. Note that consumption of biomass by small settled animals does not go beyond natural biomass fluctuations. In contrast, large nomadic animals substantially destroy the community's biomass and do not return to the same community until the biomass of all the parts of the community is completely restored. Note, however, that in both cases the community spends most time in the nonperturbed state. This is ensured by ecological limitations imposed on population densities of both large and small animals (cf. Figure 6.19 in Section 6.8.5).

consumption share of net primary production β_l:

$$\beta_l = \Lambda L_l / P^+, \tag{4.4.2}$$

where $\Lambda \equiv Q/l^3$ is the rate of food consumption per unit volume of the animal's body.

The relationship in Eq. (4.4.1) makes possible quantitative differentiation between settled and nomadic life styles. If an animal traverses its feeding territory in time τ_S equal to vegetation reproduction period τ, it would return to a given area within its feeding territory in exactly the time needed for the vegetation to reproduce the part of biomass eaten by that animal during the previous attendance of that area. In that case the share of consumed plant biomass and the plant production would coincide with each other: $\beta_L = \beta_l$.

In reality, most small-sized animals go round their feeding territories in a shorter time ($\tau_S \ll \tau$), each time eating away an amount of biomass significantly smaller than the permissible share ($\beta_L \ll \beta_l$). This permits the animal to visit any part of its feeding territory at practically any time and also reduces fluctuations of vegetation biomass on that territory. Small-sized animals may therefore exist in conditions of abundance of food and metabolic energy. Fluctuations of vegetation biomass due to its consumption by small animals do not exceed the natural fluctuations of that biomass in the absence of animals (Figure 4.4a). The presence of such animals does not leave any noticeable trace and does not disturb the natural distribution of the vegetation.

A large animal has to consume a very substantial part of vegetation biomass ($\beta_L \gg \beta_l$). Vegetation and the whole community around the animal are then destroyed, and the closed matter cycle in a local range where consumption took place is disrupted. Then the animal leaves the destroyed area and returns to it only after a time τ_S, when the steady state distribution of vegetation, community structure and the closed character of matter cycles are restored. During that succession time (which is much longer than the vegetation reproduction period, $\tau_S \gg \tau$) the net primary production is consumed by different species (species-repairers, see Sections 4.7 and 6.7) which act to close the matter cycle again. As a result, the effective share of net primary production consumed by large animals remains within the ecologically permissible norm when averaged over large time period of the order of τ_S (see Figure 3.3). However, large animals leave a noticeable trace in the observable distribution of vegetation, visible, e.g. from on board an aircraft (Figure 4.4b). At any given moment of time, only a tiny part of the enormous feeding territory appears to be fit for life of a large animal. This part is equal to the ratio $\tau/\tau_S = \beta_l/\beta_L \ll 1$, see Eq. (4.4.1). All the rest of the territory must be closely guarded against the intervention of competitors. A large animal remains constantly in a state of food and energy deficit (see Figure 4.5).

The strategy of existence following which a round of the feeding territory is completed during a time shorter than that of reproduction of vegetation, so that it does not leave any traces in the natural distribution of vegetation, corresponds to a *settled* lifestyle. The strategy following which such a round takes more time than the reproduction period for vegetation, and leaves a noticeable trace in the natural distribution of vegetation, corresponds to the *nomadic* lifestyle.

If the consumption share of vegetation production β_l is fixed, all the large animals, having $\beta_{L_{\min}} > \beta_l$, may only exist in the nomadic regime (Figure 4.5). At anomalously low values of β_l even small animals may fall into an obligatory nomadic lifestyle, if their $\beta_{L_{\min}} > \beta_l$. Small animals may also turn sporadically to a nomadic lifestyle if they reach extremely high population density numbers and consume an extremely high share of vegetation biomass, while the relationship $\beta_L > \beta_l > \beta_{L_{\min}}$ holds (Figure 4.5). Such facultative nomadic behaviour apparently arises in communities disturbed by man (e.g. arable lands invaded by locusts, managed forests destroyed by insect pests, see Section 6.8.3).

In a settled regime any animal in its natural community consumes less than 20% of all the eatable biomass in any given spot of its feeding territory (Golley, 1973), which guarantees sustainability of the whole community under any fluctuations. In a nomadic regime the share of biomass consumption may even reach unity (all the edible biomass is completely destroyed), and, consequently, the whole community perishes. Following that, a long restoration period is needed during which the biomass and the community as a whole are generated anew and then regain their most competitive steady state. Such rare cases of nomadic behaviour are only encountered in the human population (slash-and-burn agriculture and modern

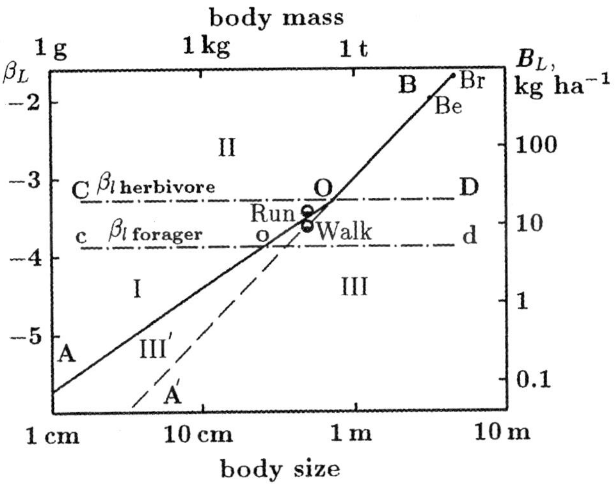

Figure 4.5. The share of vegetation biomass consumption by herbivores and foragers vs. their body size (mass). The solid line shows the minimum value of edible biomass $B_{l_{\min}}$ (and the corresponding share of the total edible plant biomass consumption $B_{l_{\min}}$ which still covers energy expenditure of animals of a given body size associated with their movement across their feeding ground. The circles Run and Walk correspond to Ho—*Homo sapiens*, Be—*Beluchitherium*, Br—*Brachiosaurus*. The breaking point of the solid line corresponds to the breaking point in Figure 4.2, which corresponds to the efficiency of movement reaching its maximum. Movement efficiency is at maximum to the right of point O (line OB). Line AO describes the observed decrease in efficiency for lower body sizes (see Figure 4.2). The dash-dot line CD corresponds to the share β_l of consumption of net primary *production* by herbivorous mammals in natural terrestrial ecosystems (see Figure 3.5b). Note that line COD crosses line AOB at the latter's breaking point, which is presumably not a random coincidence. Line cod gives the admissible share of consumption of plant production by foraging mammals (four times less than the herbivores' share). Range III to the right of line A′OB is prohibited energetically. Range III′ limited by lines A′OA is prohibited physiologically. Range I, limited by lines AOC for herbivores (Aoc—for foragers) is open for them in a settled regime, $\beta_L/\beta_l < 1$. Range II, limited by lines COB (coOB—for foragers) is open for nomadic animals, $\beta_L/\beta_l > 1$. Existence of animals of body size falling to the right of point O for herbivores (and to right of point o for foragers) is possible only in a nomadic regime.

clear cutting of wood) and also in certain insects in human-disturbed ecological communities (Holing, 1978; Isayev *et al.*, 1984).

Under pressure of anthropogenic activities, natural ecological communities are permanently destroyed every few tens of years ($\tau_S/\tau \sim 10$) in the majority of the continental areas, so that they are never able to reach their steady state corresponding to the maximum stability of community and ecosystem as a whole. The steady state of most natural communities might only be restored if the frequency of anthropogenic disturbances were reduced by at least an order of magnitude on a global scale.

4.5 CARNIVORES

As we have seen above, the stability of organisation of ecological communities is mainly dictated by the character of interaction between autotrophic plants and heterotrophic herbivores. Let us now consider the internal structure of the heterotrophic part of the ecological community in more detail, subdividing it into herbivores and carnivores.

Animal biomass production by herbivorous animals is 10 times less than the production of vegetation they consume, i.e. it is less by more than an order of magnitude than the net primary production of the ecosystem. Therefore, in order to be able to live on animal biomass, a carnivore must be locomotive. Their feeding territory must be 10 times larger than the feeding territory of their prey and, respectively, their population density number should be 10 times lower (given a metabolism level similar to that of their prey).

Predators should not only be locomotive animals themselves, but they may only feed off locomotive prey. Live biomass produced by locomotive prey is concentrated into their projection area and presents a 'source' of productivity with characteristic density exceeding net primary plant productivity by several orders of magnitude. In contrast to a herbivore, a carnivore does not need to collect the evenly-spread food products by moving over the whole feeding territory. For carnivores, it is enough to move from one 'source' (prey) to another or to keep attached to the 'source' itself (the phenomenon of parasitism). All the carnivores, despite their enormous feeding territories, always remain in the state of energy abundance.

Herbivores consume only a small part of the net primary production of plants. Due to this fact herbivores could, in principle, increase their population number at constant biomass and productivity of feeding plants. However, this is ecologically forbidden, being incompatible with the stable existence of the community (see Section 3.7). By contrast, predators consume a major part of the 'net' production of their prey. It follows that under natural conditions, predators cannot increase their population number if the population number of their prey remains constant. Therefore, natural predators of any body size are incapable of disrupting the ecological equilibrium in the community. (However, under the artificial conditions of suburban areas, dogs and cats living off man may completely extinguish their natural prey.)

The most important ecological function of predators in the community is to eliminate decay (weak, ill, injured, etc.) individuals from the prey species of the same community (see, e.g. Vorisek *et al.*, 1998). In strongly disturbed external conditions, such decay polymorphism of all individuals sharply increases (see Chapter 9). This brings about a sharp increase in the population number of the respective predator species. On the contrary, when the decay polymorphism of the prey species tends to zero, the predator species becomes less important and its population number diminishes. The same reasoning is valid for the correlation between plants and herbivorous consumers, mainly insects. A sharp increase of the decay polymorphism in plants precipitates a respective sharp increase in the population number of herbivorous insects. Such processes are observed after fires, forest clear cutting and other strong perturbations of natural communities (Holing, 1978; Isayev *et al.*, 1984; Morneau and Payette, 1989). Under stable natural conditions when the decay polymorphism in all the plant species of the community is kept at its minimum, population numbers of all the plant consumers are also very low, leaving plant biomass to be decomposed by reducers in the form of dead bodies of plants.

Correlated associations of the type of 'predator–prey', as well as all the diverse forms of symbiosis between various species, may only propagate and be further supported if ecological communities which include such associations become more competitive than communities without them. If one considers the interaction between the predator and its prey at a species level, it appears extremely difficult to reach even a dynamic stability of the system, and remains absolutely impossible to explain the sustained genetic stability of correlation of such kind.

One may similarly treat the presence of parasites in the bodies of higher living beings. Naturally, the malfunctioning individual that contains parasites in its body features lower competitive capacity as compared to the healthy one. Thus, decay of the level of organisation in parasites, and decrease of their functional activity, should have led to higher competitive capacity of their hosts and, hence, to their higher numbers. However, that is not actually the case, and no on-going degradation and extinction of parasites is observed. Even when losing some of their inner organisation, parasites only do so in the direction of increasing the efficiency of their functioning in the host body. The loss of their own respiration and distribution systems by some parasites only works to increase their competitive capacity within the host body, as compared to those parasite species that have retained such morphological features.

If a sufficiently large population of competitively-interacting parasites exists either within the host body or on its surface, their functional organisation may be supported due to stabilising selection among them, taking place within the body limits of each separate host. However, the inevitable loss of competitive capacity of a host affected by parasites should result in decreasing the number of such deficient individuals, and in their further exclusion from the population together with their parasites. In the opposite case, when the body of a single host is not large enough for a population of parasites to live in, the process of expulsion of the host affected by parasites from its own population might only be retarded via the decay of the organisation of parasites, that is via the extinction of the phenomenon of parasitism itself. In both cases, one appears unable to explain the observed stability of the phenomenon of parasitism, considering it at the level of host–parasite interactions.

The only possible way to explain the phenomenon of parasitism is to tackle the problem at the level of the ecological community to which the host species belongs. The complex nature of correlation between individuals of different species in the community results in that the community, in which a sufficient number of individuals from a given species contain parasites, appears to be more competitive than communities in which the individuals from the same species host no such parasites. Competitive interaction between communities in a population of communities results in the support of those communities which favour parasites.

The cause for higher competitive capacity of communities with parasites may be related to resulting limitations upon and control of population densities of host species in these communities. For example, an uncontrolled growth of the population density of large animals could disrupt the closed biochemical cycles of matter within the community, that are a prerequisite of environmental stability (see Section 3.7). A spontaneous increase in population density of hosts accelerates the spread of parasites, that in most cases, entails an epidemic and subsequent fall in population density down to the initial optimal value. Higher degree of correlation between the host and its parasites should also work to improve the competitive capacity of a community. That may be the reason for the observed complicated patterns of successive changes of hosts during the life cycle of many parasites (Raven and Johnson, 1988).

4.6 DIFFUSION OF EXCRETA

Let us now consider an additional, very peculiar, limitation on consumption by locomotive animals.

Fall-off of dead parts of plants provides for stability of concentration of nutrients in a local ecosystem, only if these nutrients, randomly distributed in the fall-off, have time to diffuse back to their initial position before they are again used by autotrophs. This may happen in two ways: short-lived leaves are small enough to be distributed evenly after falling off, while large tree trunks grow slowly and just as slowly decompose after falling.

Locomotive herbivores collect nutrients from their feeding territories, transport biogenic elements within their bodies and concentrate them in their excreta, of which 80% fall to urine and 20% to faeces (Kleiber, 1961; Kendeigh, 1974). As follows from the data presented in Figure 3.3, small locomotive animals (like dungbeetles) feeding off the faeces consume no more than 10% of them. The principal part of the excreta is destroyed by bacteria and fungi. Therefore, a stationary state is only possible when the nutrients, transported and concentrated in the excreta, spread again over the territory they had been collected from in the course of diffusion or with help of external matter fluxes; this process should not be longer than the period of vegetation reproduction τ.

Let L_{ex} be the average distance between two separate excreta of an animal (Figure 4.6). The average amount of excreta, proportional to the animal's body size l, may be denoted as δl^3, where $\delta \approx 0.001$ (Kleiber, 1961). Moving along the distance L_{ex} the animal scans a band of width l and eats the amount (volume) of vegetation equal to $\beta_L L_{\text{ex}} l L$, where $L = B/\rho$ is the average thickness of the layer of edible biomass. The excretion distance L_{ex} is determined from the equality $\delta l^3 = \beta_L L_{\text{ex}} l L$, from which we deduce, assuming that $\beta_L \sim \delta \sim 10^{-3} \div 10^{-4}$:

$$L_{\text{ex}} = l^2/L$$

The time τ_D that excreted nutrients take to diffuse along the distance from which they had been concentrated is of the order of:

$$\tau_D = L_{\text{ex}}^2/D = \left(\frac{1}{DL^2}\right) l^4, \tag{4.6.1}$$

where D is the diffusion coefficient with dimension $\text{cm}^2\,\text{s}^{-1}$ (that is why the term

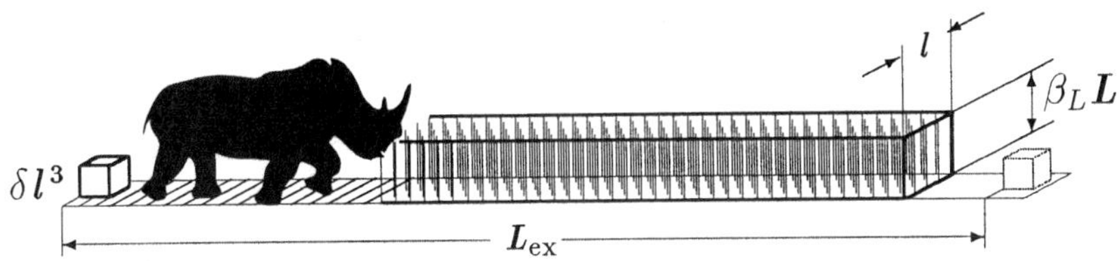

Figure 4.6. Diffusion of excreta of animals. A moving animal eats away biomass inside a band of approximate width l (l is the body size). The effective thickness of the layer of consumed biomass is equal to $\beta_L L$, where β_L is the share of biomass consumption by that animal, L is the thickness of edible plant biomass. The average distance between two excreta is equal to L_{ex}. The average volume of excreta is δl^3, where δ is a small constant coefficient of the order of β_L. In a stationary case the biogens contained in the excreta should spread over the territory where they have been collected. The time of diffusive spread $\tau_D = (l/l_D)^4 \tau$, where $\tau \sim 1$ year is the time of vegetation reproduction, $l_D = (\tau D L^2)^{1/4} \approx 1$ cm, see (4.6.2). A stationary state on the basis of diffusion is only possible at $l \leq l_D$. At $l \sim 1$ m the time of diffusion spread of excreta exceeds the time of vegetation reproduction by a factor of 10^8.

L_{ex}^2/D has the dimension of time). Spreading of excreta occurs mostly due to their molecular diffusion in water solutions (including molecular diffusion in plant roots). The molecular diffusion coefficient in water is of the order of magnitude $D \approx 10^5\ \text{cm}^2\ \text{s}^{-1} = 10^{-2}\ \text{m}^2\ \text{year}^{-1}$. The condition $\tau_D \leq \tau \approx 1$ year imposes a limit on the value of l:

$$l \leq l_D \equiv (\tau D)^{1/4} L^{1/2} \tag{4.6.2}$$

Assuming the average thickness of the metabolically-active edible live biomass layer of vegetation $L \approx 4$ mm (Section 4.3) we obtain $l_D \approx 1$ cm. As seen from Eq. (4.6.2) this value is hardly sensitive to even significant variations in any of the input variables.

Therefore, there is no problem of the excreta spread for very small animals with $l \leq l_D \approx 1$ cm. Even if this spread occurs by means of the slowest possible process, i.e. molecular diffusion in liquid, it takes less time than reproduction of vegetation. In contrast, for animals with body sizes $l \geq l_D$ the time of diffusive spread of excreta τ_D sharply increases with growing body size l, see Eq. (4.6.1). The time τ_D increases by four orders of magnitude, while body size l increases by only one. At $l \approx 10^{-1}$ m, the time τ_D is already equal to 10^4 years, while at $l \approx 1$ m we have $\tau_D \approx 10^8$ years. These estimates do not change considerably even if the diffusion coefficient D changes by a few orders of magnitude. Therefore, nutrients excreted from the bodies of large animals cannot be returned by the process of diffusion to those places from which they had been collected. Thus, processes of diffusion cannot account for the observed sustainable distribution of nutrients within natural undisturbed ecosystems.

Ecosystems inhabited by large animals may only remain stable due to external natural fluxes of matter, which mix and transport the animal excreta across the land surface. The principal role in mixing the excreta of large animals is played by the surface runoff of precipitated water. The amount of nutrients mixed by surface water runoff may be assessed from the level of ionic runoff of carbon, which amounts to about 10^{-3} of the organic carbon production (Watts, 1982; Schlesinger, 1990). In a stationary case, the amount of nutrients concentrated in the excreta of large animals must not exceed the amount of nutrients returned on average to their initial location by the surface ionic runoff. **Consequently, the global average share of consumption by all the large animals must not exceed several tenths of a percent of net primary production**, which corresponds to the distribution presented in Figure 3.3. In the areas of low surface runoff large animals may comparatively quickly turn the land surface into desert. Note that this conclusion represents an independent estimate of the large animals' share of consumption of net primary production of the biosphere compatible with environmental stability.

In coastal areas, estuaries, river floodlands and also in rivers and lakes themselves, where the ionic runoff locally approaches the level of net primary production, the share of consumption by large animals may be significantly larger than the global average. In such areas, large animals (both predators and prey) may develop very high population densities as compared to those on land.

After a substantial disturbance of an ecosystem that entails destruction of soil cover and of the natural homogeneous distribution of biogens (e.g. fire, volcanic eruption, etc.), locomotive animals may play a significant role as species-repairers in spreading biogens across the affected territory and transporting them to places where physical water fluxes are of no help (e.g. transporting biogens up a hill). Therefore, the structure of the pioneer vegetation during the early stages of a community's recovery after perturbations offers a larger share of consumption of its production to locomotive animals than in a stationary case (Section 6.7.2).

4.7 CONCLUSIONS

We will now sum up the results for the energetics of ecological communities obtained in Chapters 3 and 4.

The source of external energy supporting the life of communities of the Earth's biota is solar radiation. Due to the zero mass of photons, solar energy cannot be stored in the environment in the form of short-wave radiation. The Earth's plants, for which solar radiation is the primary source of energy, are therefore incapable of increasing their energy consumption by locomotion. That is why green plants are immobile and form a continuous vegetation cover. A continuous cover of immobile vegetation is typical for all the land ecosystems and the oceanic shelf.[1]

The fact that plants do not need to move results in the possibility of their existence in the form of weakly-correlated (modular) multicellular individuals of large size. The mode of functioning of plants is similar to that of the functioning of a completely non-correlated set of single-cell individuals of equal metabolically-active mass. That feature provides for lowering the fluctuations of photosynthesis

[1] Mobility of single-cell individuals in certain species of oceanic phytoplankton is not related to an increased energy consumption. Rather, it is determined by the fact that phytoplankton biomass follows a certain non-random distribution with depth and that a vertical movement is needed to keep that phytoplakton within the euphotic layer (Sieburth, 1976; Gorshkov, 1980).

of organic matter within the local ecosystem, as based on the action of the law of large numbers.

Destruction of synthesised organic matter in the community is also performed by the immobile, weakly-correlated individuals—bacteria and fungi. (Mobility of certain forms of bacteria, similar to that of certain forms of phytoplankton, pursues the aim of vertical movement through soil layers, but not the increase of energy consumption.) Taking into account the respiration of plants themselves, immobile weakly-correlated individuals account for decomposition of more than 95% of the photosynthesised organic matter (Section 3.7). Similar to the synthesis of organic matter, such organisation of the destruction process makes it possible to reduce fluctuations in the destruction of organic matter within the local ecosystem, as based on the action of the law of large numbers.

Organisation of the community on the basis of a large number of completely uncorrelated or modular, weakly-correlated parts, makes it possible to control both the rate of synthesis and of destruction of organic matter to a high degree of accuracy. That, in its turn, makes it possible to keep the matter cycles strictly closed, and the environment steady provided there are no external disturbances. Also, such organisation enables the community to perform an adequate reaction directed towards compensation of any external disturbances of the ecosystem. Epilithic lichens represent an example of a simplest type of community composed solely of immobile individuals (of algae and fungi).

The existence of a universal minimum admissible value of energy consumption per unit volume (or mass) in all the living beings results in a strict limitation upon the admissible effective vertical size of an individual, provided the flux of energy incident upon the unit projection area of that individual upon the Earth's surface is constant. That size is controlled by the flux of solar radiation and by the photosynthesis efficiency. For the biosphere, on average, the solid cover of the metabolically-active biomass for both synthesisers of the organic matter (green plants), and their reducers, heterotrophs (bacteria and fungi), may only reach a thickness of no more than 3 mm. The observed extremely large body size of woody plants is explained by the fact that the overwhelming part of space occupied by those plants is empty and most of the biomass of such plants is metabolically inactive (effectively dead).

In order to support life of large individuals with a metabolically-active layer much thicker than that formed by plants and other immobile organisms, it is necessary to employ energy fluxes, which by far exceed the flux of solar energy, even in the hypothetical case when the biochemical efficiency of the solar energy consumption is equal to unity. Hence, large individuals cannot form a solid continuous cover (a 'stationary crowd'). The feeding energy for these individuals should be collected from a large surface, significantly exceeding the projection area of such individuals upon the Earth's surface. Since, apart solar radiation, there are no other forms of primary energy available, while solar radiation cannot be accumulated, such individuals cannot feed off the solar energy immediately. Hence, large individuals may only feed off the secondary energy of synthesised organic matter which, unlike solar photons, has non-zero mass and can be accumulated locally. In other words, large individuals may only be heterotrophic, and have to participate in destroying organic matter.

Feeding off the locally accumulated plant biomass is only possible via movement of large individuals over their feeding territory. Movement, on the other hand, demands that the body of a large individual is a rigidly-correlated formation (unlike weakly-correlated bodies of immobile plants).

The community consisting of immobile and, hence, of effectively small-sized modular individuals can afford strict equality between the synthesis and destruction over any arbitrarily small time interval, i.e. of strict constancy of the organic and inorganic mass in each local ecosystem. This may be assured due to small relative fluctuations characterising any system comprising a large number of uncorrelated components.

The introduction of large mobile animals into the community results in a drastic change in the way of functioning of that community. Mobile animals feeding off the accumulated biomass inevitably bring about sharp fluctuations of that biomass in any local area of the community. The animal very quickly eats away that biomass, after which it is regenerated very slowly. That, in its turn, results in fluctuations of inorganic matter excreted by the animal into the environment after consumption of organic matter. Hence, the state of the environment ceases to be stationary, and suffers significant perturbations. Stationary state of the environment may only be reached after averaging over a long enough time period. That result imposes strict limitations upon the possible species composition and the behaviour of the animals entering that community.

Random oscillations of biomass and of destruction rate of organic matter in a community being strong, it becomes impossible to close matter cycles and to ensure the long-term stability of the ecosystem. Therefore, the average consumption quota of plant products by mobile animals should not exceed the natural fluctuations of plant production. When that condition is met, the presence of animals in the community does not leave any apparent traces in plant biomass. Fluctuations in plant biomass introduced by large animals grow with the body size of those animals. Hence the quota of consumption of plant production by those animals should drop for larger body size of the consumers. That conclusion agrees with the observed distribution of plant production over consumers of different body size (Figure 3.3). Reducing the number of species of large animals slows down the rate at which that consumption quota drops for a separate species with their body size increasing.

The ubiquity of large animals (i.e. the fact that they are present in the overwhelming majority of ecosystems) means that the net impact of large animals to the community's stability is positive. Communities where large animals are present appear to be more stable and competitive than communities without them. There may be several reasons for that.

After external perturbations the population density of species-repairers reaches its maximum value, while where no external disturbances are present, the ecosystem contains very few individuals of species-repairers (Sections 6.7 and 6.8.2). For example, in a climax spruce, forest birch and aspen (species-repairers) are present in very low numbers and may not even form a population. The genome of every species-repairer should contain information about optimal population density and behaviour of its individuals that corresponds best to the task of rapid recovery of the ecosystem after external disturbances. That information is only put into operation in a disturbed ecosystem. Therefore, the appropriateness of that information can be tested and maintained only in a disturbed environment, when species-repairers form large populations and there opens a possibility for stabilising selection to operate.

Under normal environmental conditions that information is continuously decaying. Natural disturbances are of a rare and non-regular nature. If the time interval between two successive disturbances becomes too long, the genetic information of species-repairers may substantially degrade. As a result, natural communities would not be able to cope with natural disturbances and the stability of their organisation would be undermined.

There are two ways to settle the problem. Firstly, stabilising selection of the normal genome of the species-repairers can be performed by the dominant species in the community. In that sense it is a kind of artificial selection: as soon as individuals of one species are unable to form its own population, they are selected by individuals of another species. The mechanisms of such artificial selection may be very complex and difficult to reveal. One may only speculate about it. For example, the dominant plant species (e.g. spruce) may modify the biochemical composition of soil in such a way that only normal birches will grow best in the ecosystem, while birches with decay genetic information will lose the competition. Also, dominant plant species may support the existence of locomotive animals that would eat up decay individuals of species-repairers (e.g. eliminate young birches that grow too slowly or too rapidly, etc.). The same function can also be performed by large animals under disturbed conditions, the efficiency of intraspecific stabilising selection of species-repairers. The observed increase in population numbers of large animals (e.g. elks) during processes of forest succession after fires, clear cutting and other disturbances testifies for such a role of large animals (Section 6.8.2).

Secondly, it is possible to introduce regular disturbances of biotic nature into the ecosystem in order to 'train' species-repairers. This task can be also performed by large animals. For example, elephants are known to make big clearings in the forest, where pioneer plant-repairers (herbs, shrubs) are given a possibility of forming relatively dense populations.

The described ecological functions of large animals (we have not mentioned here the previously discussed stabilising function of carnivores, see Section 4.5) contribute to the stability of ecological communities and thus allow large animals themselves to continue.

The obtained estimates of size of local ecosystems, that were based on the condition of low fluctuations of the community's major characteristics (Section 3.5), show that local ecosystems are much smaller in size than typical feeding territories of large animals. However, the biological community of the local ecosystem is able to control population density of large animals, e.g. by changing production of edible parts of plants. Communities that maintain population density of large animals at an optimal level become most competitive and force out all other communities. As a result, optimal population density of large animals is maintained on large territories covered by a population of local ecosystems with the most competitive communities. In that sense, large animals represent a certain component of the environment, which is kept, similarly to concentrations of the important nutrients, at a certain optimal level by the population of communities consisting of plants and microorganisms (see also Section 5.8).

Note that it is erroneously assumed sometimes that the community should include whole populations of all the species of large animals present in the ecosystem. In that case, the size of the community could not be less than the largest territory occupied by a population of large animals, and there would be no populations of communities possible at all, leaving unsolvable the task of preventing the internal correlation of communities from decay (Section 2.8). The estimated size of the community appears to be of the order of several metres (Section 3.5). We consider this result as one of the most important statements made in Chapters 3 and 4.

Easily noticeable disturbances produced by large animals in the ecosystem may be misinterpreted as the principal managing role of large animals in the community. However, as we have seen, in natural communities large animals are only allowed to consume a tiny part of the net primary production of plants, which is achieved via regulating population density of large animals by other species of the community. Thus, the community controls the large animals, but not vice versa, as it is sometimes asserted (Sher, 1990). Substantial destabilisation of ecological communities following artificial elimination of large animals (e.g. due to hunting) does not contradict this statement. There are no unnecessary species in the community, so that elimination of any of them (including large animals) would impair the community's wellbeing. However, the effect of elimination of dominant species of the community (plants, bacteria and fungi, see Figure 3.3) by far overrides disturbances that may be introduced by elimination of large animals. Elimination of dominant species leads to rapid complete degradation rather than to gradual destabilisation of the community. Thus, it is reasonable to think that during evolutionary process as well, changes in the dominant species impose on average by far a more drastic effect on the community structure than evolutionary changes in large animals.

Under natural conditions, the population density of large animals is determined not by the availability of food (the latter is always present in excess to the needs of the animals, see Section 4.3), but by the condition of maximum stability of the ecological community or its rapid recovery after disturbances. Information about the optimal population density should be written in the normal genome of a species, as well as the programme of correct interaction with all the other species in the community.

An increase of the population density of large animals above the optimal value presents a serious danger to the community's integrity due to their ability to rapidly destroy the community's biomass. Thus, most behavioural traits the animal displays are aimed at preventing excessive growth of population density of large animals, rather than to stimulate increase of population density to the limit the food resources can stand. That aim is reached via strict control of the size of the feeding territory per single individual (McNab, 1983). Such control is based on various interactions of individuals at all levels of the community organisation (voice signals of neighbours, migration of individuals with too high population density (McFarland, 1985), control of population density by parasites and carnivores, etc.). All these correlated interactions of individuals, often perceived as 'altruistic', are aimed at ensuring the maximum possible stability of the community. Communities that are able to stabilise their optimal environment in the most efficient way, survive. Species unable to perform correlated ('altruistic') interactions with other species in the community may increase their population density until resources permit, disrupt correlated organisation of the community, disturb the community's environment and finally perish together with all the communities that favoured their existence. This is what is now going on with the global environment under anthropogenic impact.

Energy spent by an animal for making the rounds of its feeding territory quickly increases with the animal's body size due to purely physical reasons (Section 4.2). According to the ecological reasons outlined above, the energy of food consumed by the animal per unit of the distance covered should reduce with the increasing body size of the animal (Sections 3.6 and 3.7). The life of an animal is physically possible when the energy spent to traverse that territory is regained with food the animal collects on that territory. This ecological condition is only met for small enough animals, their size not exceeding a certain critical value (Section 4.3). The body size of man coincides with that critical body size in its order of magnitude. With the animal body size exceeding that critical limit, the share of plant food it consumes inevitably grows. To support the ecological equilibrium, such animals may only exist in nomadic mode, at a very low average density of their population number, while at the same time saving an enormous feeding territory from intervention by competitors.

Biotic regulation of the environment is determined by a strictly specified relation between the synthesis and destruction of organic matter, which should be dependent on the current environmental situation. Possible forms of such a relationship are limited by the above enumerated strictly described laws of nature. Within the limitations set by those laws, biological relations may assume any forms, however complex. One of those forms is the trophic structure of the destruction of organic matter, that is, the way the heterotrophs are divided into reducers, destroying dead organic matter, and the consumers, who devour live individuals. Herbivores, carnivores of the first order eating away the herbivores, carnivores of the second order which feed off the carnivores of the first order, etc. are identified in natural ecosystems. Such a division into different levels forms a well-known ecological pyramid of energy fluxes, which is often envisaged as the basis of ecology (Odum, 1983). However, the pyramidal organisation of energy fluxes is only observed on condition of low ratio of production to consumption in a population of species at each level of the pyramid. For reducers like bacteria and fungi that consume the most part of the net primary production (see Figure 3.3), and for which that ratio is not low, such a pyramid loses sense as soon as the energy flux practically does not change from one species to another (Gorshkov and Dolnik, 1980; Gorshkov, 1982b).

Carnivores feature certain peculiarities (discussed in Section 4.5), that are based on the general physical laws of nature. Details of interaction between the carnivores and their prey are controlled by the correlated nature of interaction between all the species in a natural community which, in principle, may be considered as one of the forms of symbiotic interaction. Dividing the interaction between the species in the community, such as commensalism, amensalism, predation, parasitism, symbiosis, etc. (Begon *et al.*, 1986) is quite artificial and arbitrary, and in no way defines the actual sense and diversity of these interactions. All such interactions are but different forms of correlation between the species in a natural community and are similar to correlation between the different organs in a body or organelles and biological macromolecules in a cell, being aimed at maintenance of stable organisation of the ecological community.

The enormous power of information fluxes processed by the molecular structures of living beings in a natural community exceeds by many orders of magnitude the maximum achievable fluxes of information flowing through all the computers of modern civilisation (Chapter 7). This indicates that it is hopeless to try to construct any mathematical models pretending to describe the actual processes within the natural ecological communities. These processes are many orders of magnitude more complex than the processes taking place within a separate living individual, in particular those evolving within the brains of large animals and man, and will apparently never fall subject to detailed modelling. (That comment does not refer to artificial communities, devoid of a programme of stabilisation of the environment, the only aim of their construction being to provide enough food for humans.) Meanwhile, studying the physical limitations upon the processes taking place in the community yields reliable results and opens the way to unequivocal conclusions.

5

Ecological Principles of Biotic Regulation

5.1 ECOLOGICAL LIMITATIONS ON EXPANSION OF SPECIES

As noted in Section 2.1, expansion is one the most important properties of life, ensuring its stability. It is a most general characteristic of life, observed in every species; it even covers such advanced forms of claiming new territories as space ventures by man.

Expansion in its general sense may occur in two ways. First, population density of individuals may increase within the boundaries of a given territory. This type of expansion may be called 'intensive'. Second, population may expand into as yet unoccupied areas ('extensive' expansion). Genetic evolutionary changes accompanied by an increase in competitiveness become fixed in the population in the course of expansion.

The most important feature of the expansion process is that, during expansion, competitive interaction and, consequently, stabilising selection in the population weaken or even cease altogether. Individuals spend their efforts in occupying new positions instead of competing with each other. This implies that a new progressive evolutionary trait only has a chance to persist in the population if the process of expansion of its carriers takes a shorter time than it takes such a trait to decay in the absence of stabilising selection.

Were infinite territories and fluxes of energy available for life, expansion (either 'intensive' or 'extensive') would have been infinite as well. Competitive interaction would be totally switched off, so that stabilisation of the existing level of organisation of biological objects would be impossible. Therefore, life may exist sustainably only in the absence of energy and territory abundance, i.e. when all available energy fluxes are already claimed by the biota. Presence of energy or territory abundance entails biological expansion that undermines the process of competitive interaction, which is the only guarantee for the stability of life organisation (Chapter 2). Hence, evolutionary processes necessarily accompanied by expansion may only occur very infrequently, so that species spend the majority of their time in a state of evolutionary stasis (Haldane, 1954).

5.2 BIOTIC AND INORGANIC FLUXES OF MATTER IN THE BIOSPHERE

The present-day net primary production of the whole biosphere in units of mass of organic carbon is estimated as $100\,\mathrm{Gt\,C\,yr^{-1}}$. Carbon amounts to only about one-tenth of the overall live organic mass (Odum, 1983; Kendeigh, 1974). Thus the annual production of live organic matter may be estimated at $10^3\,\mathrm{Gt\,yr^{-1}}$. During the whole period of life existence, that is, about $4 \cdot 10^9$ years, that production should have reached $4 \cdot 10^{12}$ Gt. This figure practically coincides with the mass of the planet, equal to $6 \cdot 10^{12}$ Gt (Allen, 1955). The sphere available to life is only the biosphere, including the atmosphere, the ocean, and the thin soil layer on land, its overall mass being of the order of the total mass of oceanic water, which is $1.4 \cdot 10^9$ Gt (Watts, 1982). The production of biota over its lifespan exceeds that mass by a factor of several thousand.

Hence, the same atoms must have entered the synthesised organic matter many thousands of times and, for that process to be possible, all of the synthesised organic matter must have been destroyed into its inorganic components again and again. In this sense, all life-important chemical elements present in the environment may be called biogens as being many times involved in biochemical cycles. We use the term *biogen* along with the possibly more common term *nutrient*.

The necessity of keeping the environment stable in the presence of on-going, powerful, biochemical cycles of matter results in the fact that life is only possible on the basis of organic (i.e. energy-rich) substances, their energy available for use after these substances are destroyed. Since death is inevitable for any living being, the organic matter contained in all the dead bodies should necessarily be decomposed by other living individuals. Some species synthesise organic matter, directly consuming the solar energy, others destroy that organic matter, using the energy contained in it. Correlated communities of different species must necessarily be organised to support life.

For ecological communities to be able to support a stable environment, it is necessary to ensure that the biological fluxes of both production (synthesis) and destruction of organic matter significantly exceed the external abiotic flux of biogens entering the biosphere. Abiotic fluxes of biogens (e.g. filtration of inorganic carbon from the Earth's mantle) work to perturb the optimal environment suitable for life. These processes are counteracted by biological ones that are aimed at relaxation of the environment to the initial optimal state. If biological processes are too weak compared with abiotic ones, no biotic regulation of the environment is possible. Hence, the biota should tend to increase the power of biological processes to their maximum possible value, so that the difference between biotic and abiotic power is as high as possible. In the modern biosphere, fluxes of production and destruction of organic matter exceed the observed external fluxes of matter to and from the biosphere by a factor of 10^4 (see Figure 5.1 below and Section 5.4).

The external flux of biogens entering the biosphere is determined by the structure of the Earth's depths and cosmic processes. The Earth's biota is incapable of changing that flux, i.e. the values of external abiotic fluxes dictate the value of biological production compatible with stable organisation of life. This means that natural biota cannot arbitrarily change the flux of synthesis or destruction of organic matter.

5.3 EVOLUTIONARY PROGRESS AND ENVIRONMENTAL DEGRADATION

Evolutionary changes occur in the direction of increasing competitiveness of the biological object. As noted in Sections 2.4 and 2.11, increase in competitiveness does not automatically guarantee an increase in the level of organisation of the evolving object and, in particular, does not necessarily lead to increased regulatory potential of the biota.

In other words, in the course of evolution there may appear highly competitive communities that, however, do not care to maintain a stable environment. Such communities may be called 'destabilising'. If 'destabilising' communities were given a chance to dominate and to propagate throughout the whole biosphere, it would finally lead to degradation of the environment on a global scale. Thus, evolutionary expansion of such communities should be prohibited.

Let us now consider ecological limitations imposed on the evolutionary process. Let τ_κ be the time of noticeable degradation of the environment due to violation of the biotic regulation mechanism by destabilising communities. If a certain community is highly competitive in the initial environment where it actually originated, but cannot support this environment in a stable state, it will lose competitiveness when the environment which is optimal for that community is substantially destroyed. Then it can be forced out by normal communities capable of biotic regulation. 'Noticeability' of the environmental change means that such change should exceed the biotic sensitivity $\varepsilon_b \sim 10^{-4}$ (see Section 3.4). If the change in the store of biogens M is less than $\varepsilon_b M$, the community does not feel such change and does not lose competitiveness. The time τ_κ is equal to the ratio of the available store M of biogens to the rate of their change. This rate is equal to the net difference between rates of synthesis P^+ and destruction P^- of organic matter averaged over the period of observation and can be written as follows:

$$\tau_\kappa \sim \frac{\varepsilon_b M}{|P^+ - P^-|} \equiv \frac{\varepsilon_b}{|\kappa|}\tau, \qquad \kappa \equiv \frac{P^+ - P^-}{P^+}, \qquad \tau \equiv \frac{M}{P^+} \tag{5.3.1}$$

where τ is the time of turnover of the biogens in the biosphere. Normal communities performing biotic regulation of the environment keep the relative difference between average rates of synthesis and destruction of organic matter with an accuracy not exceeding ε_b when averaged over the turnover time τ (see Section 3.4). It is reasonable to assume that a destabilising community will not be able to attain a higher accuracy, so that $\kappa \geq \varepsilon_b$. Hence, the time τ_κ does not exceed the turnover time τ,

$$\tau_\kappa \leq \tau = M/P^+ \tag{5.3.2}$$

Some components of a destabilising community, e.g. large animals, may preserve high competitiveness until complete degradation of the environment (e.g. complete extinction of plant biomass) occurs, which corresponds to $\varepsilon_b \sim 1$. However, in such a case the correlation between synthesis and destruction also appears completely violated, $\kappa \sim 1$, so that relationship (5.2.2) holds anyway.

Let T be the average time of appearance and expansion of a new type of ecological community in the course of evolution. It means that during time T since their appearance, the new more competitive communities completely force out the former less competitive communities. In order to support environmental stability, the time τ_κ of noticeable environmental degradation after which destabilising communities lose competitiveness should be much less than the time T of evolutionary expansion of newly-arising communities. Using (5.3.2) that condition can be written as

$$\tau \ll T, \qquad \tau = M/P^+ \tag{5.3.3}$$

It means that the destabilising communities destroy their environment (and, hence, lose their competitiveness) well before they manage to expel normal communities from the biosphere. As soon as they lose their competitiveness, normal communities that are able to maintain their local environment in an optimal state begin to dominate again. Normal communities force out the destabilising communities preventing them from ubiquitous spread and, by doing so, protect the global environment.

In the opposite case, i.e. when $\tau \gg T$, destabilising communities remain highly competitive during the whole time of their expansion T, because the environment remains optimal for them during the whole period of expansion. In such a case, nothing can prevent destabilising communities from spreading over the whole biosphere. After that is done and all the normal communities are expelled from the biosphere, the destabilising communities remain prosperous during the time τ. However, being unable to support their optimal environment, such communities cannot counteract the inevitable degradation of the environment later on. After the degree of environmental degradation becomes substantial, these communities die out, which means an end to life as a whole.

The discussed problem is tightly linked to the problem of the amount of nutrients that can be stored in the environment so as not to violate stability of life organisation.

When absolutely no stores of nutrients are available outside the bodies of the living individuals i.e. when $M = 0$, see Eq. (5.3.3), stable life is only possible, independent of the rate of evolution, if matter cycles are completely closed. Any community violating that closure of matter cycles would then immediately disintegrate and lose its competitive capacity. Such a situation corresponds to the condition $\tau = 0$ and gives absolutely no chance for destabilising communities to appear and persist. Such communities (with their M close to zero) are likely to appear on territories affected by some natural catastrophe, such as a volcanic eruption, glaciation, etc., when the biota has not yet had enough time to accumulate a substantial amount of nutrients. Communities with the value of M close to zero do exist. These are epilithic lichens, existing at bare rock surface and organised as symbiotic individual algae and fungi (Farrar, 1976; see also Morneau and Payette, 1989), see the book cover.

However, such a situation has a reverse side. The amount of nutrients stored in the biota represents a kind of buffer that helps the community to compensate adverse changes of the environment due to abiotic fluxes. This can be illustrated as follows. It is a well-known fact that terrestrial communities contain a very substantial part of their organic carbon in soil (Melillo *et al.*, 1996). If the concentration of carbon in the atmosphere becomes too low compared to the community's optimum, communities could increase the rate of destruction of organic matter and thus compensate the atmospheric depletion of carbon at the expense of soil organic carbon store. In the opposite, more real case of today, when the atmospheric concentration of carbon becomes too high, communities may sequester excessive carbon in the same refractory biological reservoir. The larger the characteristic fluctuations of the environment, the larger buffer (i.e. the larger value of M) the community should maintain.

Thus, a certain non-zero store of biogens M enables the community to cope with adverse environmental changes and contributes to its competitiveness. On the other hand, a large store of nutrients M increases the time τ (5.3.3) weakening the inequality $\tau \ll T$ and thus leaving more opportunities for destabilising communities. **It means that very large values of M (i.e. the situation of matter abundance) are not compatible with stable organisation of life.**

Under the conditions of matter abundance it is impossible for communities that perform environmental stabilisation to persist. Until the biota exists in a state of matter abundance, destabilising communities remain most competitive. Normal communities should spend a substantial part of their power to regulation of the environment, while destabilising communities exist at the expense of abundant resources and spend all their efforts on competitive interaction. Normal communities can be compared to waste-free (closed) technologies, that are very expensive and, hence, noncompetitive on the technological market. Their competitiveness may be increased only artificially through environmental laws. However, were civilisation to exist in a state of matter shortage, so that all the wastes would have to be re-involved in the technological progress, no other technologies except for waste-free ones could dominate the market.

A very important conclusion that follows from the above statement about matter abundance is that, in a stable state of equality between synthesis and destruction, biospheric stores of nutrients in organic and inorganic form (e.g. store of organic and inorganic carbon available to biota) should coincide in their order of magnitude (i.e. be approximately equal). If it were not the case, either the synthesising individuals (if inorganic matter is in excess) or the reducers of the organic matters (when the organic matter is overabundant) would find themselves in conditions of relative matter abundance. Such a situation would lead to expansion of either the synthesisers or reducers, and the correlation between the synthesis and the destruction would be inevitably disrupted.

Let us now examine the real situation observed in parts of modern biosphere slightly (if at all) disturbed by humans.

We find that, indeed, the stores of nutrients in their organic and inorganic forms in the biosphere are of one and the same order of magnitude (Chapter 6). For example, both the organic and inorganic store of carbon (one of the most important biogens) are of the order of $M \sim 10^3$ Gt C. As noted above, the rate of production of organic carbon amounts to about $P^+ \sim 100$ Gt C yr^{-1}. Thus, we have

$$\tau \sim M/P^+ \sim 10 \text{ years} \tag{5.3.4}$$

The available data on speciation mode suggest that most species evolve rather rapidly and then remain stable during the whole period of the species' existence (Jackson, 1994; Gould and Eldridge, 1993; see also Chapter 11). This phenomenon is known as punctuated equilibrium evolution. According to the available data, the actual period of species formation is of the order of 10^4 years. This value is retrieved from the observation that even when the resolution of paleodata is as high as 10 000 years, no noticeable changes in species' morphology can be discerned. Thus, the evolutionary expansion of a newly-arisen species most probably takes about $T \sim 10^4$ years. Any new species modifies the community and imparts new properties to it. In this sense a community with a new species is a new community. Thus, the above value of T should be used as an estimate of the characteristic time of expansion of a new type of community appearing in the course of evolution.

Thus, for the modern non-perturbed biota the inequality $\tau \ll T$ (5.3.3) indispensable for a stable organisation of life assumes the form 10 yr $\ll 10^4$ yr. It means that the necessary stabilising condition is met in the non-perturbed natural biota to the strength of at least three orders of magnitude, strongly prohibiting the appearance of communities that could break the closure of matter cycles and destroy the environment.

The value of the biospheric carbon store ($M \sim 10^3$ Gt C) used in the above calculations includes only the biologically active carbon that is immediately available to the biota. It includes inorganic carbon of the atmosphere and dissolved inorganic carbon of the ocean (available to the oceanic biota) and organic carbon of soil and dissolved organic carbon of the ocean.

There is also another reservoir of carbon, though biologically inactive. This is the sedimentary organic matter that has been dispersed over the upper several kilometres of the Earth's core during the period of life existence. This store of organic carbon is of the order 10^7 Gt C, i.e. it exceeds the store of available organic carbon by a factor of ten thousand. However, it is unavailable for the biota and cannot be re-involved into the contemporary biochemical cycles.

Were the sedimentary organic matter biologically active, the time of its biologic turnover τ (5.3.1) would be four orders of magnitude longer than the present turnover time of environmental biogens, i.e. would be of the order of $\tau \sim 10^5$ instead of 10 years violating the condition $\tau \ll T$. Thus, use of sedimentary organic matter by living individuals brings about a situation of matter abundance and should be therefore strongly prohibited under natural environmental conditions.

Economic growth of the last two centuries accompanied by an explosive increase in the world human population (Figure 1.2) has been possible thanks to the ability of people to use so-called non-renewable resources of the fossil fuels—oil, coal and natural gas. The characteristic time of environmental change that will cause loss of competitiveness of the modern industrial society is of the order of 100 years ($\tau_{\text{hum}} \sim 10^2$ yr). It is determined as the time of complete extinction of the available store of natural resources, which is equal to the value of the store ($M \sim 3 \cdot 10^3$ Gt C, see Figure 5.1) divided by the rate of its exploitation, which is of the order of 6 Gt C yr^{-1} and corresponds to the power of 10^{13} W (Table 7.1, Section 7.2). The characteristic time T_{hum} of progress of the modern civilisation is determined by the average time of renovation of major technologies, which does not at present exceed 10 years ($T_{\text{hum}} \sim 10$ yr). Thus, modern humanity exists in a situation of matter abundance, $\tau_{\text{hum}} \gg T_{\text{hum}}$.

As a result, those human societies that are organised on the basis of closed matter cycles not involving non-renewable resources (e.g. indigenous societies) are absolutely noncompetitive as compared to modern industrial societies and exist on the verge of extinction. On the other hand, if tomorrow the global fossil fuel store suddenly disappeared, the indigenous societies would not even notice, while the majority of the modern human population would simply perish.

Rapid development of civilisation under conditions of matter and energy abundance makes an illusion of the possibility to go over to alternative, so-called renewable, resources of energy when the fossil fuel is exhausted. In reality, however, the renewable energy resources of humanity (the available hydraulic power of rivers, tides, wind, biota and self solar radiation) altogether can provide the humanity with a power of no more than 10^{12} W (see Table 7.1), which is an order of magnitude lower than the energy consumption of modern civilisation. When the non-renewable resources are exhausted, civilisation will not be able to support the hypertrophied economies and the huge global population. If people manage to completely destroy the biotic regulation mechanism before the fossil fuel is exhausted, the inevitable decay of the modern fossil fuel-based civilisation will be accompanied by irreversible degradation of the global environment and its transition to a state unfit for any life.

Let us now analyse what can be done by the natural biota itself to keep the condition $\tau \ll T$ satisfied. As noted in Section 5.2, the value of biological production P^+ is dictated by the value of external abiotic fluxes of biogens to and from the biosphere. The value of P^+ is determined from the condition that the fluxes of synthesis (and, consequently, destruction) of organic matter developed by the biota should significantly exceed any abiotic fluxes. That means that natural biota cannot arbitrarily change the level of its production P^+.

Similarly, the characteristic time of evolutionary changes T is also beyond the biotic control and cannot be increased infinitely to satisfy the condition $\tau \ll T$ at any values of τ. Increasing the time T would mean slowing down the evolutionary process. This can be slowed down by reducing the amount of mutations (the primary driving cause of the evolution) per replication event. Despite various mechanisms created by life to increase the stability of the DNA molecules, e.g. proof-reading and mismatch repair (Lewin, 1987), there is apparently a limit to such restrictions and, consequently, a limit to the value of T. That time is ultimately determined by the quantum characteristics of the molecules used by life, which cannot be changed by the biota.

Thus both the biotic production P^+ and the characteristic time of community's evolution T are determined by external abiotic conditions, and may not be affected by the biota. Hence, only the stores of biogens M and, consequently, their concentrations in the environment may be prescribed by Earth's biota and supported by it at a certain level to satisfy the condition (5.3.3). Biotic control of stores of biogens M can be only performed on the basis of negative feedback between the external disturbances of such stores and biotic reaction to such disturbances, which is discussed in the following sections.

5.4 MATTER CYCLES IN THE BIOSPHERE

In the course of biochemical reactions of synthesis and destruction of organic matter the biologically active chemical elements (biogens) are synthesised or decomposed in certain specific proportions known as stoichiometric ratios. Using these ratios, masses and concentrations of particular biogens may be retrieved from the known masses of other biogens. In what is to follow we set down all the quantitative relations for carbon, one of the most important biogens.

We denote respective masses of carbon stored per unit surface area in a biologically active reservoir (e.g. atmosphere, ocean, land) in the organic and inorganic forms as M^+ and M^-, respectively. Fluxes of carbon during synthesis (production) and destruction of the organic matters in the reservoir will then be P^+ and P^-, respectively. Finally, we denote net fluxes of the organic matter evacuated from the reservoir as F^+, and that of inorganic matter imported into the reservoir as F^-. Note that the net fluxes $F^\pm$ are equal to differences between the gross import and export fluxes, $F^\pm_{\text{in}}$ and $F^\pm_{\text{out}}$. The law of matter conservation for any reservoir is expressed by the equations:

$$\dot{M}^+ = P^+ - P^- - F^+, \qquad \dot{M}^- = P^- - P^+ + F^- \tag{5.4.1}$$

Here $\dot{M}^+$ and $\dot{M}^-$ stand for the rates of change of the organic and inorganic carbon stores, respectively. As far as masses M^+ and M^- characterise the state of environment, the equalities $\dot{M}^+ = \dot{M}^- = 0$ mean that the environment does not change with time.

The relative value of external fluxes $F^\pm$ (or $F^\pm_{\text{in}}$) as compared to P^+ describe how open the reservoir is with respect to external forcing. The ratios ν and ν_{in}

$$\nu^\pm \equiv \frac{F^\pm}{P^+}, \qquad \nu^\pm_{\text{in}} \equiv \frac{F^\pm_{\text{in}}}{P^+} \tag{5.4.2}$$

may be called the net (ν) and the gross (ν_{in}) openness. When $\nu \ll 1$ and $\nu_{\text{in}} \ll 1$, external fluxes remain small as compared to the synthesis P^+. In such a reservoir biological processes completely determine the state of environment. Such a reservoir may be called closed.[1] A situation when $\nu \ll 1$ while $\nu_{\text{in}} > 1$ means that though there are large fluxes both in and out of the reservoir ($\nu_{\text{in}} > 1$), these fluxes are practically

[1] It remains open only for solar radiation and thermal radiation of the Earth.

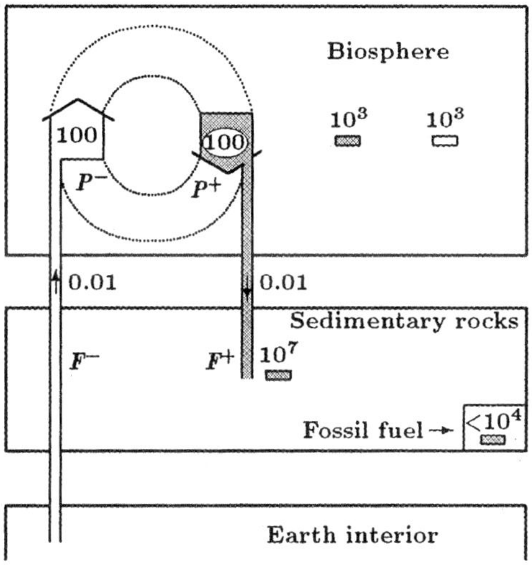

Figure 5.1. Stores (Gt C) and annual fluxes (Gt C yr^{-1}) of carbon in the biosphere shown to the accuracy of the order of magnitude. Stores of carbon are given by figures above rectangles. Fluxes of carbon are given by figures near arrows. Fluxes and stores of organic carbon are shaded. Fluxes and stores of inorganic carbon are represented as empty areas. The flux of organic carbon going to deposits in sedimentary rocks is equal to the difference between its synthesis and destruction in the biosphere. That flux coincides with the net flux of inorganic flux entering the biosphere, to a relative accuracy of about 10^{-4} when averaged over the Phanerozoi (about $6 \cdot 10^8$ yr). Fluxes of synthesis and destruction coincide with each other to about the same level of accuracy when averaged over the last several hundred years. That situation works to support stores of carbon in its organic and inorganic form in the biosphere in a stable state.

equal ($\nu \ll 1$), so that they cannot bring about considerable changes in the environment. In such a case biological processes within the reservoir still play a decisive role in determining its environment. Finally, when $\nu \gg 1$ and $\nu_{in} \gg 1$, the reservoir is completely open, and the biological processes in it are insignificant. In such a reservoir biota cannot support its optimal environment (see Section 5.6). The value of ν is small for the biosphere of the planet Earth as a whole, so the modern biosphere represents a closed reservoir.

Stores of organic and inorganic carbon in the biosphere coincide in their order of magnitude (Figure 5.1). The ratio of these stores to productivity P^+ of the global biota yields the time period of biological turnover of the biogenic store in the biosphere τ, which is of the order of tens of years, see (5.3.4). Hence, were only synthesis of organic matter to take place in the biosphere with no decomposition to accompany it, all the inorganic carbon in the biosphere would be used up and transformed into organic substances within a few decades. Similarly, were only decomposition to take place, all the organic carbon in the biosphere would vanish in decades, while the atmospheric CO_2 concentration would increase by about two times.

To escape sharp fluctuations of the environment, rates of synthesis and decomposition of organic matter should accurately coincide with each other. By measuring the concentration of carbon dioxide in air bubbles entrapped in ice cores of different ages from Antarctica and Greenland, it is found that the concentration of carbon dioxide in the atmosphere remained constant within error of measurement during the last 1000 years (Oeschger and Stauffer, 1986; IPCC, 1996). It sustained its order of magnitude during time periods of several hundred thousand years, that is, for time periods exceeding the turnover time by the factor of 10^4 (Barnola *et al.*, 1991). It follows quite unequivocally from these data that the global mean fluxes of biological synthesis P^+ and destruction P^- averaged over the last 1000 years coincide with each other to an accuracy of four digits, that is, compensate each other to the relative accuracy of 10^{-4}.

Beside biotic impact, the environment of Earth is exposed to directional physical impacts, although much less powerful. Inorganic carbon is released into the atmosphere in the process of degassing (including volcanic activity, filtration from the mantle, etc.) and is stored in sedimentary rocks leaving the biosphere in the processes of weathering. The biota has practically no control over the carbon emission from the Earth's interior. Land biota can only slightly change the rate of weathering (Schwartzman and Volk, 1989). The difference between emission F^-_{in} and deposition F^-_{out} yields the net flux F^- of inorganic carbon into the atmosphere. It appears that this flux is positive and is of the same order of magnitude as the emission and deposition themselves, $F^-_{\text{in}} - F^-_{\text{out}} \equiv F^- \sim F^-_{\text{in}} \sim F^-_{\text{out}} > 0$. It means that physical fluxes of emission and deposition of inorganic carbon do not compensate each other. The ratio of the present-day store of inorganic carbon in the atmosphere $M^-_{\text{a}} \sim 10^3\,\text{Gt C}$ to its net geophysical flux $F^- \sim 10^{-2}\,\text{Gt C yr}^{-1}$ (Degens *et al.*, 1984; Figure 5.1) corresponds to a time scale of around 100 000 years. In other words, the store of inorganic carbon in the atmosphere should have increased by a factor of ten thousand during a time period of about a billion years. That would have brought about a catastrophic greenhouse effect.

However, that never actually happened. Hence, some compensating process must function, and this process is the flux of storage of organic carbon in sedimentary rocks, F^+. Excessive inorganic carbon of the atmosphere is transformed to organic carbon by the biota and leaves the biosphere in inactive form of sediments. Direct studies have demonstrated that the stores of organic carbon, accumulated during approximately one billion years and dispersed through the sedimentary layer about two kilometres thick, exceed the stores of both the organic and inorganic carbon available for life in the biosphere by about four orders of magnitude (Figure 5.1; Budyko *et al.*, 1987).

It follows from the above that the net geophysical flux F^- of inorganic carbon into the biosphere and the flux F^+ of organic carbon buried into sedimentary rocks (which is equal to the difference between production and destruction) have, on average, coincided to an accuracy of four digits, that is, to a relative accuracy of 10^{-4}. As far as a random coincidence of two independent fluxes with such an accuracy is improbable, the obtained result may be regarded as an independent argument for the existence and precise character of the biotic regulation of the environment. It is especially worth noting that the biota is able to perform very refined regulation of the environment, because physical flux of inorganic carbon into the atmosphere is about 10 000 times less powerful than the average fluxes of synthesis and destruction of organic matter by the biota (see Figure 5.1). The biota balances fluxes P^+ and P^- that are of the order of 100 Gt C yr^{-1} so that the net difference between them accurately compensates a physical flux F^- of about 0.01 Gt C yr^{-1}, which is indeed a very precise process.

5.5 ENVIRONMENTAL HOMEOSTASIS AND INTERPRETATION OF THE BIOTIC LE CHATELIER PRINCIPLE

Let us call the chemical element X either organic or inorganic biogen (= nutrient), depending on whether it enters an organic or an inorganic substance (e.g. organic carbon enters the organic molecules within a living cell). We denote fluxes of synthesis (the net primary productivity) and of destruction (destructivity) of the organic nutrient X as P_X^+ and P_X^-, respectively. The dimension of these values is kg X m^{-2} yr^{-1}. We denote the environmental density of mass of organic and inorganic nutrient X per unit land surface as M_X^+ and M_X^-, respectively (their dimension is kg X m^{-2}).

The phenomenon that may be called *environmental homeostasis* consists in the fact that any change ΔM_X^- in the environment is accompanied by the appearance of an adequate non-zero difference $P_X^+ - P_X^-$ directed to compensate precisely for that change.

The environment is characterised by particular values of concentrations of both organic and inorganic substances. The diversity of organic substances used by the biota far exceeds the diversity of inorganic substances involved in biochemical cycles. Owing to the law of matter conservation, the biota may change concentrations of inorganic substances, transforming them into inorganic ones at constant total stores of chemical elements in the environment. Changing the character of organic substances and their localisation in the biosphere from highly active organic matter of live biomass to relatively inactive organic matter of humus and dissolved organic carbon of the ocean, the biota is able to keep environmental impact of organic substances under control.

Inorganic substances used by life are characterised by a limited set of physico-chemical properties and predictable environmental impact. For example, water vapour and carbon dioxide in the atmosphere determine the greenhouse effect of the planet, while ice and snow control the albedo of the Earth's surface and, hence, impose impact on the global surface temperature (Chapter 8). Hence, the change ΔM_X^- of inorganic substances in the environment is both an important and relatively easily parameterised variable to be used in description of the biotic regulation mechanism.

Variables ΔM_X^- and $P_X^+ - P_X^+$ have different dimensions. Correlation between them cannot be provided by any fundamental physical or chemical constant of a transitional dimension. Owing to the extreme complexity of interactions between living organisms in a natural community, such constants are lost in a huge variety of existing types of correlation. One may say that the biota 'forgets' values of such constants. Thus, the only possible way of formalising correlation between those variables is a scale-invariant proportionality between the dimensionless ratios of increments of these two variables to their initial values. Such type of correlation known as *allometric* (Peters, 1983) is most often discovered in biological and ecological observations. It may be presented in the following form:

$$\frac{P_X^+ - P_X^-}{P_{X0}^+} = \beta_X \frac{\Delta M_X^-}{M_{X0}^-} \equiv \beta_X \frac{\Delta[X]}{[X_0]},$$
$$\Delta M_X^- \equiv M_X^- - M_{X0}^-, \qquad \Delta[X] \equiv [X] - [X_0], \text{ or :} \tag{5.5.1}$$
$$P_X^+ - P_X^- = \beta_X \frac{\Delta[X]}{R_{iX}}, \qquad R_{iX} \equiv \frac{[X_0]}{P_{X0}^+},$$

where M_X^-, M_{X0}^-, $[X]$, $[X_0]$, P_X^+ and P_{X0}^+ are the perturbed and the initial nonperturbed (index '0') values of the mass (M) of nutrient X in the environment, its concentration [X] and of the net primary productivity (P_X^+) of the biota, respectively; P_X^- is destructivity. The variable R_{iX} has the meaning of internal resistance of the corresponding inorganic store of nutrient X to its synthesis by the biota. The dimensionless coefficient β_X describes correlation between the respective variables. Below, the 'X' indices are omitted in all general formulae for simplicity's sake.

The internal resistance R_i represents an essential characteristic of the plant components of the community and has a definite biochemical meaning. If R_i remains constant and independent of the concentration of nutrient [X], the productivity of organic form of X grows linearly with its concentration, $P^+ = [X]/R_i$. Due to the catalytic nature of synthesis of organic matter by the biota, when the concentration [X] climbs to a certain saturating value, the rate of synthesis of organic X becomes saturated, i.e. it ceases to grow linearly with [X]. Further increase of [X] leads to a proportional increase of R_i at nearly constant value of P^+.

If in response to external perturbations it is the productivity P^+ that changes, while the destructivity P^- remains constant, $P^- = P_0^- = P_0^+$, the relations (5.5.1) assume the form

$$\frac{\Delta P^+}{P_0^+} = \beta \frac{\Delta[X]}{[X_0]} \tag{5.5.2}$$

The equation (5.5.2) is a linear relationship between logarithmic derivatives of P^+ and [X]. Its solution has the form

$$\ln \frac{P^+}{P_0^+} = \beta \ln \frac{[X]}{[X_0]} \qquad \text{or} \qquad \frac{P^+}{P_0^+} = \left(\frac{[X]}{[X_0]}\right)^\beta \tag{5.5.3}$$

If productivity P^+ is expressed via internal resistance R_i (5.5.1) we have from (5.5.3):

$$P^+ \equiv \frac{[X]}{R_i}, \qquad R_i = [X]^{\beta-1} \frac{P_0^+}{[X_0]^\beta}. \tag{5.5.4}$$

Thus, the internal resistance R_i remains constant and independent of [X] only when $\beta = 1$.

The natural background fluxes F^- in any biologically important reservoir (atmosphere, ocean, land) do not, on average, change with time, so that the openness $\nu^- \equiv F^-/P_0^+$ remain constant. Thus when considering significant perturbations of the environment, these values may be considered constant in Eqs. (5.4.1) that describe the law of matter conservation for each reservoir.

It can be easily checked that introducing a designation

$$Z^- = \frac{\Delta M^-}{M_0^-} + \nu^- \beta^{-1}$$

we may rewrite the law of matter conservation for the store of inorganic nutrient (the first equation in (5.4.1)) in the following form:

$$\dot{Z}^- = -kZ^-, \qquad \text{where } k \equiv \beta/\tau \quad \text{and} \quad \tau \equiv M_0^-/P_0^+ \tag{5.5.5}$$

The time τ is the time of biological turnover or residence time of the respective inorganic nutrient in the environment.

Solution of Eq. (5.5.5) at constant k > 0 (i.e. when $\beta > 0$) assumes the form $Z^- = Z_0^- e^{-kt}$, where t is time. In such a case any perturbation of the environment exponentially dies out with time. It means that the biota compensates the perturbation and the environment relaxes to the initial stable state. When the value of k is negative (i.e. when $\beta < 0$) any external perturbation is exponentially enhanced by the biota making impossible the maintenance of a stable environment. Note also that when $k > 0$ productivity P^+ grows (or destructivity P^- decreases) with increasing concentration [X] of inorganic nutrient, which means that the excessive inorganic matter is turned into organic matter. At $k < 0$ the opposite destabilising trend takes place.

The condition of biotic stability thus assumes the form

$$\beta > 0, \qquad k > 0 \tag{5.5.6}$$

At $k > 0$ the value $k^{-1} = \tau/\beta$ has the meaning of characteristic time of relaxation back to the normal state of the environment. An important point is that after the perturbation ceases, the environment returns to its former (instead of a new) stable state. For example, after excessive CO_2 enters the atmosphere from external sources, the natural nonperturbed biota should transfer it into organic forms of low activity (such as soil humus, peat, and the dissolved organic matter in the ocean), thus restoring the former atmospheric concentration of CO_2 optimal for the biota.

When the organic matter of living individuals is synthesised, certain ratios should be followed between the fluxes of production P^+ (and, respectively, of destruction) of the organic nutrients, which correspond to the ratios between those nutrients in living cells. For example, phytoplankton cells are synthesised and oxygen is released by them in the ocean following the so-called Redfield molar ratio (Redfield, 1958; Redfield *et al.*, 1967; Broecker, 1982; Takahashi *et al.*, 1985; Chen *et al.*, 1996):

$$P_C^+/P_N^+/P_P^+/P_{O_2} = 106/16/1/138, \tag{5.5.7}$$

where C denotes carbon, N—nitrogen, P—phosphorus, O_2—oxygen, and the productivity P_X^+ have the dimension of mole X m^{-2} yr^{-1}. A significant part of single-cell algae have $CaCO_3$ shells, which increases the share of carbon in Eq. (5.5.7) by approximately 20% over the world ocean (Broecker, 1982; Neshyba, 1987).

Random perturbations of the environment may happen independently with respect to each nutrient. Thus, to ensure biological control of the environment, the natural biota should be capable of producing the needed differences $P_X^+ - P_X^-$ in arbitrary ratios between the nutrients (Peng *et al.*, 1987) to compensate for their arbitrary independent perturbations. That effect may be achieved via synthesis of extracellular excretions of organic matter (Khailov, 1971; Fogg, 1975; Platt and Rao, 1975; Gorshkov, 1982c) and of the metabolically inactive parts of plants (such as wood trunks or shells of the marine organisms), in which the ratios between the nutrients are different from those in living cells. Another way of ensuring a diverse response can be selective destruction of the dead organic matter (Prinn, 1994).

Equations (5.5.1) and (5.5.5) with constant coefficients k and β hold true for the situation when a separate component of the environment X is independently perturbed in each reservoir. Formally, Eqs. (5.5.5) may be written for various mutually-related reservoirs and stoichiometrically interrelated nutrients X_i. Then the X-specific values of β and k in Eqs. (5.5.1) and (5.5.5) may, as a general case, be presented as functions of all the variables, characterising the concentrations of nutrients X_i in each reservoir. When each of such nutrients suffers a relatively small perturbation, these functions may be expanded into a Taylor series, from which we then exclude the non-linear terms. As a result, one arrives at a system of linear equations, relating the time derivatives $\dot{Z}_{X_i}$ of relative perturbations of nutrients X in reservoirs i, to the increments Z_{Y_j} of nutrients Y in reservoirs j. Matrix of the obtained system of equations may be diagonalised via a linear transformation (a linear substitution of the variables), and then finding the eigenvalues (Lotka, 1925) the system may be presented in the form of a noncohesive set of equations of the type of Eq. (5.5.5). In that case, all the statements set forth above hold for each equation. Below, such an account of relations between the reservoirs is demonstrated for the example of the global cycle of carbon (see Sections 6.3 and 6.4). If all the eigenvalues of the system of linear equations are positive and for them the conditions $\beta > 0$ and $k > 0$ hold, the environment remains stable with respect to all possible perturbations of all nutrients.

One may formulate the well-known Le Chatelier principle with respect to the biota. The Le Chatelier principle is usually applied to physicochemical systems in the state of thermodynamic equilibrium. It states that in a stable system any external perturbation leads to the appearance of compensating processes that would tend to return the system to the initial state. For example, in chemical buffer systems large external perturbations lead to small deviations from the equilibrium pH values. The remarkable difference between stability of physicochemical systems and the biospheric environment lies in the fact that stable state of equilibrium in physicochemical systems is totally determined by external conditions. When the latter change, the system transits to another stable state, although the change may remain relatively small owing to compensating processes. In the process of biotic regulation of the environment all external perturbations are *fully compensated*, so that no deviation from the equilibrium state occurs irrespective of the value of external perturbations. In that sense, biological systems are equivalent to physicochemical systems with an infinitely large buffer effect. As noted above, such property of the environment in the biosphere may be called environmental homeostasis. Homeostasis characterises both separate individuals of a certain species (e.g. maintenance of constant body temperature in endothermic (warm-blooded) animals) and whole ecological communities of species and corresponding local ecosystems.

Thus, interpretation of the biotic Le Chatelier principle may sound as follows: **in the nonperturbed biosphere, any external perturbation of the optimal environment leads to appearance of compensating processes that completely quench the perturbation and relax the environment to the initial stable state**. That is, in physicochemical systems, compensating processes only *tend* to return the system to the initial state, only diminishing possible effects of perturbation; meanwhile, biotic processes in the biosphere actually fully compensate the perturbation.

The particular validity of the Le Chatelier principle is that its quantitative presentation does not require an intimate understanding of the system's structure. Due to the extreme complexity of the entangled interactions between the diversity of species in any ecological community, the values β and k may only be obtained empirically. The values of k_X and β_X are, together with P_X^+, P_X^- and $M_X^{\pm}$, the principal characteristics of the nonperturbed biota. They remain constant, and the relations in Eqs. (5.5.1) and (5.5.5) are linear at low values of relative perturbations, Z_X, provided all the other characteristics of the biota and the environment remain only weakly perturbed as well.

From the point of view of one particular species in the community all the other surrounding species may be treated as components of the environment. Perturbation of the community structure may then be described by the same equations, Eqs. (5.5.1) and (5.5.5), treating the values Z as the relative changes in the measurable characteristics of the community. One may consider the population density of individuals of a separate species as such a variable, or the fluxes of matter and energy, consumed by that species, which are uniquely related to the former.

Synthesis and destruction of organic matter, i.e. the consumption of energy and matter by living individuals, may be treated as part of the useful work performed by a given species within the community, which is aimed at stabilising both the biota and the environment. A universal characteristic for all natural communities is the distribution of the consumed fluxes of matter and energy vs. individual body sizes of the consumers independent of their taxonomic properties (Figure 3.3). Deviations from that distribution within each interval of the histogram of Figure 3.3 may be regarded as a perturbation of the community and be described by Eqs. (5.5.1) and (5.5.5).

It is apparent that compensation of perturbations of environment and biota itself may only be pursued up to a certain threshold level, i.e. the Le Chatelier principle may only be sustained that far. The most vivid illustration of that statement is that when the natural biota is completely destroyed on a given territory, biotic regulation of the environment will naturally stop there. The same result will elicit from substitution of natural communities by artificial agricultural or silvicultural systems.

Nonperturbed ('healthy') biota should be characterised by the maximum possible values of β, which means that it should ensure the quickest possible recovery to the initial stable state after external perturbation. With that perturbation growing in scope, the value of β ceases to be constant at some moment, and starts to decrease. Eqs. (5.5.1) and (5.5.5) then become non-linear with respect to the perturbation. The value of that perturbation, at which β turns to zero, may be considered the threshold beyond which the stability of the system 'biota plus its environment' is destroyed.

Threshold values of perturbations still compatible with biotic and environmental stability represent the most important characteristics of the natural biota together with the optimal concentrations of substances in the environment, M_0, relaxation times, $k_0^{-1} \equiv \tau_0/\beta_0$, and the fluxes of production, P_0^+, and destruction, P_0^-. **Finding those thresholds becomes especially important because of the growing anthropogenic forcing of the biota and its environment.**

As follows from the distribution in Figure 3.3, the principal work on environmental stabilisation is carried out by microscopic individuals. After averaging over periodic seasonal oscillations, that part of the distribution in all communities apparently suffers only small fluctuations (Gorshkov, 1981; Kamenir and Khailov, 1987; Kamenir, 1991). No more than 1% of the total flux of energy and matter in the community falls on consumption by the largest individuals of the community, e.g. mammals, despite the fact that biomass of mammals may constitute a significant part of the total biomass of all the animals in the community. Hence, mammals under natural conditions only form a superfine structure of the community. Mammals, as well as dinosaurs in the past, have never been the lords of the biosphere, as they are sometimes considered (Alwarez and Asaro, 1990; Courtillot, 1990). Extinction of the dinosaurs at the C-T border 65 million years ago (the last mass extinction) (Raup and Sepkoski, 1982) could not have driven the biosphere outside its stability threshold, and had not practically affected its capability to compensate for adverse fluctuations of the environment.

If the stability threshold is exceeded on a local scale, local perturbations of the biota remain uncompensated, which leads to accumulation of various types of pollutants in the environment. There appear diffusion fluxes of these pollutants into the adjacent regions. Size of the perturbed domain is defined by the effective distance over which the concentrations of pollutants fall to the stability threshold of the nonperturbed biota. If the rate of perturbation is fixed, a stationary state sets in with a maximum perturbation of the environment in the centre of that local perturbation. Concentration of pollutants may be lowered in that local centre either proceeding to a partially closed technological cycle (the perturbed state of the biota then persists on a local scale), or increasing the evacuation of the pollutants outside the local area. Both ways of coping with local pollution are widely used in modern civilisation. Modern technologies display a tendency to become closed, and more and more efforts are spent on evacuation of wastes from densely-populated areas.

As a rule, the value of the maximum admissible concentration of pollutants is determined as the threshold effect upon human health, which has nothing to do with the stability threshold of the biota. Within the locally perturbed area the biota ceases to execute its stabilising functions and may be substituted by any artificial community or even totally exterminated.

When the perturbation begins to grow, the affected area expands. A stationary state of the environment on a global scale may only be sustained so far as the remaining nonperturbed part of the natural biota is still capable of compensating for all the anthropogenic perturbations, i.e. so long as the stability threshold is not exceeded globally.

According to the available data, within any geological period the biota and the environment retained their stationary stable state during thousands of years (Oeschger and Stauffer, 1986; Hansen and Lacis, 1990; Lorius *et al.*, 1990). By all appearances, periods of glaciations, that are often referred to as periods drastically unfavourable for life on the planet, represent fluctuations of the stationary stable state of the global environment, that might have been triggered by some external (e.g. astronomical) factors. Were there no biotic regulation of the environment, such external perturbations could drive the global environment outside the life-compatible interval of conditions (see Chapter 8). The natural biota of Earth damps those and all other perturbations of the environment and prevents the environmental characteristics from deviating away from their optimal values. The presently-observed global changes of the environment (Hansen and Lacis, 1990; Lorius *et al.*, 1990; IPCC, 1994, 1996) occurring over the period of several decades and possibly hundreds of years, unambiguously suggests that the stability threshold of the natural biota has been surpassed on a global scale.

5.6 BIOTIC REGULATION OF MATTER CYCLES

In the previous chapters we discussed the indispensability of limited spatial size of ecological communities (Section 2.8) and found it to vary from 1–10 m not exceeding the overall body size of higher plants (Section 3.5). **An ecological community together with its environment forms a local ecosystem, that, apparently, also features finite size.** Ecological communities maintain environmental homeostasis within local ecosystems. It means that concentrations of nutrients within the space domain inhabited by living beings (which is called *environment* in the ecological science) may differ drastically from concentrations of those nutrients in the external abiotic milieu where no life exists.

The existence of such a difference in concentrations, $\Delta[X] \equiv [X_{in}] - [X_{out}]$, results in appearance of diffusion fluxes, which tend to even the concentrations of biogens between the environment ($[X_{in}]$) and abiotic milieu ($[X_{out}]$). This process may only be counteracted and the necessary difference in concentrations supported, if the physical diffusion fluxes are compensated by biological transport of biogens, which occurs in the process of synthesis, P^+, and subsequent destruction, P^-, of the organic matter. Note that the physical diffusion processes are spontaneous and increase the entropy (chaos) of the environment tending to the state of thermodynamic equilibrium (maximum chaos), whereas biological transport of biogens is a highly ordered process which is ultimately based on consumption of highly ordered solar energy by living organisms and which increases the orderliness of the environment.

Biotic regulation of the environment is characterised by some finite sensitivity threshold, $\Delta[X]_{min}$, which means that the biota does not notice too small changes in concentrations of biogens, $\Delta[X] < \Delta[X]_{min}$, and, hence, does not react to them (see Section 3.4). The value $\varepsilon_b \equiv \Delta[X]_{min}/[X]$ is to be called the sensitivity (resolution) of the biota with respect to nutrient X. The biota may only react to the relative change $\varepsilon \equiv \Delta[X]/[X]$ in the concentration of X such that $\varepsilon > \varepsilon_b$. The less the value of ε_b, the better the sensitivity (resolution) of the biota.

Borders of local ecosystems are prescribed by those surfaces at which net biological transport of biogens turns to zero to the relative accuracy of about ε_b. The principal biological transport of biogens in land biota is performed by higher plants. Thus the size of the local ecosystem should be of the order of the largest plants in the community (to be more exact, of the order of maximal distance between the correlated parts of higher plants). Within the areas occupied by separate higher plants, autonomic local ecosystems of smaller size may be found (e.g. lichens on trees).

Due to competitive interaction between ecological communities, the neighbouring local ecosystems occupying the same geographical region (biotope) should feature identical environmental characteristics (to within the sensitivity of the biota). Hence, the differences in concentrations of biogens, physical diffusion fluxes, and corresponding compensating biochemical fluxes in the horizontal, y, directions should be much less than in the vertical, z, direction: $\Delta_y[X] \ll \Delta_z[X]$. The diffusion flux in the vertical direction within a local ecosystem is proportional to the concentration gradient and is expressed as:

$$F = -D\frac{d[X]}{dz} \approx \frac{D}{H}\Delta[X] \equiv \frac{[X_{in}] - [X_{out}]}{R_e} = F_{out} - F_{in} \tag{5.6.1}$$

where D is the coefficient of either the molecular or the eddy diffusion of the biogen; H is the vertical size of the local ecosystem; $R_e \equiv H/D$ is the external resistance of the diffusion transport; $F_{out} = [X_{in}]/R_e$ and $F_{in} = [X_{out}]/R_e$ are the fluxes of the biogen out and into the ecosystem, respectively (i.e. F_{out} is the export and F_{in} is the import of the biogen, see below Figure 5.2).

While the external milieu remains intact, and the net physical diffusion fluxes of biogens into the local ecosystem are compensated by the biological transport, the concentrations of biogens inside the local ecosystem do not change either. If the external conditions change, the balance between the physical diffusion and biological transport of biogens is broken, which results in directional changes of concentrations of biogens both within and outside the local ecosystem. Stationary stable environment favourable for the ecological community may only be supported on condition that the changes in processes taking place in the biota affected by external perturbations are directed to compensate such perturbations and to return the local ecosystem to its unperturbed condition. In other words, the biotic response to external perturbations should be based on a negative feedback mechanism working in accordance with the biotic Le Chatelier principle (Section 5.4).

Within a separate local ecosystem, concentrations of biogens may only be controlled by the biota on condition that the fluxes of biological synthesis and destruction of organic matter exceed the net fluxes of physical diffusion and/or advective transport. In other words, the biota should be powerful enough to compensate spontaneous physical degradation of the optimal for life, highly-ordered environment. Let us now discuss this problem in more detail.

Introducing the internal resistance to synthesis of organic matter, see (5.5.1)

$$R_i \equiv [\mathrm{X}]/P^+, \tag{5.6.2}$$

where $[\mathrm{X}] \equiv [\mathrm{X_{in}}]$ is the concentration of X inside the local ecosystem, we obtain the following relationship between the values of the gross ν_{in} and net ν openness of the local ecosystem with respect to the biogen X:

$$\nu_{\mathrm{in}} \equiv \frac{F_{\mathrm{in}}}{P^+} = \frac{R_i}{R_e}, \qquad \nu \equiv \frac{F_{\mathrm{in}} - F_{\mathrm{out}}}{P^+} \equiv \nu_{\mathrm{in}}\varepsilon, \qquad \varepsilon = \frac{\Delta[\mathrm{X}]}{[\mathrm{X}]} \tag{5.6.3}$$

Resistances R_e and R_i defined by (5.6.1) and (5.6.2) can be found empirically.

As soon as the biota's ability to control the concentration of the biogen X is concerned, three distinctive situations are possible:

(1) Biological fluxes of synthesis and destruction of organic matter are more powerful than the physical fluxes of import and export of the biogen into and out of the local ecosystem.

For simplicity let us consider the situation when the physical flux of import F_{in} is greater than the flux of export from the ecosystem F_{out}. Then condition (1) can be written as $F_{\mathrm{in}} < P^+$ or, using the gross openness $\nu_{\mathrm{in}} \equiv F_{\mathrm{in}}/P^+$, as

$$\nu_{\mathrm{in}} < 1 \tag{5.6.4}$$

In such a case the difference between the internal and external concentrations of the biogen with respect to the local ecosystem may be of the order of the internal concentration of the biogen itself, $\varepsilon \equiv \Delta[\mathrm{X}]/[\mathrm{X}] \sim 1$. In other words, the biota is able to gradually enrich the local ecosystem with such biogen up to an arbitrary optimal level, irrespective of whatever low concentration of the biogen is in the external abiotic milieu. Quantitatively, such a situation may be described as follows, using (5.6.1), (5.6.2), (5.6.3) and (5.6.4):

$$\nu_{\mathrm{in}} = \frac{R_i}{R_e} = \frac{[\mathrm{X}]D}{P^+H} < 1. \tag{5.6.5}$$

Such biogens may be called *locally regulated*, because even a single local ecosystem is able to control their concentrations. All the biologically-active substances of land soil enter that group and for them the relationship (5.6.5) holds true. This can be demonstrated on the example of phosphorus, X = P. Soil biogens exist in the form of water solutions and are transported in soil by molecular diffusion. In other words, the net physical flux F in soil represents the flux of biogens leaving the local ecosystem in the process of molecular diffusion. Phosphorus is characterised by the diffusion coefficient $D \leq 10^{-2}\,\mathrm{m}^2\,\mathrm{yr}^{-1}$ (Broecker and Peng, 1974; Fedorov and Gilmanov, 1980). Thickness of the soil layer is equal to $H \sim 0.3$ m, production of phosphorus by terrestrial biota is about $P^+ \sim 0.3$ mole P $\mathrm{m}^{-2}\,\mathrm{yr}^{-1}$ (Whittaker and Likens, 1975), concentration of phosphorus in soil is of the order of $[\mathrm{P}] \leq 10\,\mathrm{mole\,P\,m}^{-3}$ (Fedorov and Gilmanov, 1980). Using these values we obtain that for phosphorus $\nu_\mathrm{P} \sim 0.1 \leq 1$, as in (5.6.5).

The second situation is as follows.

(2) Physical fluxes of import and export into and from the local ecosystem are much larger than the biological fluxes of synthesis and destruction of organic matter. However, the relative difference between the internal and external concentrations, ε, still exceeds the biotic sensitivity threshold ε_b.

In such a situation, the difference between the concentrations of the corresponding biogen inside and outside the local ecosystem is much less than its concentration inside the local ecosystem, the latter practically coinciding with the concentration in the external abiotic milieu, $\varepsilon \equiv \Delta[\mathrm{X}]/[\mathrm{X}] \ll 1$.

Quantitatively this situation can be described as

$$\nu_\mathrm{in} \gg 1, \qquad \varepsilon > \varepsilon_b \tag{5.6.6}$$

For example, physical mixing of biogens in the atmosphere and ocean is so strong that it is very difficult to discriminate between the local ecosystem's environment and the external abiotic milieu. In such a situation, concentrations of biogens in the local ecosystem cannot be controlled by that local ecosystem alone, contrary to the previous case. Rather, the optimal concentrations of biogens are maintained by a large number of uniform local ecosystems occupying large territories of the Earth's surface. It is natural to call such biogens *globally regulated*. The process of biotic regulation of globally-regulated biogens is organised as follows.

If the external concentration of a globally regulated biogen differs from the community's optimum, the community activates processes aimed at compensation of that difference. Compensating processes can be based on increasing productivity as compared to destructivity, or vice versa. For example, if the global atmospheric concentration of CO_2 is larger than the optimal concentration for a terrestrial ecological community, the latter may increase the productivity as compared to destructivity and begin to deposit excessive CO_2 in organic form in humus and other relatively inactive reservoirs. As a result, the internal concentration of CO_2 in the local ecosystem becomes slightly lower than its concentration in the atmosphere outside the ecosystem (e.g. above the canopy layer). This slight difference, $\varepsilon \ll 1$, will induce a local physical influx of CO_2 into the local ecosystem (Figure 5.2).

If such slight change can be discerned by the biota, $\varepsilon > \varepsilon_b$, it gives the community advantage, i.e. it makes the community more competitive. Such a community can thus force out other communities that cannot perform such change. As a result, all neighbouring equally-competitive communities will ensure the same small change of

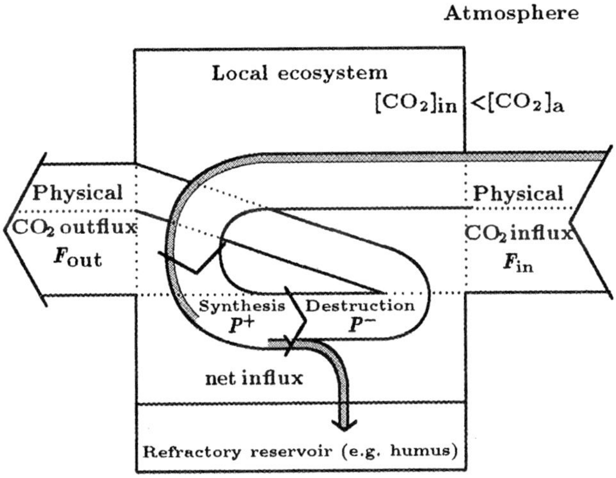

Figure 5.2. Biotic control of globally-regulated biogens on the example of atmospheric CO_2, see (5.6.6). If the atmospheric concentration of CO_2 is higher than the community's optimum, the ecological community changes the balance between synthesis P^+ and decomposition P^- of organic matter tending to decrease the CO_2 concentration within the local ecosystem. There appears a net influx of inorganic carbon from the atmosphere into the local ecosystem where it is stored in organic form in some refractory reservoir, e.g. soil humus. The appearing slight difference between CO_2 concentrations inside and outside the local ecosystem gives the community advantage and increases its competitiveness. If equally competitive communities occupy a large territory, the absolute flux of CO_2 leaving the atmosphere becomes globally important. The process goes on until the atmospheric concentration of CO_2 reduces down to the optimal for community value.

CO_2 concentration in their local environment. Provided that the total area occupied by such communities is large, there will be a globally significant influx of CO_2 into the biota. Such flux will proceed until the global atmospheric CO_2 concentration becomes equal to the community's optimum, or, to be more exact, until the relative difference ε between the internal concentration of the local ecosystem and the global atmosphere becomes less than the biota's sensitivity, ε_b. (Further decrease of the internal concentration of CO_2 as compared to the atmospheric one will not give the community any competitive advantage, so that the global process will be stopped when $\varepsilon \leq \varepsilon_b$.)

So, small relative changes in concentrations of globally-regulated biogens performed by local communities may lead to large absolute changes in the global environment.

The situation with globally-regulated biogens can be illustrated by the following economic analogy. Each normal community that ensures the needed fluxes can be compared to an employee who performs some amount of work in a factory. The employee will only work if he is interested, i.e. if he receives a certain per cent of the total profit. This per cent should be sufficient to considerably increase the person's living standard and social status ($\varepsilon > \varepsilon_b$). The more the employee works, the more he receives, which makes him work better, even though the overwhelming part of the profit goes away from the workers to the factory owner (this corresponds to huge physical fluxes). In contrast, when those who work and those who do not get nearly equal salary and are equally competitive in the society ($\varepsilon < \varepsilon_b$), it becomes disadvantageous to work, so that the working process disintegrates. Such a situation took place in the socialistic countries where people had no financial stimulus to work, while various types of social reward (e.g. medals for labour) proved to be insufficient to increase considerably the social status of productive workers as compared to lazybones.

Examples of globally-regulated biogens are atmospheric water vapour, fresh water of the continents (see Section 6.6 for more details), water dissolved oxygen (see below) and atmospheric carbon. Consider the last example in more detail.

Vertical physical transport of atmospheric carbon in forests is characterised by eddy diffusion coefficient, which depends on wind speed and constitutes on average $D \sim 10^7 \mathrm{m}^2\,\mathrm{yr}^{-1}$ (Fedorov and Gilmanov, 1980). The average vertical size of forest ecosystems is of the order of $H \sim 30$ m. Maximum productivity of forest communities in carbon units is about $P^+ \sim 10^2$ mole C m^2 yr^{-1} (Whittaker and Likens, 1975). The global average atmospheric carbon concentration is equal to $[CO_2]_a \sim 10^{-2}$ mole C m^{-3} (Watts, 1982). From this we obtain using (5.6.5) that for carbon $\nu_{in} \sim 30 \gg 1$. According to (5.6.6) this means that **if atmospheric carbon can be regulated by the terrestrial biota at all, it may only be globally regulated**, i.e. its concentration can be only maintained at an optimal level provided that forest communities occupy significant areas of the Earth's surface. In simple words, winds that destroy an ecosystem's environment are stronger than biological productivity which creates it, so that a separate local ecosystem cannot control atmospheric concentration of carbon on a local scale. Compare this situation to the above example of phosphorus and other biogens in soil, where molecular diffusion that destroys the environment is less powerful than biological productivity.

To find out whether atmospheric carbon can be regulated by the terrestrial biota, one needs to test the second condition in (5.6.6). Highly productive vegetation of forest communities form their internal atmosphere, where concentration of CO_2 may differ from that in the atmosphere above the canopy layer by about one per cent (Fedorov and Gilmanov, 1980), i.e. $\varepsilon \sim 0.01 > \varepsilon_b \sim 10^{-4}$ (Section 3.4). Thus, with respect to carbon in the terrestrial ecosystems we have $\varepsilon > \varepsilon_b$, which, according to (5.6.6), shows that atmospheric carbon can indeed be regulated by the terrestrial biota.

The difference in concentrations of biogens between the local ecosystem and external abiotic milieu, $\Delta[X]$, that can be supported by the biota, is directly proportional to the vertical size H of the local ecosystem, all other parameters remaining the same, see (5.6.1). Thus, the relative difference in concentrations $\varepsilon \equiv \Delta[X]/[X]$ that can be supported by local ecosystems of grasslands with $H \leq 1$ m should be at least 30 times lower than that for forest ecosystems and may approach ε_b. Thus, it is very likely that grasslands and similar low-sized terrestrial ecosystems are not able to control the atmospheric concentration of carbon.

We have seen that the mechanism of biotic regulation is based on redistribution of the regulated biogen between active (more important), and refractory (less important), reservoirs. For example, atmosphere is the active reservoir with respect to carbon, because the atmospheric concentration of CO_2 is critically important for photosynthesis. Meanwhile, the amount of organic carbon in humus layers is less important for the ecosystem, which is demonstrated by the fact that on previously glaciated territories organic matter of humus accumulates very slowly, which does not hinder normal functioning of forest communities. Forest communities may use the inactive humus reservoir of organic carbon to regulate the atmospheric concentration of inorganic carbon. Organic carbon accumulated in soil humus imposes no considerable impact on the major environmental characteristics critically important for the biota. So the organic carbon of humus represents a buffer store which may be used by the biota to control the atmospheric CO_2 concentration.

An important conclusion that follows from the above consideration is that if biogen X is globally regulated, its cumulative mass in inorganic form in the external milieu, M_X^-, should be of the same order of magnitude as its cumulative mass M_X^+ in organic form in the biosphere, $M_X^+ \sim M_X^-$ (cf. Section 5.3 where another reason for such relationship is considered). If the mass of the biogen in the biosphere is considerably smaller than its mass in the external abiotic milieu, $M_X^+ \ll M_X^-$, the biota is not able to compensate considerable environmental fluctuations accompanied by spontaneous decrease of the inorganic biogen.

For example, in the extreme case when the whole atmospheric CO_2 somehow vanished from the atmosphere, the biota would be able to completely compensate this fluctuation only if the cumulative stock of carbon in the inactive organic reservoirs of the biosphere were not less than the stock of atmospheric CO_2. In such a case it is possible to decompose the organic carbon of these reservoirs and, by doing so, fully replenish the optimal atmospheric stock of CO_2. Thus, for a globally-regulated biogen, M_X^+ cannot be considerably less than M_X^-. On the other hand, M_X^+ cannot be considerably larger than M_X^-, either. The organic matter may only be accumulated due to functioning of the biota itself, and that process stops when the optimal concentrations of both organic and inorganic nutrient are reached in the environment. The excessive mass of the organic biogen, not needed for regulation of M_X^-, will be inevitably withdrawn from the biochemical cycle and from the biosphere via storage in inactive reservoirs completely unavailable for the biota, e.g. in sedimentary rocks. Thus, for a globally-regulated biogen

$$M_X^+ \geq M_X^- \qquad (\text{not } M_X^+ \gg M_X^- \text{ or } M_X^+ \ll M_X^-) \tag{5.6.7}$$

For example, the cumulative mass of inorganic carbon in the atmosphere is about $M_C^- \sim 750$ Gt C, whereas the cumulative mass of organic carbon in the terrestrial part of the biosphere is about $M_C^+ \sim 1000$ Gt C (Watts, 1982; Schlesinger, 1990), so that the condition (5.6.7) holds true for atmospheric carbon.

(3) If the physical fluxes of a biogen are so large that the biota is incapable of producing any considerable difference in the concentrations of the biogen between the local ecosystem and the external abiotic milieu that would exceed the biota's sensitivity, such biogens cannot be regulated by the biota at all.

This situation can be described as

$$\nu_{\text{in}} \gg 1, \qquad \varepsilon < \varepsilon_b \tag{5.6.6}$$

For example, free oxygen and nitrogen of the atmosphere cannot be regulated by the terrestrial biota. For atmospheric oxygen in forest ecosystems we have $D \sim 10^7\,\text{m}^2\,\text{yr}^{-1}$, $H \sim 30\,\text{m}$, maximum productivity in oxygen units is about $P^+ \sim 10^2\,\text{mole}\,O_2\,\text{m}^2\,\text{yr}^{-1}$, average concentration of oxygen in the atmosphere $[O_2]_a \sim 8\,\text{mole}\,\text{C}\,\text{m}^{-3}$ (Whittaker and Likens, 1975). From this we obtain using (5.6.5) that for oxygen $\nu_{\text{in}} \sim 3 \cdot 10^4 \gg 1$.

The relative difference in concentrations of oxygen, ε, can be estimated from relationship (5.6.3) as $\varepsilon = \nu/\nu_{\text{in}}$, where ν is the net openness of the local ecosystem with respect to oxygen, $\nu \equiv (F_{\text{in}} - F_{\text{out}})/P^+$. The maximum value of ν is equal to unity, which corresponds to the case when all the oxygen produced within the local ecosystem, P^+ is carried away with the physical flux F_{out}. Thus, $\varepsilon = (\nu_{\text{in}})^{-1} \sim 3 \cdot 10^{-5}$ is the upper limit for oxygen. The resolution of the terrestrial biota ε_b with respect to oxygen remains unknown. Under the reasonable assumption that it is not significantly lower than that for carbon, $\varepsilon_b \sim 10^{-4}$, we obtain that even the upper estimate of ε is significantly lower than ε_b. According to (5.6.6) this means that atmospheric oxygen cannot be regulated by the terrestrial biota.

Another indication of the fact that atmospheric oxygen cannot be regulated by the terrestrial biota is that the cumulative mass of oxygen available for the biota reservoirs (living and dead biomass, organic carbon of humus, fossil fuel), $M^+_{O_2} \sim M^+_{C} \sim 4 \cdot 10^3\,\text{Gt}\,\text{C}$, is more than two orders of magnitude less than the cumulative mass of atmospheric oxygen, $M^-_{O_2} \sim 10^6\,\text{Gt}\,O_2$ (Schlesinger, 1990; Martin and Fitzwater, 1992). Thus for oxygen $M^+_{O_2} \ll M^-_{O_2}$, which prohibits its regulation, see (5.6.7). Among the non-regulated biogens are also the biogens in flowing waters (rivers) and in the upper oceanic layer, where eddy diffusion at average sea roughness is extremely high.

Concentrations of biotically non-regulated biogens still may be under biotic control, if the reservoirs of these biogens are in either chemical or physical equilibrium with those of the biotically-regulated biogens. We illustrate this seemingly controversial statement with the example of oxygen.

As shown above, atmospheric oxygen cannot be regulated by terrestrial biota. However, atmospheric oxygen is in physical equilibrium with oxygen dissolved in soil water and in the ocean. The equilibrium concentration of water dissolved oxygen is 40 times less than that in the air. Furthermore, the coefficient of diffusion of oxygen in water is four orders of magnitude less than in the air. Thus, physical fluxes of oxygen in water are six orders of magnitude less than the physical fluxes of O_2 in air. As a result, the water dissolved oxygen is locally biotically regulated by the aquatic biota. Due to physical equilibrium between concentrations of oxygen in the atmosphere and oxygen dissolved in water, the former also appears to be subject to biotic regulation, although the resulting concentration of oxygen in the atmosphere may appear not to be optimal for the terrestrial overground biota. Thus, atmospheric oxygen is ultimately regulated by oceanic biota and terrestrial underground biota inhabiting water capillaries in soil. For these organisms the concentration of oxygen in their environments is optimal. Terrestrial plants are unable to control atmospheric oxygen. Thus the biota functioning in air has to be adapted to the existing concentration of that gas in the atmosphere.

5.7 LIMITING BIOGENS. RESOURCES: RENEWABLE AND NON-RENEWABLE

It is evident that biological processes in natural ecological communities may only be limited by the biotically non-regulated biogens. Concentrations of all the regulated biogens is, by definition, supported by the biota at an optimal level, so that regulated biogens can by no means take part in limiting the functioning of the biota. In other words, when the biota chooses and supports the optimal concentrations itself, neither of them may limit its functioning. In the absence of biotically non-regulated biogens, the biotic productivity may be limited only by the available maximum rates of chemical reactions and by the power of solar radiation. In natural ecological communities, productivity of many plants actually reaches those limits. That testifies in favour of lack of limitations upon the biotic productivity from non-regulated biogens. Hence, if such non-regulated biogens do exist, their concentrations should remain within those optimal (or, at least, permissible) limits to which the biochemistry of the contemporary biota is actually tuned.

It is, however, easy to see that such a situation is improbable. Concentrations of biotically non-regulated biogens are not controlled by the biota. Biochemical transformations of such biogens cannot follow closed cycles, since the ecological communities with closed cycles of non-regulated biogens are in no way advantageous, as compared to those with their cycles open, see (5.6.6).[2] As a result, concentrations of non-regulated biogens appear to be unstable and cannot remain within the limits optimal for functioning of the biota for a long period of time. Instead, concentrations of biotically non-regulated biogens should drift randomly and finally get outside the margin fit for life.

Thus, we come to the conclusion that concentrations of all chemical substances used in life cycles must be subject to biotic regulation. Evolution favours ecological communities that are the most efficient in regulating the environment. Hence, concentrations of biogens in the natural environment of the non-perturbed biosphere correspond to the maximum efficiency of the biotic regulation. Being based on complex processes of synthesis and decomposition of organic matter into which all the species of the community are involved the biotic regulation of the environment is performed by the ecological community as a whole. The maximum efficiency of functioning of the community as a whole does not necessarily correspond to the maximum productivity of individuals of all its species. In very much the same manner, the state of absolute health of an organism as a whole does

[2] The closed character of cycles of non-regulated biogens might be associated with the closed character of cycles in biotically-regulated biogens, if certain stoichiometric relations between the former and the latter hold. However, such stoichiometric relations are themselves subject to unpredictable change in the process of evolution.

not correspond to equal rates of divisions of all cells of its organs. Thus, if we consider a single organism in a natural ecological community, e.g. a plant, it may well happen that its productivity under natural conditions does not reach the biochemical maximum. One might call productivity of that plant limited by some biogen. However, there would be no ecological sense in that statement, because the organisation of the biota as a whole does not follow the principle of maximisation of productivity of all species. Instead, the principle of maximisation of the efficiency of biotic regulation is observed in the biosphere.

Limiting biogens make sense only for strongly perturbed or artificial communities (e.g. agricultural systems) created by people. Consider this problem in more detail.

Fluxes of biogens (i.e. values of productivity $P^+_{X_i}$ expressed in units of concentration of the respective biogen X_i) in the process of synthesis of organic matter should agree with stoichiometric ratios that characterise the relative content of those biogens in the synthesised organic matter, (X_i):

$$P^+_{X_1}/P^+_{X_2}/\dots/P^+_{X_n} = (X_1)/(X_2)/\dots/(X_n) \tag{5.7.1}$$

On the other hand, productivity $P^+_{X_i}$ is related to concentration of the respective inorganic biogen in the environment, $[X_i]$, via internal resistance R_{iX_i}, see (5.6.2):

$$P^+_{X_1}/P^+_{X_2}/\dots/P^+_{X_n} = \frac{[X_1]}{R_{iX_1}} \Big/ \frac{[X_2]}{R_{iX_2}} \Big/ \dots \Big/ \frac{[X_n}{R_{iX_1}} \tag{5.7.2}$$

Therefore ratios of concentrations of biogens in the organic matter, (X_i), are related to ratios of concentrations of biogens in the environment, $[X_i]$, through values of internal resistances R_{iX_i}, which characterise the process of consumption by the biota of inorganic biogens from the environment:

$$[X_1]/[X_2]/\dots/[X_n] = (X_1)R_{iX_1}/(X_2)R_{iX_2}/\dots/(X_n)R_{iX_n} \tag{5.7.3}$$

It is evident from (5.7.3) that the ratios of concentrations of biogens in the organic matter, $(X_1)/(X_2)/\dots/(X_n)$, and in the environment, $[X_1]/[X_2]/\dots/[X_n]$, may only coincide if all the internal resistances R_{iX_i} are equal to each other, $R_{iX_i} = R_i$ for any X_i.

If the ratio between environmental concentrations of biogens, $[X_i]$, deviate from (5.7.3), then the biogen that is present in the environment in the relatively smallest amount will limit production of organic matter. In ecology, this evident statement is known as the Liebeg principle (Lotka, 1925; Odum, 1983). The Liebeg principle is most clearly manifested in agricultural systems. Actually, the very discovery of this principle owes itself largely to land-use practices. When this principle was realised it became possible to find out what biogens are limiting, and artificially increase productivity by introducing the limiting biogens into the agrosystem in the form of chemical and organic fertilisers.

The notion of limiting biogens is close to the notion of resources, which are divided into renewable and non-renewable. When productivity is limited by something, a store of that something represents a resource. Generally, non-renewable resources can be defined as stocks of matter, energy or territory that inevitably exhaust when being used, e.g. fossil fuel. Examples of renewable resources are solar radiation, geothermal and tidal power, biological production of wood, river run-off, etc. Thus, while non-renewable resources are characterised by their stores, renewable resources are best characterised by their fluxes. Fluxes of renewable resources represent ever-cycling processes that do not exhaust, in contrast to fluxes of anthropogenic consumption of non-renewable resources. Therefore people perceive fluxes of matter and energy that are cycling in the biosphere as being at their disposal. For example, as soon as the flux of productivity of the global biota is equal to about 100 Gt C, namely this figure is used when estimating the upper limit of food resources of humanity.

Such consideration is dangerously misleading. The very notion of renewable resource implies that there exists a strict correlation between the production and consumption of that resource, i.e. a rigid correlation between its producer and its consumer. When the correlation is broken, the renewable resource is no longer renewable and rapidly exhausts. For example, consider a city with a factory owned by its workers who produce cars, sell them, buy new materials and are again able to produce cars. There is a stationary flux of car productivity in that city. The anthropocentric view on resources outlined above corresponds to a situation when somebody decides to take away all the cars from the workers while at the same time expecting that the flux of car production will remain the same. It will not. As soon as the workers are not able to receive payment for their cars, they are not able to buy new materials and the factory is stopped. Hence, simple existence of fluxes of matter and energy in the environment where people live does not mean that these fluxes are designed to cover human needs. In a stationary and sustainable biosphere there cannot be such a notion as a resource that may or may not be consumed by people, at their disposal.

Consumption of energy and matter fluxes by any living organism of the nonperturbed biota represents an integral part of the closed matter cycle of matter taking place in the biosphere. In other words, in the nonperturbed biosphere all organisms have enough of everything. Thus, in a stationary state of a population the competitive interaction of individuals, which forms the basis for life stability (Chapter 2), cannot depend on availability of some limiting resource. In this sense the competitive interaction is purely aggressive, i.e. it is a competition for its own sake. However, during the non-stationary periods of population expansion, living individuals both consume non-renewable resources, e.g. territory, and destroy stationary correlation between consumers and producers of renewable resources, e.g. biomass. When expansion is stopped and a stationary stable state sets in, the destroyed correlation links between producers and consumers are formed anew, so that the notion of resource again loses any sense. Thus existence of resources is an indication of a non-stationary unstable state.

For compensation of all the external perturbations of the environment to be the quickest, primary productivity developed by the ecological community as a whole should on average be at its maximum.[3] That would correspond to the minimal

[3] This does not mean that the productivity of all species in the community precisely reaches its biochemical maximum. Provided that there are many species in the community, the task of maximisation of productivity in all of them may be in principle insoluble.

possible internal resistance (5.7.2), which exclusively depends on the internal structure of the living bodies of organisms inhabiting the community. Hence, the ratios between concentrations of different biogens in the environment are prescribed by the right-hand part of Eq. (5.7.3), i.e. by values of internal resistance for each biogen and their stoichiometric ratios in the organic matter. Now, what is the controlling factor for the absolute values of biogen concentrations in that environment? Evidently, there exist certain minimal concentrations $[X]_{min}$ for each biogen X, below which life becomes impossible. The maximum possible concentration $[X]_{max}$ of that biogen may be related to the maximum possible primary productivity of the biota at a prescribed incoming flux of solar radiation: $[X]_{max} = R_{iX} P^+_{X\,max}$. However, maximum concentrations of biogens may not necessarily correspond to the optimal values of other characteristics of the environment, such as its temperature, atmospheric pressure, etc. Therefore, concentrations of all biogens are supported by the biota at levels intermediate between those maxima and minima.

If one of the biogens is limiting, all the other biogens are present in the environment in excess. It means that their consumption by the biota goes in the regime of saturation, when both the productivity and destructivity of the biota become independent of changes in concentration of the consumed biogen, while the internal resistance R_i changes proportionally to that concentration, $P^+ = [X]/R_i = \text{const}$. In such a case, biotic regulation of non-limiting biogens becomes impossible, because whatever changes their concentrations undergo, the biota does not react to them. Thus, if one assumes the concept of biological regulation from the start, the notion of limiting biogens becomes meaningless. None of the biogens can be limiting in the non-perturbed biosphere and none of the biogens can be consumed by the biota in the saturation regime. It is only in this case that the natural biota is able to react to external perturbations of any biogen in accordance with the biotic Le Chatelier principle, compensating those perturbations by directed changes in productivity and destructivity. The internal resistance R_i should then change non-linearly with concentration following Eq. (5.5.4). In particular, the internal resistance may remain constant and productivity change linearly with concentration of the biogen, see (5.5.4).

However, when a species is introduced into an unnatural environment, the ratios (5.7.3) may noticeably deviate from the respective ratios typical for the natural conditions of the natural ecological niche of that species. In that case biogens with the lowest concentrations may appear to play the limiting role. Concentration of the limiting biogen prescribes then the productivity of the species. Changes of other concentrations do not affect that productivity, and the respective internal resistances change in proportion to those concentrations. All the cultivated agricultural species exist in such conditions.

The principal condition for biological regulation to operate is the capability of the biota to react to external perturbations of the environment. That means that such perturbations should enter the sensitivity margin of the biota. Concentrations of biogens (that is, of the environmental determinants regulated by the biota) are kept at levels intermediate between the maximum and the minimum possible. Information on such intermediate concentrations, which are optimal for the biota, is written into the normal genomes of all the species combining into natural ecological communities. Such information (that is, the stabilising programme for keeping environmental conditions in a state fit for life) may suffer decay, similar to any other programme. The process of spontaneous decay of the stabilising genetic programme is counteracted by competitive interaction between independent ecological communities within an ecosystem, which essentially represents a population of homologous communities. Communities with impaired (decayed) stabilising programmes (decay communities) deteriorate their environmental conditions, lose their competitive capacity, and are ousted by the normal communities, which are capable of supporting, and do support, the optimal environment.

It is the existence of the population of competitively interacting, internally-correlated communities that is the basis for sustainable existence of life. Competitive interaction and stabilising selection among independent, internally-correlated ecological communities is similar to selection among any internally-correlated living structures, e.g. multicellular organisms. It has nothing to do with the repeatedly discussed group selection of populations and species (Sect. 2.6), which is indeed impossible under natural conditions. Competitive interaction among separate communities does not go in the direction of the quickest consumption of biogens (environmental resources) by 'normal' communities, thus depriving the less operative 'decay' ones of their resources. Such a strategy of competitive interaction would have led to quick degradation of the environment and to extinction of the ecosystem as a whole. Thus, the true normal communities are those which are capable of balanced use of environmental stores and fluxes of matter and energy, so that the optimal life conditions are preserved. Conversely, communities capable of the maximally quick consumption of environmental resources are the decay ones, featuring lower competitiveness. Thus, the principles of extremes (Parker and Maynard Smith, 1990) cannot provide an insight into the way in which a natural ecological community functions.

Conservation of the genetic stabilising programme of each species is based on intraspecific competitive interaction of individuals of that species. To preserve the stabilising genetic programme of each species it is necessary to ensure that normal individuals carrying this programme in their genomes have the maximum competitiveness. Only species predominantly composed of normal individuals may maximise the competitiveness of the whole community. Thus, in the overwhelming majority of cases, competitive interaction of individuals in a population does not result in increased consumption of a limiting biogen or in absolute maximisation of the population number. Instead, competitive interaction serves to keep the relative number of normal individuals (those carrying in their genomes the programme for sustaining an optimum population density of the given species) at maximum.

Competitiveness of normal individuals of each species depends on the presence of other species in the community. One can say that the wide variety of complex interrelationships between different species in the community provides life with a variety of different ways to maintain genetic stability of species. In some specific cases, an artificial selection of individuals of one species by individuals of some other, controlling, species is also possible. In such a case the competitive interaction between individuals of the controlled species may be significantly weakened or even completely switched off. Such a situation is apparently realised with respect to some herbivores and the plants they consume, and with respect to some predators and their prey. Herbivores, in the first case, and predators, in the second, play then the role of controlling species (see the next section for more details). See also Section 4.5 about predators.

Competition for the limiting biogens and energy resources inevitably springs up among individuals of a species introduced into an alien ecological community. Within an alien community the stabilising programme of such species loses its meaning. Decay individuals, devoid of the stabilising programme but capable of using the limiting environmental resources at the highest rate and reaching the highest possible population density, happen to be the most competitive ones. The extreme principles may correctly describe the non-stationary ecology of populations of such individuals picked out from their natural communities (Parker and Maynard Smith, 1990).

5.8 IMMIGRATION IN THE ECOLOGICAL COMMUNITY

For the ecological community to remain stable and perform efficient regulation of the environment in accordance with the stabilising programme of its species, the relative number of decay individuals, i.e. those with impaired genetic programmes, should be low in all the species. To provide for that condition, immigration of decay individuals into the community and emigration of the normal individuals from the community should remain less powerful than both elimination of decay individuals and reproduction of normal individuals within the community. Immigration of individuals into the community occurs from neighbouring communities, among which there may be decay communities, where population density of decay individuals is high. Thus, the only way to support the level of organisation of any species in the community is to keep immigration small, as compared to reproduction of individuals within the community itself.

Consider for a species α its productivity P_α^+, which corresponds to the birth rate of its individuals, destructivity P_α^-, which corresponds to the death rate of its individuals, immigration $F_{\alpha\,\mathrm{in}}$ and emigration $F_{\alpha\,\mathrm{out}}$ of individuals of that species into and from the community, respectively. These parameters are related to the change in the biomass of that species, $\dot{M}_\alpha$, as follows:

$$\dot{M}_\alpha = P_\alpha^+ - P_\alpha^- + F_{\alpha\,\mathrm{in}} - F_{\alpha\,\mathrm{out}} \tag{5.8.1}$$

If we measure biomass, productivity and destructivity in units of the average body mass of individuals (i.e. if we put the average body mass of individuals equal to unity), these parameters are exactly equal to the population density, birth rate and death rate of individuals, respectively.

The condition that immigration remains low as compared with reproduction of normal individuals in the community is equivalent to the condition for locally regulated biogens, see Eq. (5.6.4), and has the form

$$\nu_{\alpha\,\mathrm{in}} \equiv F_{\alpha\,\mathrm{in}}/P^{+}_{\alpha} < 1 \tag{5.8.2}$$

Denote the volume density of individuals of the species α as $[\alpha]$. Then the biomass (or the density of the number of individuals, which is proportional to it) of individuals of the species α in the community is $M_\alpha = [\alpha]H_\alpha$, where H_α is the vertical dimension occupied by the species α in the community. The productivity P^{+}_{α} can be expressed as $P^{+}_{\alpha} = [\alpha]H_\alpha/\tau_\alpha$, where τ_α is the characteristic reproduction period, which coincides with the average time lapse between the two successive generations in a stationary population. Immigration $F_{\alpha\,\mathrm{in}}$ occurs in the horizontal dimension and can be characterised by the horizontal diffusivity D_α as $F_{\alpha\,\mathrm{in}} = D_\alpha[\alpha]/L$, where L is the horizontal size of the community. Then the relationship (5.8.2) assumes the form

$$\nu_{\alpha\,\mathrm{in}} = \frac{D_\alpha \tau_\alpha}{LH_\alpha} \ll 1 \tag{5.8.3}$$

For free-floating organisms incapable of independent movement in water or air streams (e.g. plankton), D_α coincides with the coefficient of eddy diffusion which characterises the environment in the local ecosystem. In other words, such organisms move as fast as they are carried by the water or air currents present in the ecosystem. For locomotive animals capable of independent movement, the diffusivity D_α is determined by the process of random wandering and can be described as $D_\alpha \sim u_\alpha L_\alpha$, where u_α is the average available speed of movement and L_α is the linear size of its feeding territory (home range) of the locomotive animal.

In normal communities the relationship (5.8.2) holds true for all species that maintain their genetic stability on the basis of intraspecific competitive interaction. The relationship (5.8.2) is only broken in decay communities. Being expelled by normal individuals, decay individuals emigrating from a normal community may only settle down in a decay community. Thus, if a decay community appears, decay individuals from all the adjacent normal communities begin to immigrate into it. On the other hand, only normal, highly-competitive individuals may leave the decay community and immigrate into a normal one where they have to stand the competition with aboriginal normal individuals. Under such conditions the decay community rapidly loses competitiveness and dies out, being substituted by a normal, highly-competitive one. After that the emigration-immigration fluxes fall down to the level determined by relationships (5.8.2) and (5.8.3).

The conditions in Eqs. (5.8.2) and (5.8.3) should be met for the principal species in the community. In other words, species that control the principal part of both energy and matter fluxes in the community should maintain their stability on the basis of intraspecific competitive interaction. Along with the task of elimination of decay individuals of its own species, the principal species in some communities are also able to control the number of decay individuals in other species, which may be called subordinate. The competitive interaction within subordinate species may be then weakened or even completely switched off. In that case the principal species in the community conduct 'artificial' selection of the subordinate species, leaving the normal individuals in them (that is, individuals carrying the necessary stabilising programme) and eliminating the decay ones (those devoid of such programme). Such subordinate species are apparently incapable of maintaining genetic stability outside their natural community. Their genetic programme may be completely devoid of information on the necessity of competitive interaction. In other words, individuals of those species may be incapable of discriminating between normal and decay individuals of their own species. The information on such differences is then contained in the normal genomes of the principal species, for which the 'artificial selection' of the subordinate species is one of the natural ecological functions in the community.

In some environments, organisation of ecological communities without division into subordinate and controlling species is simply impossible. Below we discuss one of the most vivid examples of such a situation.

In the ocean, photosynthesis of organic matter is only possible in the upper euphotic oceanic layer, which can be penetrated by the solar light. Even in the most transparent oceanic waters solar light appears to be completely absorbed at depths greater than 100 m, where photosynthesis becomes impossible. The most active photosynthesis is restricted to the upper 25 m of oceanic depth, where up to 90% of solar light is absorbed (Neshyba, 1987). Decomposition of organic matter is not light-dependent and may go on at arbitrary oceanic depths. The aerobic decomposition may only occur at those depths to which oxygen may penetrate. The anaerobic decomposition may occur independently of the availability of oxygen.

What is the possible structure of ecological communities in such conditions? Positive buoyancy makes it physically possible for the aquatic organisms to keep close to the oceanic surface. For example, multicellular algae contain special organs with air bubbles in them, which may keep them floating near the water surface. One might imagine an ecological community consisting of individuals retaining positive buoyancy of their bodies and of all their excretions. In that case decomposition of the photosynthesised organic matter could also be concentrated within the euphotic layer. In such a hypothetical ocean, life would be absent below the euphotic layer.

However, due to the extremely strong wind mixing of the euphotic oceanic layer, all the components of the environment of such community would appear to be biotically non-regulated (Sect. 5.6). The eddy diffusion coefficient in the surface layer approximately 100 m deep averages $D \sim 10^6\ \mathrm{m^2\,yr^{-1}}$, reaching as high as $10^{11}\ \mathrm{m^2\,yr^{-1}}$ during storms (Ivanoff, 1972, 1975). The reproduction time constant for phytoplankton is $\tau_\alpha > 10^{-2}$ yr. Assuming $L \sim 10$ m and $H_\alpha \sim 100$ m we find from (5.8.3) that for phytoplankton $\nu_{\alpha\,\mathrm{in}} \sim 10 \gg 1$. Hence, immigration of the individuals of phytoplankton exceeds their reproduction in the surface layer, and no ecological community with a genetically-constant composition of phytoplankton individuals can be formed there.

Another way to look at the situation is as follows. The extremely high physical mixing of the euphotic layer makes it impossible to arrange effective competitive interaction of phytoplankton individuals. Competitive interaction essentially means discrimination of decay individuals in one way or another. The power of the process of discrimination is naturally proportional to the productivity P^+. In other words, normal individuals in the community create unfavourable for decay individuals conditions with power P^+. Meanwhile physical mixing does not know anything about decay and normal individuals and evens out environmental conditions for all the individuals in the community. This significantly hinders suppression of decay individuals. If the 'equalling' physical mixing is more powerful than the 'discriminating' productivity, competitive interaction of individuals becomes meaningless, because normal and decay individuals will all the time find themselves in equal conditions. This renders the corresponding community genetically and environmentally unstable. Decay individuals that accumulate in the community finally lose their stabilising programme and deteriorate the community's environment.

In the ocean, effective competitive interaction of individuals can be only organised at depths exceeding 100 m, where the eddy diffusion coefficient drops by three orders of magnitude as compared to that at the surface and is equal to about $10^3\ m^2\ yr^{-1}$ (Broecker *et al.*, 1985; Lewis *et al.*, 1986). However, photosynthesis is impossible at those depths, so that only heterotrophic organisms, i.e. bacteria and animals feeding off the organic matter, are able to exist there. Thus, no ecological community could be arranged at depths either.

Life found an elegant way out of the situation. Heterotrophs existing at depths maintain their genetic stability via intraspecific competitive interaction. For them the condition (5.8.2) holds true. These heterotrophs should perform 'artificial selection' with respect to photosynthesising phytoplankton individuals, predominantly eating away decay individuals among them. Thus genetic stability of phytoplankton species inhabiting the euphotic layer is ensured by the heterotrophic species inhabiting deeper layers. To have a possibility to eliminate live decay phytoplankton individuals, the corresponding heterotrophs should migrate in the vertical dimension from the depths to the surface.

In other words, heterotrophic individuals compete with each other at depth, where it is possible, and eliminate decay individuals from their own population. The genetic programme of the remaining normal heterotrophic individuals contains information about characteristics of normal phytoplankton individuals, that inhabit well-mixed surface waters where intraspecific competitive interaction is impossible. Normal heterotrophic individuals migrate then to the surface, recognise decay phytoplankton individuals and eat them away. They then return to depth to take part in the intraspecific competitive interaction, and so on. This phenomenon is indeed observed in the ocean. Individuals of oceanic zooplankton species perform cyclic vertical migrations with daily or seasonal periodicity (Ehrhardt and Seguin, 1978; Hays *et al.*, 1997), which can be hardly explained outside the proposed ecological pattern.

In heterotrophs living at depths, the condition in (5.8.3) is satisfied due to the fact that their reproduction time, τ_α, is increased an order of magnitude, as compared to the reproduction time for algae (Whittaker and Likens, 1975). Also, heterotrophs spend much time at depths where the eddy diffusion coefficient D is low, and little time at the surface where it is high. Furthermore, heterotrophs may choose the time of their vertical migration and rise to the surface in calm weather, when the surface eddy diffusivity decreases as well.

Destruction of living biomass of decay phytoplankton individuals necessarily constitutes the essence of stabilising selection of phytoplankton. Hence, in the ocean, a considerable part of matter cycles must necessarily be closed within the euphotic layer, particularly via eating away of live individuals, instead of dead individuals and their excretions. Indeed, according to the observed data, the share of oceanic primary production eaten up by the microconsumers (which eat live cells of phytoplankton) amounts to about 10%, while that share does not exceed 1% on land (Whittaker and Likens, 1975).[4]

However, the major part (about 90%) of all the primary production of the ocean is still absorbed in the detritus channel, i.e. in the form of dead particles and dissolved organic matter (Khailov, 1971; Toggweiler, 1990). Hence, to ensure genetic stability of the phytoplankton, the rate of genome decay μ (see Section 9.1), of these species should remain below 0.1, which means that only one-tenth of the total number of offspring in each generation may contain new mutations absent in the parent generation. These decay individuals are then eaten away by the controlling species of zooplankton and other heterotrophs. If the proportion of decay individuals is higher than 10%, the controlling heterotrophic species will not be able to cope with all the decay phytoplankton individuals and the genetic stability of phytoplankton would be threatened. The rate of genome decay, μ, is directly proportional to the genome size of the species, see (9.1.3). Single cell organisms have generally smaller genomes than multicellular ones (Lewin, 1983). It follows then that oceanic phytoplankton should consist mainly of single cell algae, which agrees with observations.

In those few areas of the ocean, where the algae consist mainly of multicellular individuals (e.g. in the Sargasso Sea) and feature larger genomes and, hence, higher rates of genome decay, μ, the share of algae eaten alive by the small-sized heterotrophs should be significantly larger in the absence of intraspecific competitive interaction between the algae individuals themselves. However, the high density of free-floating multicellular algae significantly reduces physical mixing of the surface waters. This results in a significant drop of the diffusion coefficient and may open a possibility for intraspecific competitive interaction between the algae individuals themselves, which may lessen the ecological burden imposed on heterotrophs with respect to their controlling function.

Decay individuals of any species have impaired genetic programmes. Hence, their functioning does not contribute to the environmental stability of the community to which they belong. Left without control, the environment spontaneously deteriorates, the corresponding community loses competitiveness and becomes extinct.

[4] The major consumers of live biomass on land are large animals, which are allowed to consume not more than 1% of the primary production. In the ocean, the major part of organisms consuming live biomass is composed of the smallest invertibrate individuals, so that the natural distribution of consumption (Figure 3.3), inherent to non-perturbed terrestrial ecosystems, is not violated in the ocean (Neshyba, 1987; Kamenir, 1991).

Thus communities without decay individuals are more competitive than those with them. This opened a possibility for the complex vertically-arranged oceanic communities to establish, where heterotrophs migrate upwards to eat away predominantly decay individuals of the algae. The population density of decay phytoplankton individuals is controlled by the community in very much the same manner as environmental concentrations of locally and globally regulated biogens (Section 5.6). To find out whether decay phytoplankton individuals can be regulated by a separate community (i.e. whether phytoplankton is locally regulated), one needs to compare the flux of decay individuals through the community and the community's ability to control them.

The flux of decay phytoplankton individuals, $F_{d\,\text{in}}$, entering the community, is equal to $F_{d\,\text{in}} = \beta F_{\text{in}}$, where F_{in} is the total incoming flux of phytoplankton individuals, β is the share of decay individuals among them. In a stationary case the rate at which heterotrophs eliminate decay individuals, P^d, is proportional to the phytoplankton productivity P^+, $P^d = \gamma P^+$, where γ is the share of decay individuals in the primary production. Note that γ cannot be greater than 10%, because it is the amount of live biomass consumed by heterotrophs. Using these expressions we may write for the gross openness $\nu_{d\,\text{in}}$ of the community with respect to decay phytoplankton individuals:

$$\nu_{d\,\text{in}} \equiv F_{d\,\text{in}}/P^d = (\beta F_{\text{in}})/(\gamma P^+) = (\beta/\gamma)\nu_{\text{in}}, \tag{5.8.4}$$

where $\nu_{\text{in}} \equiv F_{\text{in}}/P^+$ is the gross openness of the community with respect to the total flux of phytoplankton. According to (5.8.3), $\nu_{\alpha\,\text{in}}$ is about 10 for a community with a horizontal size L of the order of 10 m, see above. Using $\nu_{\alpha\,\text{in}} \sim 10$ and $\gamma < 0.1$, we obtain that

$$\nu_{d\,\text{in}} > 100\,\beta \tag{5.8.5}$$

If the proportion of decay individuals, β, is less than 1%, the value of $\nu_{d\,\text{in}}$ could be in principle less than unity, so that the population density of decay phytoplankton individuals could be regulated locally, see (5.6.4). However, it seems to be unlikely that β is significantly lower than 0.01. Furthermore, the proportion of decay individuals among the biomass eaten alive by the heterotrophs is most likely lower than 100%, so that γ is actually less than 0.1, which additionally increases the value of $\nu_{d\,\text{in}}$. Thus, the conclusion that decay phytoplankton individuals are globally instead of locally regulated seems to be justified, see (5.6.6).

At large values of $\nu_{d\,\text{in}}$ the heterotrophs cannot change the population density of decay phytoplankton individuals to such an extent that it would differ by a factor of several units within and outside the community. However, if a small difference in density that can be provided by the heterotrophs still gives their community advantage and makes it the most competitive, such communities will form the majority on a given territory. As a result, each normal community would perform a net flux of normal individuals from the corresponding local ecosystem. In other words, each normal community will partly clean decay phytoplankton individuals from the phytoplankton flux flowing through the community, so that the global average density of decay individuals will be kept at a low level.

Thus phytoplankton individuals do not belong to any definite community and do not stick to any particular space domain occupied by a particular local ecosystem during their lifespan. Intraspecific competitive interaction and, hence, intraspecific stabilising selection, are completely switched off within each species of phytoplankton. The remaining physical selection eliminating lethal mutations from the phytoplankton population is by far insufficient to support correlated interaction of the phytoplankton with other species in the community and ensure stability of the environment and community itself. Individuals of zooplankton and other heterotrophic species that do belong to a concrete community purify from decay individuals the phytoplankton flux passing through the local ecosystem. As a result, each local ecosystem is a source of normal individuals of phytoplankton, which is the factor supporting the level of organisation of the global population of algae in the ocean. Decay communities, incapable of eliminating decay individuals from the phytoplankton population, locally perturb the environment, lose their competitive capacity and become extinct being ousted by normal communities.

The same principle is apparently implemented in supporting genetic stability of large animals. Large animals have feeding territories exceeding by far the characteristic size of local ecosystems. Thus a large animal cannot belong to any particular community. Organisation of large animals is supported along the same principles which govern the concentrations of the globally-regulated biogens, and the organisation of phytoplankton in the open ocean. The major difference lies in the fact that, unlike phytoplankton individuals, large animals do possess intraspecific competitive interaction which prevents many important features of their organisation from decay. In other words, while phytoplankton organisation is completely controlled by 'artificial selection' performed by the heterotrophic species, in large animals some part of the stabilising work is done by the intraspecific competitive interaction between the conspecific animals, leaving a smaller amount of work to the community. This is possible because, unlike phytoplankton, large animals depend on energy and matter fluxes that are not subject to strong physical mixing, so that large animals possess power enough to discriminate their decay conspecifics. In this sense large animals can be compared to the multicellular algae of the Sargasso Sea that were discussed above.

6

Biotic Regulation in Action

Concrete examples of processes of biotic regulation of the environment are discussed in detail, in particular: the biological pump; regulation of atmospheric carbon by the oceanic biota using the dissolved organic carbon reservoir; biotic regulation of water regime on land; and processes of recovery of boreal forest communities after perturbations.

6.1 THE BIOLOGICAL PUMP OF ATMOSPHERIC CARBON

As noted in Section 5.8, a considerable part of organic matter synthesised in the euphotic layer is consumed by heterotrophic organisms (zooplankton and nekton) inhabiting deeper oceanic layers. Such structure of oceanic ecosystems provides for genetic stability of life and environment in the ocean. Thus, the two domains—that of synthesis of organic matter and that of its decomposition—are spatially separated in the vertical, which results in typical vertical profiles of all the biogens in the ocean (Figure 6.1).

To keep the matter cycles closed and the concentration profiles stable in both organic and inorganic matter, the flux of organic matter precipitating into oceanic depths should be compensated by a back flux of inorganic matter brought up into the euphotic zone. The part of primary production which originates as a result of consumption of biogens entering the euphotic zone from below, is called *new primary production* (Dugdale and Goering, 1967). The rate of total ('gross') photosynthesis minus the rate of respiration of the phytoplankton itself is called the *net primary production*. The ratio of new primary production to the total (gross) primary production is called the f-ratio.

The upflux of inorganic biogens (= nutrients) into the euphotic zone is mainly supported by eddy diffusion, which is dependent on the vertical concentration gradients of the biogens (Figure 6.1). Concentrations of inorganic nitrogen N and phosphorus P upwelling into the euphotic zone from below, increase downward. In

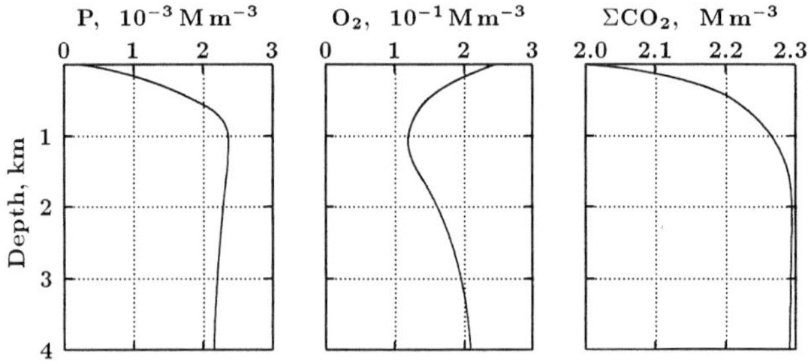

Figure 6.1. The observed change in oceanic concentrations of dissolved inorganic phosphorus P, oxygen O_2 and inorganic carbon $\sum CO_2$ (see 6.1.1) with depth. Observational data averaged over the world ocean are given (Takahashi *et al.*, 1981; Levitus, 1982; Bolin *et al.*, 1983; Sarmiento *et al.*, 1988).

contrast, the concentration of oxygen that penetrates the zone of oxidation from above drops off with depth (Figure 6.1).

In the absence of life, all the biogens would have been evenly distributed within the whole oceanic depth, and their surface and deep concentrations would have eventually evened out.[1] Life thus operates as a biological pump, which creates a considerable difference in surface and depth concentrations of inorganic biogens. Functioning of the biological pump is related to the non-zero new primary production, i.e. $f > 0$. Were the whole oceanic community (both synthesisers and reducers) located within the well-mixed euphotic zone, the action of that biological pump would have stopped. As demonstrated in Section 5.8, such community structure would be incompatible with the stable existence of oceanic environment and oceanic life as a whole.

Non-volatile dissolved inorganic nitrogen in the form of NO_3^- and NH_4^+ and phosphorus in the form of PO_4^{3-} are redistributed by the biota within the ocean only. Meanwhile, the inorganic carbon dissolved in the oceanic surface layer is at physicochemical equilibrium with the atmospheric CO_2. A depletion of CO_2 dissolved in the surface water results in a depletion of the concentration of atmospheric CO_2 (Henry's Law). Thus, operation of the biological pump keeps the atmospheric concentration of carbon at a significantly lower level compared to what would have formed in the atmosphere if the ocean remained lifeless (Gorshkov, 1979; 1982c; 1983b).

Inorganic carbon dissolved in the ocean mainly exists in the form of bicarbonate HCO_3^- and carbonate CO_3^{2-} ions. Surface concentration of the dissolved carbon dioxide reaches only 0.6% of the total concentration of dissolved inorganic carbon, which is denoted as $\sum CO_2$ (Ivanoff, 1972, 1975; Keeling, 1973):

$$[\Sigma CO_2] \equiv [HCO_3^-] + [CO_3^{2-}] + [CO_2] \tag{6.1.1}$$

$$2.0 \qquad 1.8 \qquad 0.2 \qquad \sim 0.01 \quad \text{mol C m}^{-3}.$$

[1] Only very slight concentration gradients would be preserved then, namely those due to temperature gradients and different solubility of gases and salts at different temperatures and due to the global thermohaline overturning of oceanic waters. However, these gradients would have been at least an order of magnitude weaker than those resulting from the presence of life in the ocean.

Typical values of the respective concentrations are shown here for the surface oceanic layer. All the compounds of the dissolved inorganic carbon are in the state of chemical equilibrium. The ratios between concentrations in the right-hand part of (6.1.1) are determined by the respective chemical equilibrium constants. If the concentration of the dissolved carbon dioxide, $[CO_2]$, changes, the concentrations of HCO_3^- and CO_3^{2-} change as well. In the absence of external perturbations the surface concentration of dissolved carbon dioxide, $[CO_2]$, is stationary and in physical equilibrium with the atmospheric concentration of that gas, $[CO_2]_a$. Since the solubility of CO_2 (i.e. the ratio between equilibrium concentrations of the gas in water and in the atmosphere) equals unity at 15°C (the global mean surface temperature), the surface concentration of dissolved carbon dioxide approximately coincides with its atmospheric concentration (Broecker and Peng, 1974).

Small relative changes of $[\sum CO_2]$, $d[\sum CO_2]$, are proportional to the corresponding small relative changes of $[CO_2]$, $d[CO_2]$. The proportionality coefficient ζ is called the buffer factor (Keeling, 1973):

$$\frac{d[CO_2]}{[CO_2]} = \zeta \frac{d[\sum CO_2]}{[\sum CO_2]}$$

This relationship corresponds to direct proportionality between logarithms of $[CO_2]$ and $[\sum CO_2]$. According to direct measurements, the dimensionless buffer factor ζ varies from about 9 to 15 in various regions of the world ocean, dependent on water temperature, deviating on average by no more than 30% from the average global value of $\zeta = 10$ (Broecker *et al.*, 1979). Hence, a 10% change in $[CO_2]$ entails a 1% change in $[\sum CO_2]$ only.

The observed concentration of $[\sum CO_2]$ at large depths is 15% larger than in the surface oceanic layer (Figure 6.1). Integrating the above equation in the approximation of constant buffer factor ζ we have:

$$\frac{[CO_2]}{[CO_2]_s} = \left(\frac{[\sum CO_2]}{[\sum CO_2]_s}\right)^{\zeta}, \tag{6.1.2}$$

where surface concentrations are indexed with 's'. Hence, with the increase of $[\sum CO_2]$ by 15% with depth the concentration $[CO_2]$ increases by a factor of $(1.15)^{10} \approx 4$. The dependence of the buffer factor ζ on CO_2 concentration somewhat increases this figure. Thus, the deep layer concentration of the dissolved carbon dioxide is at least four times higher than the surface one. At the same time the surface concentration of CO_2 is at equilibrium with the atmospheric CO_2 concentration. If life in the ocean ceased, the CO_2 concentrations in both the surface and deep layers would even out. Then the concentration of CO_2 in both the surface oceanic layer and the atmosphere would increase by a factor of four! That could bring about an unfavourable change of the greenhouse effect and, consequently, of climate within the mixing time of the oceanic layer in which the corresponding gradients of biogens are observed (Figure 6.1). This mixing time is of the order of several hundreds of years (Degens *et al.*, 1984). Hence, the oceanic biota keeps both the atmospheric concentration of CO_2 and the mean global surface temperature at a level fit for life.

In the ocean, consumption of the principal inorganic biogens (C, N, and P) and release of O_2 during photosynthesis, as well as release of C, N and P and consumption of O_2 in the reverse processes of destruction of organic matter, all occur in proportions determined by the Redfield ratio, see (5.5.7):

$$P^+_C/P^+_N/P^+_P/P_{O_2} = 106/16/1/138$$

The ratio between concentrations of inorganic nitrogen and phosphorus, the N/P ratio, in marine water coincides with the Redfield ratio. The cumulative store of dissolved inorganic carbon is such that the C/P ratio in marine water exceeds the Redfield ratio by a factor of five approximately. It is well known that the concentration of nitrogen compounds that are consumed by phytoplankton may be changed by the biota via both fixation of free nitrogen by bacteria and denitrification of bound nitrogen not used by phytoplankton. Thus one may assume that oceanic nitrogen is a biotically-regulated biogen. Biotic capabilities with respect to regulating the total phosphorus in the ocean remain as yet unknown. It is assumed therefore that phosphorus cannot be regulated by the biota and is a factor limiting biological productivity in the ocean (Sarmiento *et al.*, 1988). Other candidates for the limiting biogens are those minor biogens that are used by the biota in relatively small concentrations, e.g. iron (Martin *et al.*, 1990; Peng and Broecker, 1991; Timmermans *et al.*, 1998).

The general conclusion runs as follows. The total nitrogen in the ocean is accumulated by the biota following the Redfield ratio and is limited by the available amounts of phosphorus (or, possibly, iron) not subject to the biotic control. The total dissolved inorganic carbon, which is present in the ocean in excess of the Redfield ratio, therefore cannot be regulated by the biota either.

In what is to follow, we show that such a position logically contradicts the available empirical data. In reality, concentrations of all the biogens used by the biota are subject to biotic control and are maintained at certain definite levels (see also Section 5.7). In response to external perturbations of the environment, these concentrations can be changed by the natural biota (and are actually changed by it) following the principle of negative feedback. As a result, all the natural and many of the anthropogenic environmental perturbations are damped by the oceanic biota, so that the oceanic environment is supported at the optimal for the biota level.

The gross, P^+_{gX}, and new, P^+_{nX}, primary productivity of the ocean with respect to biogen X can be written as follows:

$$P^+_{gX} = \frac{[X]_s}{R_{iX}}, \qquad P^+_{nX} = \frac{[X]_s}{R_{iX}/f} \tag{6.1.3}$$

where $[X]_s$ is the surface concentration of inorganic biogen X = C, N, P; R_{iX} is the internal resistance to synthesis, see (5.5.4). The internal resistance may depend on the concentration of the biogen $[X]_s$. At small values of $[X]_s$ the internal resistance is constant, so that the gross primary productivity increases linearly with growing surface concentration of the respective biogen. At very large concentrations exceeding the saturation one, the gross primary productivity is no longer dependent on $[X]_s$, which means that R_{iX} changes in proportion to $[X]_s$.

In a stationary case the new primary productivity P^+_{nX} coincides with the diffusion flux of biogen X from the oxidation zone at depth up to the synthesis zone at the surface:

$$P^+_{\mathrm{nX}} = \frac{[\mathrm{X}]_{\mathrm{d}} - [\mathrm{X}]_{\mathrm{s}}}{R_e}, \qquad R_e = \frac{L_e}{D}, \tag{6.1.4}$$

where $[\mathrm{X}]_{\mathrm{d}}$ is the concentration of the biogen at depth in the oxidation zone, L_e is the average depth of the oxidation zone, D is the eddy diffusion coefficient, R_e is the external resistance to synthesis, which is the same for all the biogens.

Equating expressions (6.1.3) and (6.1.4) for the new primary productivity, we obtain:

$$P^+_{\mathrm{gX}} = \frac{[\mathrm{X}]_{\mathrm{d}}}{fR_e + R_{i\mathrm{X}}}, \qquad [\mathrm{X}]_{\mathrm{s}} = \frac{R_{i\mathrm{X}}}{fR_e + R_{i\mathrm{X}}}[\mathrm{X}]_{\mathrm{d}}. \tag{6.1.5}$$

Gross and new primary productivity are characterised by the same stoichiometric ratios of the principal biogens:

$$P^+_{\mathrm{gC}}/P^+_{\mathrm{gN}}/P^+_{\mathrm{gP}} = P^+_{\mathrm{nC}}/P^+_{\mathrm{nN}}/P^+_{\mathrm{nP}} \equiv \mathrm{C/N/P} = 106/16/1 \tag{6.1.6}$$

Due to this fact the ratio $f \equiv P^+_{\mathrm{nX}}/P^+_{\mathrm{gX}}$ is the same for all the biogens.

According to the observations (Figure 6.1), the concentrations of nitrogen N and phosphorus P at depth exceed the respective surface concentrations by about an order of magnitude. It means, see (6.1.5), that the value of fR_e is about 10 times greater than the internal resistance $R_{i\mathrm{X}}$.

Neglecting the small value of $R_{i\mathrm{X}}$ in the first expression in (6.1.5), we obtain that concentrations of nitrogen and phosphorus at depth, $[\mathrm{N}]_{\mathrm{d}}$ and $[\mathrm{P}]_{\mathrm{d}}$, should follow the stoichiometric Redfield ratio (6.1.6):

$$P^+_{\mathrm{gN}}/P^+_{\mathrm{gP}} = [\mathrm{N}]_{\mathrm{d}}/[\mathrm{P}]_{\mathrm{d}} = 16/1 \tag{6.1.7}$$

The relationship (6.1.7) holds true in all cases when the deep concentrations of N and P are significantly greater than those at the surface, irrespective of the actual ratio between the surface concentrations. In other words, in conditions when $[\mathrm{X}]_{\mathrm{s}} \ll [\mathrm{X}]_{\mathrm{d}}$ for N and P, the Redfield ratio between deep concentrations of nitrogen and phosphorus, $[\mathrm{N}]_{\mathrm{d}}/[\mathrm{P}]_{\mathrm{d}} = 16/1$, follows unambiguously from the very existence of new primary production. Meanwhile, surface concentrations of N and P only follow the same ratio if the internal resistances $R_{i\mathrm{N}}$ and $R_{i\mathrm{P}}$ are equal to each other, which is an additional and dispensable condition.

As follows from (6.1.5), the oceanic biota is able to control the values of gross primary productivity of nitrogen and phosphorus and their surface concentrations changing the ratio f between the gross and new primary productivity. This can be done by different ways of organising the oceanic ecological community with a different distribution of synthesis and destruction of organic matter over depth, i.e. with different localisation of home ranges of synthesisers and reducers over depth.

The minimum possible gross primary productivity and the minimum possible surface concentrations of nitrogen and phosphorus correspond to the maximum possible value of $f = 1$, when all the gross primary productivity is converted to the new primary productivity, i.e. all the organic matter synthesised at the surface is decomposed at depth. In other words, oceanic regions that are characterised by relatively large values of f are oligotrophic, i.e., they display low concentrations of biogens at the surface and low values of gross primary productivity. In particular, such a situation is observed in the tropical regions that account for the major part of the world ocean. As discussed in Section 5.8, the maximum possible value of new primary productivity makes it possible for the ecological community to ensure the most efficient control of the phytoplankton quality and, by doing so, to maintain a high degree of stability of the community itself and its environment.

The maximum possible surface concentrations of N and P close to their deep concentrations and the maximum possible gross primary productivity are observed at small values of f. In such a situation, the new primary productivity is small, and the major part of synthesised organic matter is decomposed at the surface. This is observed in cold subpolar regions of the ocean, where the gross primary productivity of the oceanic biota reaches its maximum. Note that cold subpolar regions occupy a minor part of the world ocean. A small value of new primary productivity does not allow the community to perform effective selection of decay phytoplankton individuals, which decreases the stability of the community and its environment (see Section 5.8). (In the extreme case when there is no new primary production at all, $f = 0$, and all the organic matter is synthesised and decomposed in the close vicinity to the surface, it is absolutely impossible to arrange competitive interaction of living individuals in such a community, because the physical mixing at the surface is so high that the environment is always the same for both normal and decay individuals. Thus, genetic stability of such a community cannot be supported (see Section 5.8).)

The biotic environmental impact is generally proportional to the gross primary productivity. The greater the gross primary productivity, the more powerful the biotic potential of compensating environmental perturbations. However, the peculiar organisation of the oceanic environment makes it impossible for the oceanic biota to attain the highest possible gross primary productivity without undermining the genetic stability of species. On the other hand, high new primary productivity that guarantees the genetic stability of species is inevitably accompanied by a lower gross primary productivity. Thus, the oceanic biota has to seek for a balance, i.e. for an optimum value of the ratio f between the new and gross primary productivity. In different regions of the world ocean, this optimum may correspond to different values of f, which determines the observed differences in productivity of the tropical and subpolar regions.

The observed changes in gross primary productivity that follow changes in surface concentrations of nitrogen and phosphorus from the tropical to the subpolar regions suggest that the oceanic gross primary productivity is far from saturation with respect to N and P. When $f \to 0$ (i.e. there is no new primary production), the deep and surface concentrations of nitrogen and phosphorus even out. In such a situation, the gross primary productivity attains its maximum value, which is determined by the maximum observed intensity of photosynthesis (see Table 3.3). In that case, the biotic productivity is limited by light but not by the available surface concentrations of N and P. This is an additional testimony in favour of the statement that neither nitrogen nor phosphorus limit the productivity of the biota. Their stocks in the ocean are such that they would be sufficient for an order of magnitude higher productivity than is observed in most parts of the ocean. The decreased productivity of the oceanic biota compared to the maximum possible one is dictated by the necessity to support stable organisation of the biota and oceanic environment. This makes the ecological communities with low gross and high new primary productivity more competitive compared to those featuring the maximum possible productivity and zero new primary productivity.

Let us now consider the oceanic primary productivity with respect to carbon. The surface, and hence the atmospheric, concentration of CO_2 is uniquely related to the surface and deep concentrations of N and P, because the Redfield ratio (6.1.6) is satisfied between the values of primary productivity of C, N and P. Dividing the carbon primary productivity P^+_{gC} expressed as in (6.1.3) as $[X]_s = [CO_2]_s \approx [CO_2]_a$ by P^+_{gX} expressed as in (6.1.5) at $X = N$ or P we arrive at the corresponding Redfield ratio (C/X). Using this relationship and again expression (6.1.5) for $[X]_d$ we obtain for the atmospheric CO_2 concentration:

$$[CO_2]_a = \left(\frac{C}{X}\right) \frac{R_{iC}}{fR_e + R_{iX}} [X]_d = \left(\frac{C}{X}\right) \frac{R_{iC}}{R_{iX}} [X]_s \tag{6.1.8}$$

At average observed values of f, $f \leq 0.5$, the internal resistance R_{iX} is significantly smaller than fR_e for N and P, $R_{iX} \sim 0.1 fR_e$. Thus, changing the value of f, the oceanic biota is able to significantly change the atmospheric concentration of carbon as compared to its present value. If f changes from zero to unity, the atmospheric concentration of carbon will change by a factor of 20! According to (6.1.2), the accompanying changes of deep and surface $\sum CO_2$ concentrations would not exceed a factor of 1.3. All this means that atmospheric concentration of carbon is completely under biotic control.

Note again that all the discussed potential changes of the atmospheric CO_2 concentration by the oceanic biota are initiated by changing the value of f, which is the ratio of gross primary productivity to the new primary productivity. The value of f is determined by the inherent, strictly specific structure of the respective ecological communities of the ocean. The danger of cultivation of the oceanic biota accompanied by an arbitrary restructure of ecological communities, as is done on land, can hardly be overestimated.

Let us now show that the sensitivity of the oceanic biota with respect to changes in atmospheric concentration of CO_2, $[CO_2]_a$, exceeds the biotic sensitivity with respect to changes in surface concentrations of nitrogen and phosphorus, $[N]_s$ and $[P]_s$. In other words, it is $[CO_2]_a$ that is regulated by the oceanic biota with the highest accuracy.

As we have seen in (6.1.1), dissolved inorganic carbon is mainly present in the ocean in the form of bicarbonates and carbonates, the dissolved carbon dioxide constituting but a very small portion of the total inorganic carbon stock $\sum CO_2$. However, it is the dissolved carbon dioxide CO_2, but not carbonates and bicarbonates, that is consumed by phytoplankton and used in photosynthesis.[2] The dissolved carbon dioxide is in local chemical equilibrium with carbonates and bicarbonates. Thus, its flux from depth to the surface is completely dictated by the diffusion flux of $\sum CO_2$, which is determined by the observed concentration gradient (Figure 6.1).

Thus, when equating expressions (6.1.3) and (6.1.4) for the new primary productivity of carbon, we have to substitute X for CO_2 in (6.1.3), which means that the gross primary productivity of carbon P^+_{gC} is dependent on CO_2:

$$P^+_{gC} = \frac{[CO_2]_s}{R_{iC}}, \qquad P^+_{nC} = \frac{[CO_2]_s}{R_{iC}/f}$$

and substitute X for $\sum CO_2$ in (6.1.4), which means that it is the $\sum CO_2$ concentration gradient that dictates the influx of inorganic carbon into euphotic layer from depth and, hence, determines the new primary productivity in a stationary state:

$$P^+_{nC} = \frac{[\sum CO_2]_d - [\sum CO_2]_s}{R_e}$$

Using the observed values $[\sum CO_2]_d - [\sum CO_2]_s \approx 0.1$ mole m^{-3} and $[CO_2]_s \approx [CO_2]_a = 0.012$ mole m^{-3}, we thus obtain, equating the above two equations for P^+_{nC}, that $R_{iC} \approx fR_e/25$. Since for nitrogen and phosphorus $R_{iN} \sim R_{iP} \sim fR_e/10$, the surface concentrations of CO_2, N and P are related as

$$[CO_2]_s/[N]_s/[P]_s = 40/16/1, \tag{6.1.9}$$

i.e. the surface concentration of CO_2 is about 2.5 times lower than is required for the stoichiometric ratios (6.1.6) characterising oceanic productivity. Thus, if the internal resistances for all biogens were equal to each other, while the surface concentrations of biogens could not be controlled by the biota, namely carbon would play the role of the limiting biogen, contrary to the common opinion which ascribes the limiting function to phosphorus or nitrogen. In reality, however, the biota is able to control the surface concentrations of CO_2 changing the ratio f and internal resistances R_{iC} and R_{iX} given the constant stoichiometric ratios (C/X) and deep concentrations $[X]_d$, where X = N or P, see (6.1.8).

[2] This becomes clear from the analysis of the ratios between concentrations of carbon isotopes ^{13}C and ^{12}C in the biota, ocean and atmosphere. The atmospheric and oceanic dissolved gas CO_2 are characterised by approximately equal $^{13}C/^{12}C$ ratios. The terrestrial and oceanic biota are also characterised by approximately equal $^{13}C/^{12}C$ ratios. Meanwhile the total inorganic carbon stock in the ocean, $\sum CO_2$, is characterised by a $^{13}C/^{12}C$ ratio which is 9‰ greater than that in the atmosphere. The terrestrial biota uses atmospheric CO_2 for photosynthesis. If the oceanic biota had used $\sum CO_2$ in the process of photosynthesis, the $^{13}C/^{12}C$ ratio in the oceanic biota would have been the same 9‰ greater than that in the terrestrial biota, which is not the case (Degens *et al.*, 1968). Note that shells of marine organisms are built by the biota using biochemical reactions other than photosynthesis and with use of $\sum CO_2$. Accordingly, the $^{13}C/^{12}C$ ratio in shells and similar structures is considerably higher than that in the rest of the organic matter of the ocean (Druffel and Benavides, 1986).

6.2 CHANGING PRODUCTION OF DISSOLVED ORGANIC MATTER IN THE OCEAN

If matter cycles are closed, change in new primary production is compensated by a respective change in its destruction, so that the cumulative mass of organic matter in the ocean remains unchanged. In what is to follow, we consider another way of regulation of concentrations of biogens, which is related to conversion of excessive inorganic biogens into organic ones, and vice versa. Such a process is necessarily accompanied by changes in the stock of organic matter.

If the gross primary production in the ocean had been limited by concentrations of nitrogen and phosphorus in the surface layer, and possibly by solar radiation in the subpolar oceanic surface area, while consumption of CO_2 occurred at saturation, one would have had grounds to conclude that oceanic biota is incapable of reacting to the observed anthropogenic increase of atmospheric CO_2 caused by fossil fuel burning and degradation of the terrestrial biota. That popular opinion resulted in excluding oceanic biota from the possible candidates for the sink of atmospheric CO_2 (Degens *et al.*, 1984; Prentice and Fung, 1990; Tans *et al.*, 1990; Schlesinger, 1990; Falkowski and Wilson, 1992).

To discard the concept of limiting biogens, and to consider concentrations of all the biogens as both formed and supported by the biota at levels optimal for it, calls for a revision of such attitudes. As demonstrated in the preceding section, the sensitivity of the biota to changes in concentration of CO_2 in the surface water is higher than its sensitivity to changes of concentrations of nitrogen and phosphorus, which are believed to be the limiting ones. It may thus be expected that the biota should react more efficiently to a relative perturbation of the atmospheric CO_2 at constant concentrations of nitrogen and phosphorus, than to equal relative perturbations of nitrogen and phosphorus at a constant concentration of CO_2. Here one should understand perturbation as external forcing, similar to anthropogenic distortion of the environment, instead of seasonal and geographic variations to which the natural biota should be adapted.

An enormous mass of the dissolved organic carbon (DOC) is present in the ocean, which is 1000 times higher than the cumulative mass of all the living beings of the world ocean and which approximately (by order of magnitude) coincides with the mass of atmospheric carbon. In a stationary state, the dissolved organic carbon is very slowly destroyed to inorganic components and is as slowly produced. Its preindustrial production did not exceed several tenths of a per cent of the net primary production of the ocean (Gorshkov and Makarieva, 1998). So far the functional role of DOC in the ocean remains unclear. It is natural, however, to suggest that DOC is a reservoir controlled by oceanic biota, which the biota is capable of using to sustain optimal concentrations of inorganic nutrients in the environment.

The present-day mass of total oceanic DOC is about 700 Gt C or $0.6 \cdot 10^{17}$ mole C (Sugimura and Suzuki, 1988; Druffel *et al.*, 1989; Ogawa and Ogura, 1992; Martin and Fitzwater, 1992; Suzuki, 1993; Siegenthaler and Sarmiento, 1993; Bauer *et al.*, 1998; Wiebinga and de Baar, 1998). The oceanic stocks of inorganic phosphorus, inorganic nitrogen and dissolved inorganic carbon ($\sum CO_2$ or DIC) are as follows: ~ 100 Gt P ($\sim 4 \cdot 10^{15}$ mole P), ~ 800 Gt N ($\sim 6 \cdot 10^{16}$ mole N), and ~ 40000 Gt C ($\sim 3 \cdot 10^{18}$ mole C), (see Figure 6.1 and relationship (6.1.7)). Synthesis and decomposition of the dissolved organic carbon follows apparently the same stoichiometric Redfield ratios (5.5.7) as the rest of the oceanic organic matter. Hence, the modern DOC store binds up about 10% of the inorganic stores of nitrogen and phosphorus and about 2% of the dissolved inorganic carbon. It means that, keeping the Redfield ratios, the oceanic biota is principally capable of increasing the cumulative DOC mass by almost 10 times. The store of DIC should then decrease by about 6%. In accordance with the buffer relation, Eq. (6.1.2), that would lead to approximately a 70% decrease of the equilibrium concentration of atmospheric CO_2. Inversely, shrinking oceanic DOC the biota is capable of increasing the equilibrium atmospheric concentration of CO_2 by about 20% (equilibrium refers to the system ocean-atmosphere).

All the gross primary production by phytoplankton may take part in changing production of DOC. Thus the compensating reaction of oceanic biota based on change in production of DOC appears to be amplified by the factor of f^{-1}, as compared to the reaction of the oceanic biota based on change of the new primary production alone described in the preceding section. It gives one grounds for assuming that the change in both production and mass of the oceanic DOC pool gives the biota a chance to control the environment most efficiently (Gorshkov, 1979; 1982c; 1984b).

Analysis of the available data on modern as well as prehistoric changes in the global carbon cycle allows one to determine the major quantitative characteristics of the process of biotic regulation of the environment performed by non-perturbed ecological communities of ocean and land.

6.3 GLOBAL CARBON CYCLE CHANGE

According to measurements taken from 1958 onwards by many observatories both on land and at sea, the modern concentration of atmospheric CO_2 keeps growing. The analysis of gas composition of air bubbles from Antarctic ice cores (Friedli *et al.*, 1986, Staffelbach *et al.*, 1991; Leuenberger *et al.*, 1992; Raynaud *et al.*, 1993; Siegenthaler and Sarmiento, 1993; IPCC, 1996) yields information on the atmospheric concentration of CO_2 from the very start of the perturbation at the end of the 18th century (Figure 6.2). It follows from the ice core data that, to within the error margin, the 'preindustrial' concentration of atmospheric CO_2 had approximately been equal to 280 ppmv (parts per million volume) (Siegenthaler and Oeschger, 1987), which corresponds to the preindustrial mass of carbon in the atmosphere equal to $M_{a0} = 590$ Gt C. This concentration had remained constant for the last few thousand years (Oeschger and Stauffer, 1986; Siegenthaler and Sarmiento, 1993; Lorius and Oeschger, 1994; IPCC, 1996). Today, with respect to the ocean, the non-equilibrium atmospheric concentration of CO_2 reaches 360 ppmv (770 Gt C) (IPCC, 1996), which is about 30% higher than the preindustrial level.

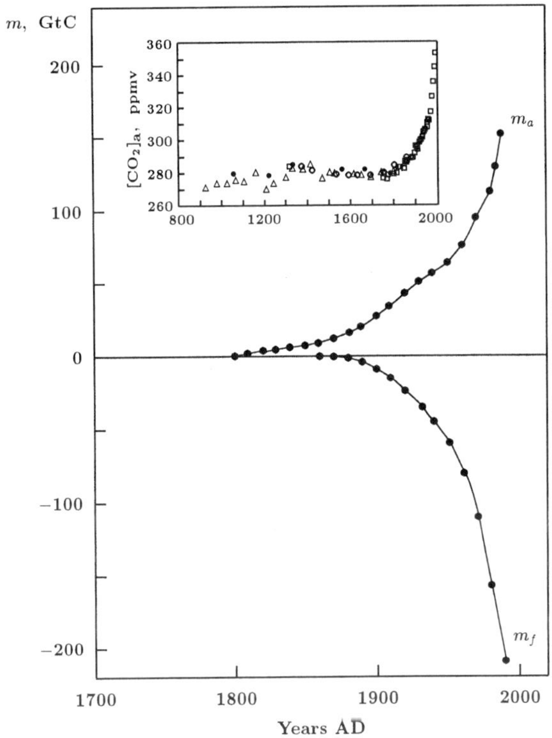

Figure 6.2. The observed global changes of carbon stores. m_a is the increase of the mass of atmospheric carbon according to measurements of CO_2 concentration $[CO_2]_a$ in the atmosphere after 1958 (Watts, 1982; Gammon *et al.*, 1986; Trivett, 1989) and in ice cores prior to 1958 (Friedli *et al.*, 1986; Oeschger and Stauffer, 1986; Leuenberger *et al.*, 1992); m_f is the depletion of fossil carbon due to combustion of coal, oil, and natural gas (Starke, 1987; 1990; la Riviere and Marton-Lefevre, 1992). The internal diagram shows the observed constancy of the preindustrial equilibrium atmospheric CO_2 concentration according to combined ice core and atmospheric data (Lorius and Oeschger, 1994; IPCC, 1996; Keeling *et al.*, 1996). According to ice core data, the global build up of the atmospheric carbon store had started before combustion of fossil fuel was initiated. That means that global changes in the environment are related to changes of carbon stored in the global biota. The state of the global environment is determined by the degree of perturbation of the global biota.

It is the enormous power of the production and destruction of organic matter developed by the natural biota which is implicitly dangerous, and may precipitate a quick disintegration of the environment should the closed matter cycles be disrupted randomly instead of in accordance with the Le Chatelier principle (Section 5.5).

We show below that the observed global change of carbon cycle is mainly due to anthropogenic perturbation of the natural land biota which had exceeded the biotic stability threshold by the middle of the 18th century. Perturbations of land biota are superimposed by direct anthropogenic perturbations of the environment, principally due to fossil fuel combustion. The oceanic biota remains stable and keeps on compensating perturbations of the environment. However, the oceanic biota already fails to cope with global anthropogenic perturbations, so that the end result is the observed global change of the environment in both the atmosphere and the ocean.

With respect to the modern carbon cycle, it means that the non-perturbed biota of the most part of the ocean and of the remaining small part of continental surface outside human activities should react to the global increase of CO_2 concentration in the atmosphere, absorbing, in accordance with the biotic Le Chatelier principle, the excessive CO_2 from the atmosphere. The perturbed biota on areas affected by anthropogenic activities by itself violates the Le Chatelier principle and emits additional CO_2 into the atmosphere.

The present-day global change in CO_2 content pertains to five principal reservoirs: the atmosphere, the fossil fuel, the oceanic dissolved inorganic carbon (DIC), the oceanic dissolved organic carbon (DOC), and the terrestrial biota. As noted above, the atmospheric concentration of carbon has been directly measured since 1958, and is also available from the ice core data. Emissions from fossil fuel are well estimated from the very start of the industrial era. In 1991–94 the emissions of carbon from fossil fuel occurred at the rate of $5.9\,\mathrm{Gt\,C\,yr^{-1}}$, and the carbon stock in the atmosphere reached 760 Gt C with about 2.2 Gt C accumulating each year (Keeling *et al.*, 1996). According to very rough estimates, the emission of carbon due to land use amounted to $1.6\,\mathrm{Gt\,C\,yr^{-1}}$ (IPCC, 1996).[3] Thus the known sum of carbon emissions into the atmosphere from the areas of active anthropogenic (industrial and agricultural) activities was $7.5\,\mathrm{Gt\,C\,yr^{-1}}$ for the 1991–94 time period. Of that amount, $2.2\,\mathrm{Gt\,C\,yr^{-1}}$ was accumulating in the atmosphere. The remaining $5.3\,\mathrm{Gt\,C\,yr^{-1}}$ could be absorbed by either the ocean or that part of the land surface biota which remains outside the scope of industrial activity. If the ocean absorbed more than $5.3\,\mathrm{Gt\,C\,yr^{-1}}$, that would mean that the remaining perturbed land surface biota not accounted for by the land use statistics also emits CO_2 into the atmosphere, that is, it violates the biotic Le Chatelier principle and is beyond the threshold of admissible perturbations.

It is not easy to discriminate between the impacts of the ocean and the terrestrial biota. With respect to the terrestrial biota, direct measurements appear of little practical help owing to their high uncertainty. The net primary production on land is estimated at $60\,\mathrm{Gt\,C\,yr^{-1}}$ (Ajtay *et al.*, 1979; Lurin *et al.*, 1994), and that in the ocean at about $40\,\mathrm{Gt\,C\,yr^{-1}}$ (Mopper and Degens, 1979; Platt *et al.*, 1989; Falkowski and Wilson, 1992; Falkowski and Woodhead, 1992; Holligan and de Boois, 1993; Lurin *et al.*, 1994). The error in both these estimates may well exceed 30% (Whittaker and Likens, 1975). Meanwhile if the global mean rate of destruction of organic matter exceeded the rate of synthesis by only 10%, that would mean an annual reduction of biomass of the terrestrial biota at a rate of $6\,\mathrm{Gt\,C\,yr^{-1}}$, which is of the order of the fossil fuel burning itself.

An indirect method of estimating the major terms in the global carbon budget is based on the analysis of changes in the concentration of atmospheric oxygen. The global carbon cycle change is due to oxidation of organic matter in processes such as fossil fuel burning and deforestation, which are accompanied by emission of CO_2 in the atmosphere and binding of the atmospheric oxygen. On the other hand, the global carbon cycle change is due to synthesis of excessive organic matter by the

[3] This figure is likely to be a gross underestimate (Houghton, 1989).

oceanic biota and the non-perturbed part of terrestrial biota, which is accompanied by depletion of atmospheric CO_2 and release of atmospheric oxygen.

The stoichiometric ratios O_2/CO_2 for all these processes are well known, which makes it possible to relate the corresponding carbon and oxygen changes to each other and to retrieve the unknown terms of the global carbon budget using the law of matter conservation. When organic matter is synthesised either on land or in the ocean, the corresponding oxygen changes mostly pertain to the atmosphere, as far as the water solubility of O_2 is small and the ocean contains two orders of magnitude less oxygen than the atmosphere. The atmospheric store of oxygen exceeds that of carbon by four orders of magnitude. Thus, a 30% increase in atmospheric CO_2 concentration entails only a 0.05% change in the oxygen content. The recently developed, highly-sensitive technique of measurements (Keeling *et al.*, 1996) makes it possible to trace such slight changes of oxygen content in the atmosphere.

The last factor that matters in the global carbon cycle change is the break of the physicochemical equilibrium of the dissolved inorganic carbon in the ocean, which is due to the present-day elevated concentration of the atmospheric CO_2 compared to the preindustrial equilibrium value. As a result, there appears a physical flux of inorganic carbon from the atmosphere to the ocean aimed at recovery of that equilibrium. Unlike the previously-mentioned biochemical processes, this process is not accompanied by any oxygen changes. Note that changes in functioning of the biological pump controlled by the oceanic biota may, as shown in Section 6.1, significantly influence (enhance) the rate of physical flux of carbon into the ocean. Hence, this rate cannot be calculated from mathematical models, since the natural biota principally does not lend itself to modelling due to huge fluxes of information processed by living organisms (see Chapter 7).

Using the available carbon and oxygen data, we now estimate the five major terms in the global carbon budget, i.e. carbon content changes in fossil fuel (f); atmosphere (a); organic carbon pool on land (terrestrial biota) (b); dissolved organic carbon pool of the ocean (sea) (s^+), and dissolved inorganic carbon pool of the ocean (s^-). Other contributions to the global carbon cycle change (volcanic activity, river run-off, etc.) either remain at the preindustrial level or are negligibly small (Degens *et al.*, 1984; Ludwig *et al.*, 1996).

According to the law of matter conservation we can represent the global budgets for atmospheric carbon and atmospheric oxygen (low index O) as

$$\dot{m}_a + \dot{m}_f + \dot{m}_{s^-} + \dot{m}_{s^+} + \dot{m}_b = 0 \tag{6.3.1a}$$

$$\dot{m}_{aO} + \dot{m}_{fO} + \dot{m}_{s^+O} + \dot{m}_{bO} = 0 \tag{6.3.1b}$$

Budget terms here are annual averaged rates of carbon and oxygen mass change in the corresponding global carbon reservoirs. Sinks of carbon enter (6.3.1a) as positive values, sources of carbon as negative values. Processes like fossil fuel burning or destruction of organic matter are accompanied by binding of atmospheric oxygen and thus represent sources of atmospheric carbon but sinks of atmospheric oxygen. Therefore the oxygen terms in (6.3.1b) will have opposite signs as compared to the corresponding carbon terms in (6.3.1a).

As noted above, the observed positive change in atmospheric carbon, $\dot{m}_a$, is measured directly. The corresponding negative change in atmospheric oxygen, $\dot{m}_{aO}$, is known from precise measurements of modern atmospheric O_2/N_2 ratio changes (Keeling and Shertz, 1992; Keeling *et al.*, 1996) for the time period 1991–94. The rate of carbon emission resulting from fossil fuel burning, $\dot{m}_f$, is well known for the whole industrial era. The corresponding rate of oxygen depletion can be calculated as $\dot{m}_{f_O} = \alpha_f \dot{m}_f$, where α_f is the stoichiometric ratio characterising the reaction of oxidation of the organic matter of fossil fuels, $\alpha_f = 1.38 \pm 0.04$ (Keeling *et al.*, 1996).

The rate of absorption of inorganic carbon by the physicochemical system of the ocean can be determined empirically on the basis of analysis of the ratio of concentrations of carbon isotopes $^{12}C/^{13}C$ in the dissolved inorganic carbon of the ocean. For the time period 1970–90 the average rate of physical carbon absorption by the ocean approached

$$\dot{m}_{s^-} = 2.0 \text{ Gt C yr}^{-1} \tag{6.3.2}$$

(Quay *et al.*, 1992; Heimann and Maier-Reimer, 1996). Note that as soon as this process is not accompanied by changes in the oxygen content, the corresponding term is absent from (6.3.1b).

It is possible to update this estimate for the 1991–94 time period, for which the atmospheric oxygen data are available. The rate at which a physicochemical system tends to the equilibrium after a perturbation is directly proportional to the perturbation itself at small values of the latter. According to the ice core data, the system atmosphere-ocean had remained in the state of physicochemical equilibrium during several thousand years before the industrial era (Figure 6.2). During this period the atmospheric mass of carbon had remained nearly constant and equalled $M_{a0} \approx 590$ Gt C ('0' stands for the preindustrial equilibrium state) (Siegenthaler and Sarmiento, 1993; IPCC, 1996). According to direct measurements, the average atmospheric carbon content during 1970–90 amounted to $\overline{M_a} = 720$ Gt C. The absolute difference, m_a, between this value and the equilibrium one is then equal to

$$m_a \equiv \overline{M_a} - M_{a0} = 130 \text{ Gt C}, \tag{6.3.3}$$

so that the relative perturbation of the preindustrial equilibrium state, m_a/M_{a0}, did not exceed 20% at that time. Thanks to this fact it is possible to use linear approximation when relating the rate of relaxation of the physicochemical system ocean-atmosphere to the equilibrium state, $\dot{m}_{s^-}$, to the atmospheric perturbation,[4] m_a:

$$\dot{m}_{s^-} = k_{s^-} m_a \tag{6.3.4}$$

[4] The fact that the perturbation m_a is measured with respect to M_{a0}, see (6.3.3), rather than with respect to some new equilibrium value of the atmospheric carbon mass, is explained by the fact that the physicochemical system of the ocean represents a powerful buffer with respect to dissolved inorganic carbon, and, hence, to the atmospheric carbon which is in the equilibrium with the latter (Section 6.1). The buffer system of the ocean is able to absorb up to 85% of excessive atmospheric carbon. An account of possible negative feedback of the biological pump will only increase this figure. Thus, if the modern perturbations of the atmospheric carbon content stopped now, the system ocean-atmosphere would return to nearly the same state of equilibrium.

The proportionality coefficient k_{s^-} is a fundamental characteristic of stability of the physicochemical system ocean-atmosphere controlled by the biological pump. It can be calculated using the data (6.3.2) and (6.3.3) for the 1970–90 time period:

$$k_{s^-} = \frac{m_{s^-}}{m_a} = \frac{2.0 \text{ Gt C yr}^{-1}}{130 \text{ Gt C}} = 0.015 \text{ yr}^{-1} \tag{6.3.5}$$

In 1991–94 the value of m_a approached 170 Gt C. Hence, according to (6.3.4), the mean rate of absorption of atmospheric carbon by the physicochemical system of the ocean during this time period was equal to

$$\dot{m}_{s^-} = 2.6 \text{ Gt C yr}^{-1} \qquad (1991\text{–}94) \tag{6.3.6}$$

Summing up, we now know the first three terms in (6.3.1a) and the first two terms in (6.3.1b). The remaining unknown terms are related to each other by the known stoichiometric ratios $\alpha \equiv O_2/CO_2$. Synthesis as well as decomposition of organic matter by terrestrial biota is on average characterised by the ratio $\alpha_b = 1.10 \pm 0.05$ (Keeling *et al.*, 1996). The corresponding ratio for the oceanic biota is determined by the Redfield ratio (5.5.7) and equals $\alpha_{s^+} = 1.30 \pm 0.03$ (Redfield, 1958; Chen *et al.*, 1996). The difference is due to the fact that the majority of the terrestrial biota consists of wood, which is decomposed following the stoichiometric ratio α close to unity. Oceanic biota does not need bearing structures like wood and, on average, features different organic matter composition and, consequently, different value of α.

Using the stoichiometric ratios α_f, α_{s^+} and α_b we may rewrite Eqs. (6.3.1) as follows:

$$\dot{m}_a + \dot{m}_f + \dot{m}_{s^-} + \dot{m}_{s^+} + \dot{m}_b = 0 \tag{6.3.7a}$$

$$\dot{m}_{aO} \frac{\alpha_f \dot{m}_f}{\dot{m}_{fO}} + \alpha_f \dot{m}_f + \alpha_{s^+} \dot{m}_{s^+} + \alpha_b \dot{m}_b = 0 \tag{6.3.7b}$$

The observed changes in the atmospheric concentration of oxygen during 1991–94 were initially measured in units of per meg, equal to the relative change $(O_2 – O_{2\,\text{ref}})/O_{2\,\text{ref}}$, where $O_{2\,\text{ref}}$ is the standard initial oxygen concentration (Keeling *et al.*, 1996). The rate of binding oxygen during fossil fuel burning, $\dot{m}_{fO}$, can also be expressed in the same units. The atmospheric term $\dot{m}_{aO}$ in (6.3.7b) is thus multiplied by transforming coefficient $(\alpha_f \dot{m}_f / \dot{m}_{fO})$, which turns units of oxygen change measurements into units of carbon change measurements (Gt C yr^{-1}).

Using the available empirical data discussed above, $\alpha_f = 1.38 \pm 0.04$; $\alpha_b = 1.1 \pm 0.05$ (Keeling *et al.*, 1996); $\alpha_{s^+} = 1.30 \pm 0.03$ (Redfield, 1958; Chen *et al.*, 1996); $\dot{m}_f = (5.9 \pm 0.3)$ Gt C yr^{-1}; $\dot{m}_a = (2.2 \pm 0.2)$ Gt C yr^{-1}; $\dot{m}_{aO} = (-42.6 \pm 0.2)$ per meg (1991–94); $\dot{m}_{fO} = (57 \pm 3)$ per meg (1991–94) (Keeling *et al.*, 1996), we arrive at the following relationship between rates of responses of the oceanic biota, $\dot{m}_{s^+}$, terrestrial biota, $\dot{m}_b$, and ocean as a physicochemical system, $\dot{m}_{s^-}$, to the anthropogenic CO_2:

$$\begin{aligned} \dot{m}_{s^+} &= -9.4\,\text{Gt C yr}^{-1} + 5.5\,\dot{m}_{s^-} \\ \dot{m}_b &= 13.1\,\text{Gt C yr}^{-1} - 6.5\,\dot{m}_{s^-} \end{aligned} \tag{6.3.8}$$

Relative uncertainties of all figures in (6.3.8) do not exceed 10%. Substituting $\dot{m}_{s^-} = 2.6\,\text{Gt C yr}^{-1}$ (6.3.6) into (6.3.8) we obtain:

$$\dot{m}_{s^+} = 4.9\,\text{Gt C yr}^{-1}, \qquad \dot{m}_b = -3.8\,\text{Gt C yr}^{-1} \qquad (1991\text{–}94) \tag{6.3.9}$$

It means that oceanic biota absorbs excessive atmospheric carbon, while terrestrial biota as a whole emits carbon into the atmosphere and adds to the direct anthropogenic pollution. The obtained estimate $\dot{m}_b = 3.8\,\text{Gt C yr}^{-1}$ exceeds direct measurements (IPCC, 1996) (that mostly pertain to the tropical zone with its intensive agriculture) more than twice and approaches earlier estimates (Houghton, 1983), which means that areas perturbed by anthropogenic activities but not accounted for in the land use statistics represent a substantial source of atmospheric carbon.

Excluding $\dot{m}_{s^-}$ from (6.3.8) yields the relation between the rates of changes in mass of organic carbon on land and in the ocean:

$$\dot{m}_b = 2.0\,\text{Gt C yr}^{-1} - 1.2\,\dot{m}_{s^+} \tag{6.3.10}$$

If we assume that oceanic biota does not react to the increasing atmospheric CO_2 concentration at all, $\dot{m}_{s^+} = 0$, we arrive at the result obtained by Keeling *et al.* (1996). They found a global terrestrial biotic sink $\dot{m}_b = 2.0\,\text{Gt C yr}^{-1}$ instead of a source (6.3.9). Neglect of the oceanic biota as a possible sink of atmospheric carbon is based of the statement that oceanic productivity is limited by concentrations of nitrogen and phosphorus, which now remain nearly constant.

We showed in Section 6.1 that such a view is erroneous and the oceanic biota is able to control concentrations of all biogens within wide margins. However, even if the concentrations of nitrogen and phosphorus do remain constant, there is still a possibility for the oceanic biota to absorb the excessive atmospheric carbon. As noted above, the equilibrium production of the refractory dissolved organic carbon (DOC) in the ocean constitutes a few tenths of a percent of the gross primary production of the oceanic biota (Gorshkov, 1995; Gorshkov and Makarieva, 1998). In response to the growing atmospheric concentration of CO_2 the biota is able to increase the refractory DOC production up to several per cent of the gross primary production, keeping the latter constant. This can be done, for instance, by increasing the share of production of extracellular excretions of phytoplankton and reducing the share of cellular production accordingly (see Figure 6.3). Then the carbon/nitrogen and carbon/phosphorus ratios in the increased production of DOC can remain at the preindustrial level.

Destruction of DOC depends on the available concentration of dissolved oxygen, which does not change with increasing atmospheric CO_2. Thus, destruction of DOC does not increase in response to growing atmospheric CO_2, so that an increase in

DOC production is inevitably accompanied by accumulation of DOC in the ocean and, consequently, depletion of CO_2 from the atmosphere.[5]

Apart from ignoring the impact of oceanic biota ($\dot{m}_{s+} = 0$), the obtained estimate (6.3.10) contradicts other empirical data as well. Modern data on land use show that exploited lands (tropics mostly) represent a source of carbon at 1.6 Gt C yr^{-1}. If the global biota as a whole represents a sink at $\dot{m}_b = 2.0$ Gt C yr^{-1} (6.3.10), then the remaining part of terrestrial biota should remove from the atmosphere 3.6 Gt C yr^{-1}, i.e. about two-thirds of what is released from burning of fossil fuels. Among terrestrial ecosystems there are no possible candidates that could ensure such a powerful sink of carbon (Ciais *et al.*, 1995; Schlesinger, 1990). Furthermore, it follows from (6.3.8) that at $\dot{m}_{s+} = 0$ and $\dot{m}_b = 2.0$ Gt C yr^{-1}, the rate of carbon absorption by the physico-chemical system of the ocean equals $\dot{m}_{s-} = 1.7$ Gt C yr^{-1} which is one and a half times less than the independent estimate of this term (6.3.6) obtained on the basis of $^{13}C/^{12}C$ measurements and updated according to formula (6.5.4).

The analysis of the carbon-storage potential of soils indicates that soils are incapable of accumulating carbon quickly, so that if quick accumulation of carbon by land biota does take place, it may only result from changes in the biomass of terrestrial vegetation (Schlesinger, 1990; Wofsy *et al.*, 1993). At the same time, destruction of soil organic matter may occur at an arbitrarily high rate. Mean loss of soil carbon following agricultural conversion reaches about 30% in different types of ecosystems. In certain tropical forest areas that value may reach the maximum of 70% (Schlesinger, 1986; Bouwman, 1989).

Expectations of an increase in biomass of the terrestrial vegetation are based on the concept of limiting biogens. It is assumed that productivity of the terrestrial biota is limited by the atmospheric concentration of CO_2 and should grow in response to its increase. This effect has never been unambiguously documented under natural conditions, so such expectations are based solely on experiments with plants in pots (Scholes, 1999). As we have shown in the previous chapter, the concept of limiting biogens may work for separate plants picked out from their natural ecological community but fails for the ecological community as a whole. Thus, there are no grounds to expect that the productivity of the terrestrial biota significantly increases in response to the increase in atmospheric carbon due to relaxation of the CO_2 limitation.

Finally, even if productivity of the terrestrial biota does increase, why should not one expect that the destructivity of organic matter (i.e. the rate of its decomposition) will also increase in proportion to productivity, so that the net effect in natural ecosystems will be close to zero? As noted before, the amount of oxygen in the atmosphere exceeds that of organic carbon by about four orders of magnitude, so that there are no obstacles for the increased productivity to be counteracted by increased destructivity. The latter can be manifested, for example, in increased population numbers of heterotrophic organisms (bacteria, fungi, animals) for which

[5] Rate of DOC decomposition may depend on the concentration of DOC itself. However, at large absolute values of DOC mass, the changes in concentration are small and cannot increase the equilibrium value of DOC destruction.

more food becomes available (food, following the same 'limiting' logic, should be the limiting factor controlling the population numbers of heterotrophs).[6] Summing up, if one discards the biotic regulation concept and logically follows the concept of limiting biogens, one cannot expect any compensating reaction of the terrestrial biota to the elevated atmospheric CO_2.

Such a reaction could be only performed by the remaining non-perturbed terrestrial biota that still functions in accordance with the biotic Le Chatelier principle. The latter forms the basis for the biotic regulation concept, which implies that rates of both productivity and destructivity are controlled by the ecological community to maximise the degree of environmental stability. However, by all appearances, the degree of anthropogenic perturbation of the terrestrial biota is so high that the compensating impact of the remaining non-perturbed biota is by far exceeded by the destabilising effect of the major part of terrestrial biota that is affected by human activities.

6.4 HISTORICAL DYNAMICS OF THE GLOBAL CHANGE

Meanwhile, the oceanic biota only slightly perturbed by human activities as yet negatively feeds back to the increasing atmospheric CO_2 in accordance with the prediction of the biotic regulation concept. As with the reaction of the physicochemical system of the ocean, the compensating reaction of oceanic biota to perturbations in the atmospheric CO_2 content should be proportional to the value of the perturbation itself. Similarly to (6.3.4), we may write for the oceanic biota:

$$\dot{m}_{s+} = k_{s+} m_a \tag{6.4.1}$$

Substituting $\dot{m}_{s+} = 4.9\,\text{Gt C yr}^{-1}$ (6.3.9) and $m_a = 170\,\text{Gt C}$ for the 1991–94 time period into (6.4.1), we obtain

$$k_{s+} = \frac{m_{s+}}{m_a} = \frac{4.9\,\text{Gt C yr}^{-1}}{170\,\text{Gt C}} = 0.029\,\text{yr}^{-1} \tag{6.4.2}$$

The obtained positive values of k_{s+} (6.4.2) and k_{s-} (6.3.5) reflect the fact that the stationary preindustrial state of the carbon cycle is stable: increase in atmospheric CO_2 is counteracted by oceanic absorption of carbon, whereas depletion of atmospheric CO_2 can be compensated by oceanic emission of carbon.

Note that the physicochemical system of the ocean alone is not able to return the system ocean-atmosphere precisely to its initial preindustrial state. When a certain amount of carbon is released into the atmosphere, some part of it is absorbed by the ocean, whereas some part will be left in the atmosphere forming a new state of

[6] It is sometimes stated that destructivity (i.e. the rate of decomposition of organic carbon) of the terrestrial biota will be increased solely as a result of an increase in the mean global surface temperature following the enhanced greenhouse effect (Scholes, 1999). Metabolic rate of cold-blooded heterotrophic organisms indeed increases with temperature (see Section 3.1). However, an increased population number of heterotrophs at constant surface temperature will produce a far more significant and rapid effect.

physicochemical equilibrium. In spite of the fact that the ocean is characterised by a large buffer effect with respect to carbon (i.e. it could absorb the overwhelming majority of the present-day excessive carbon dioxide) the effect of shifting to a new equilibrium will still be present, although very small. Functioning of the biota is not based on such simple processes of establishing physicochemical equilibrium. Genetic information of species makes it possible for the biota to use highly-ordered solar energy in order to recover precisely the initial optimal values of all the important environmental characteristics. With respect to the atmospheric CO_2 this is done by converting the inorganic atmospheric carbon into refractory dissolved organic carbon of the ocean.

The rate of change of organic matter in the ocean, $\dot{m}_{s^+}$, is equal to the difference between the change in the rate of its production $\Delta P^+_{s^+}$ and destruction $\Delta P^-_{s^+}$:

$$\dot{m}_{s^+} = \Delta P^+_{s^+} - \Delta P^-_{s^+}, \qquad \Delta P^{\pm}_{s^+} \equiv P^{\pm}_{s^+} - P^{\pm}_{s^+0}, \tag{6.4.3}$$

where $P^{\pm}_{s^+0}$ are the stationary non-perturbed preindustrial values of production and destruction in the oceanic biota, for which the equality $P^+_{s^+0} = P^-_{s^+0}$ holds true. Using (6.4.1) one may relate the relative change in productivity to the relative change of the atmospheric concentration of CO_2, which is equal to the respective relative change in the cumulative mass of atmospheric carbon, $\Delta[CO_2]_a/[CO_2]_{a0} = m_a/M_{a0}$:

$$\frac{\Delta P^+_{s^+} - \Delta P^-_{s^+}}{P^+_{s^+0}} = \frac{k_{s^+} m_a}{P^+_{s^+0}} \equiv k_{s^+} \frac{M_{a0}}{P^+_{s^+0}} \cdot \frac{m_a}{M_{a0}} \equiv k_{s^+} \tau_{s^+} \frac{\Delta[CO_2]_a}{[CO_2]_{a0}} \equiv \beta_{s^+} \frac{\Delta[CO_2]_a}{[CO_2]_{a0}} \tag{6.4.4}$$

Here $\tau_{s^+} \equiv M_{a0}/P^+_{s^+0}$ is the time of atmospheric carbon turnover through the oceanic biota which is equal to 14 years ($M_{a0} = 590$ Gt C, $P^+_{s^+0} = 40$ Gt C yr^{-1} (see Table 3.1)); $\beta_{s^+} \equiv k_{s^+} \tau_{s^+} = 0.41$ (6.4.2) is the dimensionless scaling factor (see Section 5.5). The expression (6.4.4) differs from the often-used scaling relation known as 'fertilisation' of biota by excessive CO_2 (Kohlmaier *et al.*, 1987; Scholes, 1999) in containing, in addition to the change in production, $\Delta P^+_{s^+}$, change in destruction, $\Delta P^-_{s^+}$ as well.

As repeatedly noted above, fluxes of synthesis and destruction should be balancing in such a way that the environment could remain stable both in the absence and in the presence of external perturbations. If the biota were functioning in accordance with the principle of limiting nutrients, that would be impossible. The compensating reaction of the natural biota to external perturbations in accordance with the biotic Le Chatelier principle cannot be reduced to simple increase or decrease productivity (although in some cases it is also possible). Rather, it represents a directional change in the balance between production and destruction, which may assume very complicated forms.

To find out all the details of the biotic response to external perturbation one would need to know the genetic programmes of functioning not only of each particular organism but also of the correlated interaction between different organisms in the community. As will be shown in Chapter 7, this is beyond the possibilities of modern civilisation. Thus, the questions *how* does the biota react to

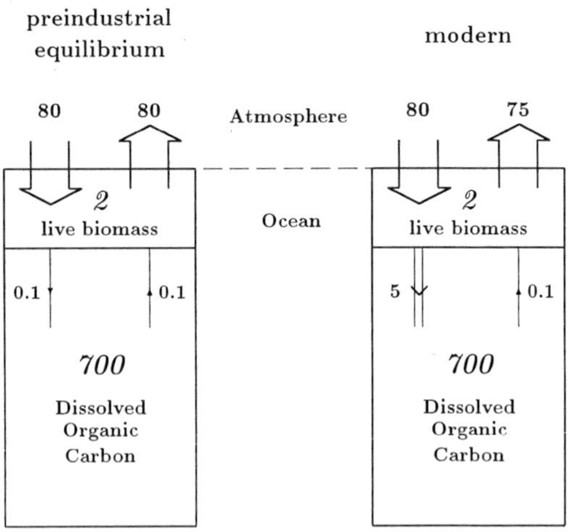

Figure 6.3. Reaction of the nonperturbed oceanic biota to increase in atmospheric CO_2 concentration. Figures near arrows represent rates of production (arrows down) and destruction (arrows up) of organic matter in $\mathrm{Gt\,C\,yr^{-1}}$. Figures in italics represent stores of organic carbon in the form of dissolved organic carbon and live biomass in Gt C. The gross primary production of the oceanic biota is taken equal to doubled net primary production, $P_g^+ = \mathrm{GPP} = 80\ \mathrm{Gt\,C\,yr^{-1}}$ (Whittaker and Likens, 1975; IPCC, 1996). The preindustrial equilibrium rate of DOC production is equal to the ratio between the DOC mass $M \approx 700$ Gt C (Druffel and Williams, 1990; Bauer *et al.*, 1998; Wiebinga and de Baar, 1998) to the time of DOC turnover $T = 8400$ years (Gorshkov and Makarieva, 1998): $M/T \approx 0.1\ \mathrm{Gt\,C\,yr^{-1}}$. Modern state is shown according to the result (6.3.9).

the perturbation and *why* does the natural biota react in this or that way, are unscientific. One can only find out *what* is going on.

In the considered case of the reaction of oceanic biota to the anthropogenic increase of atmospheric carbon, the biota increases production of the refractory dissolved organic carbon (DOC) as compared to the preindustrial value, while the gross primary production and the rate of DOC decomposition remain practically unchanged (Figure 6.3). It means that a certain part of the primary production, which used to be spent on synthesis of the short-lived biomass of living cells that was rapidly decomposed, is now spent on synthesis of refractory DOC which is decomposed very slowly. Such reaction of the biota is equivalent to an effective decrease of the rate of decomposition of organic matter at constant rate of its synthesis. This leads to accumulation of organic matter in the ocean in the form of DOC, so that the oceanic biota ensures a sink of atmospheric carbon. *How* this process is organised, via increased rate of production of phytoplankton extracellular excretions or excretions of heterotrophic organisms, or both, can hardly be investigated in detail. Besides, there is hardly any need for such an investigation.

The resulting reaction of the oceanic biota to the increased atmospheric CO_2 concentration represents, therefore, a decrease in the rate of destruction of organic matter at the constant rate of its production. This non-trivial fact cannot be either understood or predicted on the basis of the concept of limiting nutrients.

The non-perturbed terrestrial biota should have reacted to the external perturbation of the environment in a manner similar to the non-perturbed oceanic biota, which means that terrestrial biota is characterised by the same value of the scaling factor β_b, $\beta_b = \beta_{s^+} = \beta = 0.41$, which is likely to represent a fundamental characteristic of the contemporary biota as a whole and is determined by the universal biochemical organisation of life. Note that under this assumption the absolute value of biotic response, $\Delta P^+ - \Delta P^-$, as observed in different regions of the biosphere, is proportional to the non-perturbed value P_0^+ of the net primary production of the biota of those regions, see (6.4.4).

We may write for the rate of the carbon mass change in the terrestrial biota:

$$\dot{m}_b = k_b m_a \tag{6.4.5}$$

Using the relationship $\beta_{s^+} = k_{s^+}\tau_{s^+}$ we estimate the non-perturbed preindustrial value of k_b, k_{b0}, as

$$k_{b0} = k_{s^+}\tau_{s^+}/\tau_b = 0.041 \text{ yr}^{-1} \tag{6.4.6}$$

where $\tau_b \equiv M_{a0}/P_{b0}^+$ is the time of atmospheric carbon turnover through the terrestrial biota, $\tau_b = 9.8$ years ($M_{a0} = 590$ Gt C, $P_{b0}^+ = 60$ Gt C yr^{-1}, see Table 3.1). Note that there is no reservoir of inorganic carbon on land that would be in physicochemical equilibrium with the atmospheric CO_2 similar to the dissolved inorganic carbon of the ocean. Thus, the reaction of land to atmospheric carbon perturbations is described by a single parameter k_b, whereas the oceanic reaction has both biotic and physicochemical components, $k_s = k_{s^+} + k_{s^-}$.

When the degree of perturbation of the terrestrial biota grows, the value of k_b begins to drop off and turns to zero when the biotic stability on land is broken. This corresponds to a situation when the remaining non-perturbed areas compensate the destabilising impact of perturbed areas, so that the net effect is equal to zero. To analyse environmental changes related to the degree of perturbation of the terrestrial biota, it is convenient to write the law of matter conservation (6.3.7a) in terms of k_i:

$$k_a + k_s + k_f + k_b = 0, \qquad k_i \equiv \dot{m}_i/m_a \tag{6.4.7}$$

Here k_f for the fossil fuel reservoir is a formal value found from the empirical data on $\dot{m}_f$ and m_a. Variables k_s and k_b for the non-perturbed ocean and biota have the meaning of the rate of relaxation of slightly perturbed systems to the initial equilibrium non-perturbed state. Since the oceanic system has been only slightly disturbed during the industrial era, the coefficient k_s does not change with time. The behaviour of k_b can be retrieved from Eq. (6.4.7) (see Figure 6.4). In the absence of fossil fuel burning ($k_f = 0$) and perturbation of the terrestrial biota ($k_b = k_{b0}$) the rate of relaxation of the atmosphere to the preindustrial stable state after any perturbation is determined as the sum of biotic and oceanic responses, $k_{a_0} = -(k_s + k_{b0})$.

With the signs of $\dot{m}_i$ defined as they are, see (6.3.3), the atmospheric stability corresponds to $k_a < 0$, and the stability of the ocean and terrestrial biota is described

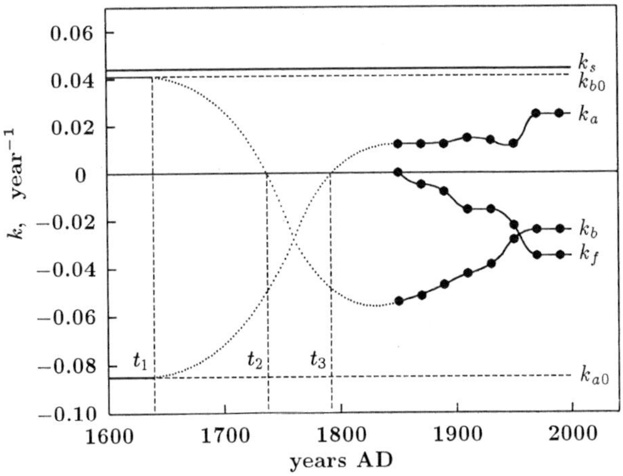

Figure 6.4. Coefficients of environmental stability (relaxation coefficients), see (6.4.7). $k_i \equiv \dot{m}_i/m_a; m_a \equiv M_a - M_{a0}; M_a$ and M_{a0} are the varying (perturbed) and the stationary (non-perturbed) mass of atmospheric carbon; $k_{f0} = 0$, k_{b0} and k_{a0} are the non-perturbed relaxation coefficients of fossil carbon, terrestrial biota and atmosphere, respectively. Points show two decade averages, plotted from ice core data up to 1958, and from atmospheric measurements after 1958 (see notes to Figure 6.2). Dashed lines interpolate between the observed and the non-perturbed values. Vertical dotted lines denote: $t = t_1$—the time margin of the highest degree of stability of the global environment (the maximum rate of relaxation after perturbation), $k_b = k_{b0}$ at $t < t_1; t = t_2$—loss of stability of the terrestrial biota, $k_b = 0; t = t_3$—loss of atmospheric stability ($k_a = 0$) provoked by further degradation of the terrestrial biota and initiation of the global change.

by $k_s > 0$ and $k_b > 0$. Therefore, the conditions for satisfying the biotic Le Chatelier principle in the biosphere as a whole have the form:

$$k_s > 0, \qquad k_b > 0 \tag{6.4.8}$$

It may be seen from Figure 6.4 that the stability of terrestrial biota and, respectively, that of the atmosphere, started to decrease, $k_b < k_{b0}$, from the middle of the 17th century, $t = t_1$. By the middle of the 18th century, $t = t_2$, the stability of land biota was completely violated, $k_b = 0$. Starting from that time and up to the beginning of the 19th century, stability of the atmosphere was supported by the oceanic biota alone. From the beginning of the 19th century, $t = t_3$, the anthropogenic perturbation of the terrestrial biota exceeded the critical level corresponding to $k_a = 0$, after which biota of the ocean already failed to cope with stabilising the atmosphere, and the process of global change of the environment had started, $k_a > 0$.

As the natural terrestrial biota is further destroyed, the relative rate of emission of carbon from it, k_b, gradually starts to drop off, although the absolute rate, $\dot{m}_b$, keeps on growing (Figures 6.2 and 6.4). Apparently, with the total destruction of natural land biota and its substitution with managed agrosystems, the stabilising potential of terrestrial biota will turn to zero, and the rates k_b and $\dot{m}_b$ will fluctuate randomly around their zero values. That tendency is already observed in Figure 6.4.

Data from Figure 6.4 make it possible to relate the stability of global environment to the global human population number (Figure 6.5). The average rate of

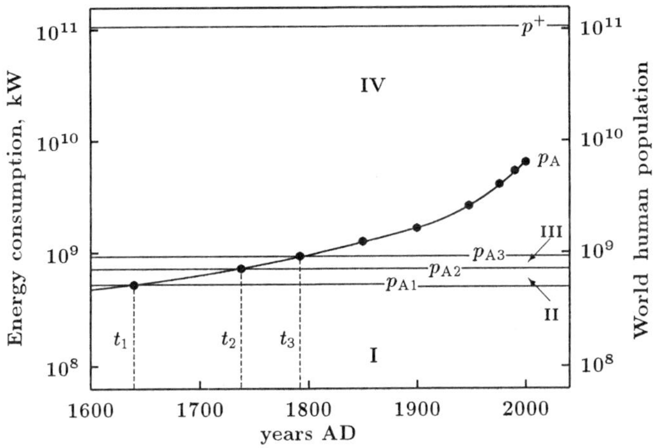

Figure 6.5. Energy thresholds for the global environmental stability. The curve p_A describes the cumulative anthropogenic consumption of the net primary production of the biosphere in power units. It includes food for people, cattle fodder and consumption of wood. The per capita consumption of the net primary production by humans is assumed to be about 1 kW (Gorshkov, 1995). Horizontal lines are as follows: line p_{A1} gives the maximum permissible value of p_A compatible with the maximum stability of the global environment; line p_{A2}—loss of stability of the terrestrial biota; line p_{A3}—loss of stability of the global environment as a whole. Horizontal line p^+ is the power of the net primary production of the total biosphere. Area I is ecologically permitted, area IV is ecologically prohibited. The transitional areas II and III are very narrow. See also notations to Figure 6.4 and text.

anthropogenic per capita consumption of the net primary production of the biosphere can be estimated in power units as 1 kW person^{-1} (Gorshkov, 1995). This estimate accounts for plant and animal food consumed by humans themselves, fodder consumed by cattle and anthropogenic consumption of wood. Using the available data on the global population growth (Figure 1.2), one may draw a curve describing the increasing rate of energy consumption of humanity, p_A, which is solely due to consumption of organic matter of the biosphere (excluding fossil fuel burning). Figure 6.5 presents this curve in the logarithmic scale. Using the data of Figure 6.4, we draw vertical lines $t = t_1, t_2, t_3$ until they cross the curve p_A and draw horizontal lines p_{A1}, p_{A2} and p_{A3} through the three points of intersections. These three lines divide the graphic plane into four areas.

Within the first area, $p_A \leq p_{A1}$ $(t \leq t_1)$, biotic regulation was characterised by the maximum possible efficiency. Consumption of the net primary production by the global population of humans did not exceed the ecologically permissible value allocated to large animals under natural ecological conditions, i.e. it did not exceed 1% of the total primary production (see Figure 3.3).

Within the second area, $p_{A1} < p_A \leq p_{A2}$ $(t_1 < t \leq t_2)$, the regulatory potential of the terrestrial biota began to degrade. However, the terrestrial biota as a whole still imposed compensating impact on the global environment. Within the third area, $p_{A2} < p_A \leq p_{A3}$ $(t_2 < t \leq t_3)$, the terrestrial biota as a whole lost its stabilising capacity and began to contribute to the environmental degradation. However, the non-perturbed oceanic biota was still able to stabilise the global environment.

Finally, within the fourth area, $p_A \geq p_{A3}$ $(t \leq t_3)$, the global environment becomes unstable. The fourth area corresponds to the share of consumption that in natural environments is allocated to the smallest heterotrophic organisms (bacteria, fungi) which, together with plants, carry out the major part of work on environmental regulation (see Section 3.4 and Figure 3.3). Anthropogenic invasion of the fourth area entails violation of the biotic regulation mechanism on a global scale and initiates global environmental changes (see Section 1.10 and Figure 1.1).

Thus, while the first area is ecologically permissible, the fourth area is ecologically prohibited. Note that the intermediate zone (areas II and III) between the states of maximum efficient biotic regulation and maximum environmental stability, on the one hand, and violation of biotic regulation on a global scale and global degradation of the environment, on the other, is very narrow (Figure 6.5). The transition from the first area, where the genus *Homo* had existed for about a million years, and the species *Homo sapiens* for at least several thousand years, to the fourth area of the global degradation of the environment, occurred nearly instantaneously.

Figure 6.5 shows that the ecologically-permissible global population number of humans should not exceed the value attained in the 16th century, which is of the order of half a billion people. Although the presented estimate is rather rough, the order of magnitude of the obtained result is unlikely to be biased.

It is sometimes stated that the destructive environmental impact of humans can be combated by decreasing energy consumption in the developed countries. The USA is often blamed for consuming the major part of world energy not proportional to their population. This is related to high living standards in the developed countries. In this respect it is important to note that only the basic need of humans, namely their food, is accounted for in Figure 6.5. It means that the current world population number still would be incompatible with environmental stability even if people in all countries stopped fossil fuel burning and producing cars and all other industrial goods. (By the way, in reality any considerable decrease of energy use is impossible at high population density, see Chapter 12 and Figure 12.1). Thus, the idea of sustainability via energy consumption equality is scientifically unsound. The only way out of the global ecological crisis is the gradual reduction of the global human population number.

6.5 STOPPING THE GLOBAL CARBON CHANGE

Using quantitative estimates obtained in Section 6.3 it is possible to calculate more accurately the minimum amount of non-perturbed biota that is sufficient to stop the present-day changes of the global carbon cycle. This means partial restoration of the regulatory biotic potential to such a level that the non-perturbed biota would be able to compensate the fossil fuel burning and carbon emission from the perturbed areas. Note that it does not mean complete restoration of the maximum possible stabilising potential of the global biota.

The net primary production of the ocean is about 40 Gt C yr^{-1} and constitutes about $s = 0.4$ (40%) of the net primary production of the whole biosphere,[7] which is about $P^+ = 100$ Gt C yr^{-1} (see Table 3.1). It leaves about 60 Gt C yr^{-1} for the terrestrial biota, $b = 0.6$ (60%). Anthropogenically perturbed land areas now amount to $b_A = 0.6$ (i.e. about 60%) of the total land area occupied by the terrestrial biota. The areas of land wilderness still remaining outside the scope of human activities constitute $b_W = 0.4$ (40%) of that area (World Resources, 1988). Thus the net primary production of the non-perturbed and perturbed terrestrial biota are about 24% and 36% of the net primary production of the whole biosphere, respectively ($b_W b = 0.24$ and $b_A b = 0.36$). As noted above, it is natural to assume that negative feedback reaction of the non-perturbed biota is proportional to the net primary production—the more the production, the more powerful the biotic response to perturbations, see (6.4.4).

Accordingly, as far as the oceanic biota accounts for 40% of the net primary production of the whole biosphere and absorbs $\dot{m}_{s^+} = 4.9$ Gt C yr^{-1} (6.3.9) from the atmosphere, we find that at present the remaining non-perturbed terrestrial biota can absorb

$$(\dot{m}_{s^+}/s) \cdot b_W b = (4.9 \text{ Gt C yr}^{-1}/0.40) \cdot 0.24 \approx 2.9 \text{ Gt C yr}^{-1}.$$

The terrestrial biota as a whole currently emits about 3.8 Gt C yr^{-1} to the atmosphere, see (6.3.9). That means that the perturbed terrestrial biota emits about $\dot{m}_{bA} \approx 6.7$ Gt C yr^{-1} (see Figure 6.6a).

Were we able to return the whole terrestrial biota and, consequently, the whole biota of the biosphere, to its non-perturbed state, it would absorb about

$$B_W = (\dot{m}_{s^+}/s) = 4.9 \text{ Gt C yr}^{-1}/0.40 \approx 12.3 \text{ Gt C yr}^{-1}$$

from the modern atmosphere, which would over-compensate anthropogenic carbon emission from the present-day fossil fuel combustion. The atmospheric CO_2 concentration would then begin to decrease at a rate of about 9 Gt C yr^{-1} (the biotic response will be helped by absorption of inorganic carbon by the ocean) and the preindustrial equilibrium state could be restored in 7–8 decades (see Figure 6.6b).

Were the biota of the whole biosphere including ocean completely perturbed to the degree that it is in exploited land areas, it would emit about

$$B_A = (\dot{m}_{bA}/b_A b) \approx 6.7 \text{ Gt C yr}^{-1}/0.36 \approx 18.7 \text{ Gt C yr}^{-1},$$

so that the atmospheric CO_2 concentration would increase at a catastrophic rate of about 22 Gt C yr^{-1} (see Figure 6.6c). Note that in such a situation even complete abandonment of fossil fuel burning would reduce this value down to only

[7] With account made for relation between the gross and net primary production (Whittaker and Likens, 1975), estimates of the net primary production of the biosphere provided by different authors (De Voogs, 1975; Fogg, 1975; Platt and Rao, 1975; Vinogradov and Shushkina, 1988; Platt *et al.*, 1989; IPCC, 1996) coincide within the uncertainty limits with this value (see Table 3.1).

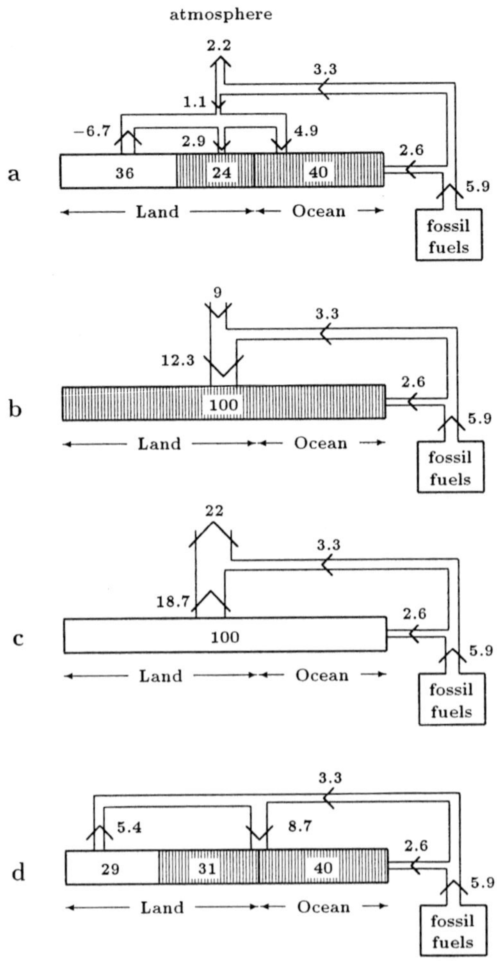

Figure 6.6. The global change and the natural biota. The global carbon cycle change: a—for the modern state of the biosphere; b—for completely unperturbed biosphere; c—for completely perturbed biosphere; d—for stopping the global change. Shaded boxes stand for natural nonperturbed biota, blank boxes stand for anthropogenically perturbed biota. Figures near arrows are carbon fluxes in Gt C yr^{-1}. Figures in shaded and blank boxes are values of net primary production of different parts of the biosphere in Gt C yr^{-1}. Carbon flux of 2.6 Gt C yr^{-1} (to the left of the fossil fuel box) stands for absorption of atmospheric carbon by the physicochemical system of the ocean, see (6.3.6).

16 Gt C yr^{-1}. Thus the state of the global environment strongly depends on the state of biota in the biosphere.

Using this fact it is possible to stop the global environmental changes of today even if the modern rate of fossil fuel burning remains the same. This can be done by increasing the area occupied by non-perturbed biota at the expense of anthropogenically perturbed territories. At present, various forms of land use account for about $b_A b \approx 0.36$ (36%) of the net primary production of the global biota. To ensure the desired effect, it is sufficient to reduce the anthropogenic consumption of the net primary production of the biosphere from the current 36% to 29% (see Figure 6.6d). In such a case the perturbed terrestrial biota would emit carbon at a rate of $0.29B_A \approx 5.4\,\text{Gt C yr}^{-1}$, while the rest of the biota (that of the ocean and of the remaining land) will find itself in a nonperturbed state and will absorb $0.71B_W \approx 8.7\,\text{Gt C yr}^{-1}$. This stabilising impact, together with that of the physico-chemical system of the ocean, will completely compensate all anthropogenic perturbations (those due to fossil fuel burning and those due to land use) (see Figure 6.6d).

Immediate reduction of the area occupied by arable lands and pastures is impossible due to the necessity to feed the current global population. Such reduction can be only achieved in a long-term perspective, when considerable changes in demographic policy of all nations already take place. About half of the 36% of net primary production of the biosphere currently consumed by man is a result of the exploitation of forests. The most real way of ensuring the desired 7% reduction in consumption of the net primary production of the biosphere needed to stop the global carbon changes seems to be a 40% reduction in exploitation of forests ($18\% \cdot 0.40 \approx 7\%$). Thus, to stop global change, humanity has to reduce the territories where forest use now takes place by about 40% and allow natural forest communities to recover in those areas.

Note that restoration of the non-perturbed biota and artificial replantation of cleared territories is absolutely not the same. The negative biotic feedback to environmental perturbations arises as the result of essentially non random interactions of particular species forming diverse ecological communities. The power of the biotic environmental impact is proportional to the net primary productivity of the natural biota, which at present exceeds by an order of magnitude the energetic power of humankind. Artificially-maintained biological systems (e.g. plantations, pastures, arable lands) represent arbitrary (from nature's point of view) sets of species placed together to cover human needs. Being deprived of the regulatory potential, such systems are not only unable to support their own stability (Soberón, 1990) but, when that is done artificially by humans, represent a major destabilising factor owing to the sustained high productivity. **In that sense complete absence of the biota on a given territory would do less harm by far to the environment than an artificially-managed ecosystem supported by humans on the same territory.**

Thus there is a principal difference between restoration of natural communities which may occur spontaneously in the course of natural succession processes (Sections 6.7 and 6.8) and artificial replantation of the cleared areas. The ultimate goal should be to restore the natural regulatory function of the biota but not merely fixing in wood a certain amount of CO_2.

6.6 THE WATER CYCLE

Water is the principal body and metabolic component in all living beings. Life on Earth exists within a narrow temperature margin corresponding to the liquid phase

Table 6.1. Stores of water in the biosphere.

Reservoir	Store, 10^{12} *t*		Source
	Salt water	Fresh water	
Ocean	1.4×10^6		Allen, 1955
Glaciers		2.0×10^4	Lvovitch, 1974
Underground waters		4.0×10^3	Lvovitch, 1974
Lakes		1.5×10^2	Lvovitch, 1974
Soil		20	Whittaker and Likens, 1975
Biota		8	Whittaker and Likens, 1975
Atmosphere		3	Mitchell, 1989
Rivers		1.4	Lvovitch, 1974

of water. Enormous stores of water in the biosphere distinguish Earth from the other planets of the terrestrial group in the solar system (Allen, 1955; Prinn, 1982). These stores are mainly concentrated in the world ocean, containing $1.4 \cdot 10^9$ Gt of H_2O (Table 6.1).

At their present rate, the processes of water outflow into the hydrosphere from the Earth's interior may be responsible for the appearance of no more than one tenth of the contemporary hydrosphere. Sedimentary rocks are on average 2 km thick (Allen, 1955). They are formed from matter fluxes rising from the Earth's depths. The relative water content in matter surfacing in rift zones and erupted by the volcanoes does not exceed 10% (Lvovitch, 1974; Degens *et al.*, 1984; Baumgartner and Reichel, 1975; Henderson-Sellers and Cogley, 1982; Shukla and Mintz, 1982). The share of water in the counterflow of matter into terrestrial depths must be at least as high. Hence, if the whole hydrosphere had been formed by the water outflow from the Earth's depths, the average depth of the hydrosphere should not have exceeded 10% of the average depth of the sedimentary layer, that is 200 m (instead of the actually observed 3 km (Allen, 1955; Watts, 1982)).

Many biochemical reactions are accompanied by the formation of free water. As follows from Table 3.1, the annual production of the modern biosphere is about $200\,\mathrm{Gt\,yr^{-1}}$ of dry organic matter, half of that amount being organic carbon. It is natural to assume that production of the paleobiosphere had been of the same order of magnitude. Assume that free water constituted 1% of paleobiological production. It could be synthesised, e.g. from CH_4 and CO_2, which could have been present in the Earth's paleoatmosphere (Allen, 1955; Budyko *et al.*, 1987). Then the whole of the hydrosphere could have been formed by the living beings in less than 10^9 years, that is within about one-quarter of the whole lifespan of Earth life. It means that, in accordance with geological data, the hydrosphere could have been formed about $3 \cdot 10^9$ years ago (Baumgartner and Reichel, 1975; Henderson-Sellers and Cogley, 1982). Biochemical reactions could have later undergone evolutionary changes to acquire their contemporary form. Thus the possibility may not be excluded that the most part of the Earth's hydrosphere is of biological origin.

The peculiar feature of atmospheric water vapour as compared to other atmospheric gases used by the biota, such as CO_2 and O_2, is that the global mean temperature of the Earth's surface and the terrestrial air pressure correspond to the liquid phase of H_2O. The CO_2 and O_2 gases (that can only be condensed at temperatures lower than $-80^\circ C$) are relatively uniformly mixed over the whole atmosphere. Their oceanic solutions are in physicochemical equilibrium with their atmospheric concentrations. In contrast, the atmospheric concentration of water vapour is only in physicochemical equilibrium with the liquid H_2O phase (oceans, seas, lakes and rivers) in the area of immediate contact of the atmosphere with open water surface. In the upper cold layers of the atmosphere, water vapour condenses in clouds and precipitates in the form of water drops.

This peculiarity of water causes the extreme unevenness of distribution of the concentration of water vapour and clouds in the atmosphere, which changes hundreds of times from region to region (Mitchell, 1989; Raval and Ramanathan, 1989; Chahine, 1992). This non-uniformity of water vapour distribution in the atmosphere leads to formation of the atmospheric water cycle. Water evaporates from open water surfaces, soil and foliage and precipitates after condensation in the upper cold layers of the atmosphere.

The present-day store of oceanic water exceeds the store of organic matter in the whole of the biosphere by many orders of magnitude (Table 6.1). In the sense defined above, water in the ocean, as well as oxygen in the atmosphere, is a biotically non-regulated substance (see Section 5.6). In contrast, atmospheric water vapour is characterised by significantly smaller concentrations and is globally regulated. In the ocean, the precipitation regime affects such environmental characteristics as salinity of the surface waters and local greenhouse effects (Chapter 8). The oceanic biota may keep these parameters close to the optimum regulating the atmospheric concentration of water vapour via control of the rate of water evaporation from the oceanic surface. This can be done by changing the surface density of algae (which may form continuous cover, e.g. during spring blooms) and by synthesis of various surface-active substances (Lebedev *et al.*, 1974). However, by all appearances, the oceanic biota's capabilities to compensate random fluctuations of the controlled characteristics are much less powerful than those of the terrestrial biota.

The time of complete latitudinal mixing of the atmosphere is of the order of a few months (Palmen and Newton, 1969). The turnover time for atmospheric moisture (that is, the ratio of the atmospheric store of moisture to the average rate of evaporation) is of the order of 10 days (Brutsaert, 1982). That results in the observed regional inhomogeneity of the atmospheric distribution of moisture, so that the moisture regime (including the atmospheric store of moisture, precipitation and evaporation) strongly differs from sea to land. Evaporation in the ocean exceeds precipitation (Lvovitch, 1974; Baumgartner and Reichel, 1975). The difference between evaporated and precipitated moisture is transported to land, where that flux is precipitated and runs off to the ocean, in agreement with the law of matter conservation, as river runoff (Lvovitch, 1974). This part of moisture fluxes on land can hardly be regulated by the terrestrial biota.

However, in addition to precipitation brought from the ocean, there exists also a land-closed cycle of evaporation and precipitation. In the absence of terrestrial biota, the additional precipitation should have been much less than precipitation brought from the ocean. In the absence of vegetation cover and soil, the precipitated moisture is quickly transferred to underground waters, which feed rivers and lakes, occupying about 2% of the land surface (see Table 3.1). Evaporation from the surface of rivers and lakes is negligibly small in comparison with precipitation brought in from the ocean. In the absence of terrestrial biota, fluctuations of oceanic precipitation would have triggered sharp changes in the river runoff.

It is only the formation of vegetation cover and soil which results in high levels of evaporation from the whole of the land and the smoothing out of random oscillations of river runoff. After vegetation is formed, part of the river runoff is spent on storing moisture in soil, in green plants, and in the continental atmosphere (Table 6.1). The increase of that mass enhances the land water cycle. According to available measurements, precipitation on land exceeds river runoff by about a factor of three (Lvovitch, 1974; Shukla and Mintz, 1982; Chahine, 1992). Hence, only one third of precipitation is brought from the ocean and more than two-thirds come from the closed water cycle on land (Figure 6.7). Hence, the principal part of the land moisture regime is formed by terrestrial biota and is subject to biotic regulation.

It is routinely assumed that the moisture regime of the continents is completely controlled by the regular circulating air flows, which depend on the latitudinal and seasonal distribution of the solar radiation, on continental relief, and on the mutual configuration of continents and oceans (Palmen and Newton, 1969). However the data on distribution of the solar energy (see Table 7.1) indicate that the power of evaporation by foliage (transpiration power) subject to biotic control exceeds the global wind power severalfold. Irregular atmospheric fluxes, such as cyclones and tornadoes, are produced by latent heat of evaporation that is released during condensation of atmospheric moisture, the most part of which is also accumulated under biotic control. Thus, from the energetic point of view, the terrestrial biota is powerful enough to completely alter the regime of irregular air flows in the atmosphere and the related characteristics of the precipitation regime on land.

Since cumulative leafage surface greatly exceeds the surface projection area occupied by corresponding plants (this ratio is called the leaf area index; it is of the order of 10 in forests, while not exceeding 5 for land surface on average (Whittaker, 1975; Whittaker and Likens, 1975; Table 3.1)), transpiration by vegetation may significantly exceed evaporation from the areas devoid of vegetation cover, and even from open water surface.

The maximum possible transpiration corresponds to a situation when all the available solar energy is used exclusively for transpiration (see Section 3.3). In virgin forests unperturbed by human activities, up to 90% of the incident solar energy is intercepted by leafage and spent on transpiration (Duvigneaud, 1974; Odum, 1983), while the same figure does not exceed 40% in human-transformed cultivated lands (Duvigneaud, 1974). Thus human forcing of the natural forest communities during the whole of human history is likely to have nearly halved the evaporating power of vegetation on the global scale. That could lead to significant changes in the

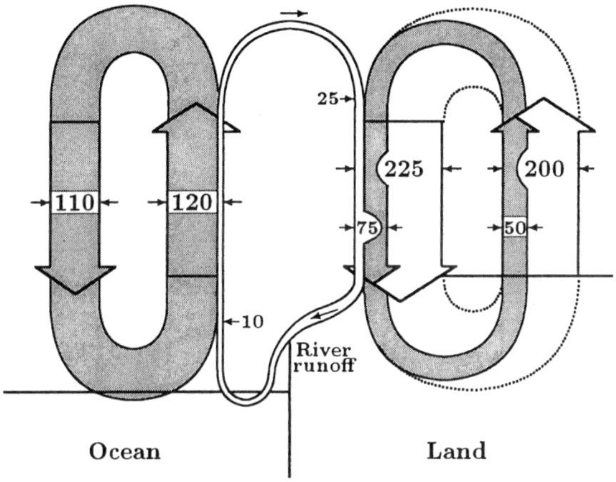

Figure 6.7. Biotic impact on water cycle. Figures indicate rates of evaporation (arrows up), precipitation (arrows down), and river runoff, cm yr^{-1} (Lvovitch, 1974; Baumgartner and Reichel, 1975; Chahine, 1992). The ratio of the rate of river runoff (cm yr^{-1}) to sea-to-land moisture transport (cm yr^{-1}) is equal to the ratio of surface areas of sea and land. Hence, the river runoff corresponds to precipitation of about 10 cm yr^{-1} in the ocean and to about 25 cm yr^{-1} on land. Evaporation and precipitation in the ocean and the sea-to-land moisture transport, equal to the river runoff, cannot be regulated by the biota. In the absence of terrestrial biota, the precipitated moisture would have rapidly transferred to underground waters. Evaporation would have mostly taken place from the surface of rivers and lakes (which constitute only 2% of the total land surface). Thus precipitation on land would have been completely controlled by moisture transport from the ocean, and would have coincided approximately with the river runoff. Plant transpiration triples the water cycle, as compared to lifeless land. If all the solar energy incident upon the Earth's surface was spent on plant transpiration, the water cycle could increase by a factor of nine (the dotted cycle). The situation close to that maximum is found in tropical forests. Shaded cycles represent the real water cycles on land and in the ocean.

circulation regime of air masses in the atmosphere and considerable reduction of precipitation on land, i.e. entail complete restructure of the moisture regime on large continental areas and enhance the process of desertification (see Figure 6.7).

Were there no wind, water would precipitate in that very location where it evaporated. In such a case, a forest community would be self-sustainable with respect to the necessary amount of precipitation. Enhancing transpiration from the foliage, the community would accordingly increase the amount of precipitation, so that the rate of water cycling could be kept by the community at an almost arbitrarily high level. In the real case, water evaporates in one location and is brought away by wind to another location, where it precipitates. Atmospheric moisture fluxes change in accordance to the direction of wind. Thus local precipitation is determined by evaporation of water by the biota of some remote, randomly-located area.

In this real case biotic regulation of the water cycle can only be performed on large areas occupied by uniform ecological communities. Self-sustainable existence of forest communities with respect to water is possible when the water evaporated in one part of the forest precipitates in another part of it, so that the total water stock in the forest cannot be reduced by external physical fluxes, i.e. by wind. Thus, the linear size of the territory occupied by forest should not be less than the average travelling distance of water vapour.

The average residence time of water in the atmosphere decreases with increasing concentration of water vapour approaching the saturated concentration. In other words, the more the atmospheric water vapour concentration, the shorter the period between two successive precipitation events. The average residence time of atmospheric water vapour over large, forest-covered territories apparently does not exceed a few days (Brutsaert, 1982). Given the global mean wind speed of about $7\,\mathrm{m\,sec^{-1}}$ (Gustavson, 1979) the average travelling distance of water from the site of evaporation to the site of precipitation is of the order of a few thousand kilometres. Thus, an inland forest can be self-sustainable with respect to water if it covers a territory of not less than 1000 kilometres in any direction. In such a case, the biotic store of moisture which is mostly contained in soil and live biomass (Table 6.1) stays within the forest communities for infinitely long periods of time irrespective of physical air fluxes. Reduction of the forest-covered territory may lead to gradual 'leakage' of water from the community through evaporation into the atmosphere and, finally, may cause complete desertification of the territory.

Thus, natural reforestation of deserts may only start from the shores of large water reservoirs—seas, large lakes and rivers, when a 1000 kilometres-wide forest front gradually penetrates into the desertified territory, creating the biotic store of moisture on already occupied areas and making possible further advancement of the forest. By all appearances, the same mechanism was applied when terrestrial communities colonised continental surfaces of the planet for the first time in the history of life. It follows that, in order to conserve the recovery potential of forests, it is necessary to conserve several kilometres-wide forest bands along all the major water reservoirs, so that these strips run continuously for at least 1000 kilometres.

Due to the inhomogeneity of moisture distribution over continental areas, biotic regulation of the water vapour concentration and water regime as a whole is possible on a regional instead of a global scale, contrary to the case of atmospheric CO_2 which is evenly mixed in the global atmosphere. Absolute humidity under the tree canopy of about 30 m in height exceeds severalfold the mean absolute humidity in the atmospheric column up to 3 km in height, where the most part of atmospheric moisture is concentrated (Whitmore, 1991). Thus, a forest community is able to change the amount of moisture in the atmospheric column by a few per cent as compared to an open area. The amount of precipitation is, on average, proportional to the concentration of water vapour in the atmospheric column. Forest communities that are able to increase humidity under their canopy receive a few per cent more rainfall than unforested territories. On the other hand, in overmoistened areas, forest communities are able to decrease the total amount of water in the community. This follows from the fact that clear-cutting in overmoistened areas is often followed by waterlogging, i.e. significant increase of the water stock per unit area (Spurr and Barnes, 1980). Since atmospheric water vapour is in local physical equilibrium with water in soil, the community is thus able to decrease atmospheric concentration of water vapour in overmoistened areas as compared to the case when biota is absent there. Thus, forest communities are able to maintain an optimal amount of moisture in the atmosphere, which determine the most part of the greenhouse effect and the temperature regime of the continents (Chapter 8).

Note that in accordance with condition (5.6.7) for biotically-regulated biogens, the amount of water stored in soil and living biomass of the global biota exceeds by nearly an order of magnitude the atmospheric store of water vapour (Table 6.1). That means that any significant fluctuations of the atmospheric water vapour concentration can be completely compensated by the biota.

Nowadays, deserts occupy about 20% of the total continental surface (Whittaker, 1975; Whittaker and Likens, 1975). Ecologically balanced flora and fauna of deserts testify by their very presence that deserts had existed even before humanity originated (Kendeigh, 1974). However, areas occupied by deserts of natural origin could have been tens or even hundreds of times smaller than those occupied by modern deserts. The natural slow process of formation of deserts on certain areas over time periods of tens and hundreds of thousands of years could not significantly weaken the biotic regulation of the environment on a global scale. In that sense modern attempts to use artificial irrigation to turn deserts into 'blooming gardens' composed of cultivated species devoid of stabilising programmes, are much more dangerous. Gardens feature high productivity, run completely open matter cycles and totally disrupt the biotic Le Chatelier principle. Only natural reforestation of modern deserts could restore biotic regulation on these areas.

6.7 FOREST SUCCESSION: RECOVERY OF FOREST COMMUNITIES AFTER PERTURBATIONS

We have so far considered biotic compensation of relatively mild environmental perturbations that, although unfavourable for the biota, do not threaten its very existence. Whether the atmospheric concentration of CO_2 becomes 30% higher or lower than the biotic optimum, it does not undermine the stability of the biota's organisation. In other words, these perturbations do not go beyond the threshold where the biota is able to exist and cope with them. However, apart from such 'mild' perturbations, there also occur 'strong' perturbations that significantly destroy the biota itself and its environment on a local area and make impossible the further existence of this type of the biota on that area. The degree of perturbation is so large that the adjacent intact ecological communities cannot cope with it and are unable to recolonise the perturbed territory immediately when the perturbation is stopped. As examples of such strong perturbations one can name fires, large-scale floods, volcanic eruptions and glaciation destroying most part of the organic matter of the affected ecosystems. Humans have added clear-cutting to the list of severe perturbations encountered in nature.

Were there no mechanism to counteract 'strong' perturbations, the whole of the terrestrial biota would finally become extinct, because any given local territory is sure to be affected by one or another strong perturbation one day. However, the biota invented a peculiar mechanism to struggle with adverse effects of such perturbations. This mechanism is known as succession, which we here consider in the example of forest communities. This section is devoted to a general theoretical discussion of succession, while the next section presents analysis of the empirical data.

Very briefly, succession can be described as follows. As soon as the dominant species of the neighbouring ecological communities are not able to recolonise the significantly perturbed areas immediately (e.g. areas previously occupied by a glacier, or burnt), the communities support existence of special species, which we call *species-repairers* (pioneer species), that are able to exist on the perturbed area. In the absence of perturbations under natural environmental conditions the population density of species-repairers is regulated by the community and kept at a low level, whereas on the perturbed areas these species develop high population numbers. The task of species-repairers is to restore the perturbed environment to its initial state and make it suitable for the dominant species of the initial communities. After that is done, the latter recolonise the repaired area and suppress population density of species-repairers down to its initial low level. As a result, the communities are again able to support their environment for infinitely long periods of time, until the next strong perturbation occurs on the same territory.[8]

A stable state of ecological community such as this is called *climax*. For example, in boreal forests of the Russian North climax communities are those dominated by spruce or pine, depending on the soil type.[9] **The principle difference between climax species (i.e. those dominating climax community) and species-repairers is that the latter change the environment in a direction that is unfavourable for themselves but favourable for the climax species, while climax species are able to keep their own optimal environment stable.**

Indeed, species-repairers are most abundant at early succession stages and become very low in numbers when the succession is finished. The very word 'succession' points to the fact that species succeed each other while the environment is being restored. Most often there are several stages of succession at which different species-repairers dominate the community until the climax species establish. Very generally, one can say that herbs (e.g. fireweed, nettle, foalfoot, dandelion) and grasses are the first to colonise the perturbed area, their exact species composition depending on the type of perturbation and properties of the area itself. Herbs and grasses are gradually replaced by shrubs (e.g. raspberry, squawbush, different species of willow) and trees (alder, aspen, birch), which finally give way to the climax tree species. The succession process brings about significant changes of the environment including changes in the

[8] Often difference is made between primary succession (i.e. colonisation of previously-unoccupied territories, such as areas appearing after retreat of glaciers) and secondary succession (i.e. restoration of the community that had occupied the affected area before the perturbation). We do not follow this discrimination because, as becomes clear from empirical evidence presented in Section 6.8, there are no considerable differences between quantitative characteristics of both types of succession processes (compare, e.g. Figures 6.10a and 6.10b, 6.12a and 6.12b, 6.14a and 6.14b).

[9] Note that in most cases a climax community can be completely identified by naming its dominant plant species, because the rest of species composition and population numbers of other species are uniquely related to it.

light, moisture and temperature regimes and concentrations of biogens. The latter are changed by the community by tens and hundreds of times (see Section 6.8.1).

The best biological analogy to succession seems to be the process of recovery of an organism from a disease. Species-repairers can be compared to various components of the immune system, which help the organism to recover. Many cells involved in the immune response are present in a healthy organism in trace numbers and develop high numbers when the organism is affected by a disease. Similarly, species-repairers are also characterised by low population densities in climax communities and become numerous after perturbation. Recovery of the organism from some common disease (e.g. influenza) can be characterised by some average time (e.g. a fortnight). Similarly, the succession process of restoration of the environment suitable for the climax community also features some characteristic time. In boreal forests complete recovery of the climax state takes about 150 years (see Table 6.4 in Section 6.8).

The inherent property of species-repairers to directionally alter their environment in an unfavourable for them direction is, as all other properties, encoded into the species' genome and cannot be changed. One may say that species-repairers act as self-murderers when they restore the environment suitable for the climax species and, by doing so, destroy the environment favourable for themselves. The existence of species-repairers is thus inherently unstable and requires a continuous external perturbation of the environment. However, external perturbations are of irregular nature. If the time interval between the two successive perturbations becomes too long, the species-repairers may completely destroy environment in all locations favourable for them and become extinct, leaving the climax community unprotected from further perturbations. To avoid it, the biota invented a mechanism of introducing regular internal perturbations into the climax community, so that species-repairers always have some territories with a favourable environment. These biotic perturbations result from the functioning of large animals that locally consume large amounts of biomass, trample down forest litter when moving and introduce other disturbances of similar kind (Section 4.4). In turn, the climax community provides large animals with shelter and food. Note that a similar function is performed by large climax trees themselves which, when fallen, give species-repairers an opportunity to achieve high population numbers on the appearing free territory.

The stability of the climax community's organisation is enhanced by such a mechanism in several ways. Firstly, it is not only that species-repairers always have a place to live, but these locations can be relatively evenly distributed over the whole territory occupied by the climax ecosystem. Thus, the perturbation in any site of it can be immediately reacted to by the nearest individuals of species-repairers.[10] Were the areas inhabited by species-repairers determined by random external perturbations of a physical nature, the space distribution of species-repairers would be

[10] A similar result is achieved owing to the creation of seed banks of species-repairers in the climax community, when seeds of species-repairers are present almost everywhere, although very few adult plants are observed (Karpov, 1969; Thompson *et al.*, 1996).

uneven, leaving some parts of the ecosystem outside easy reach. Secondly, by introducing perturbations into climax communities (e.g. making large lawns), large animals make it possible for the species-repairers to form populations and to switch on intraspecific competitive interaction in order to prevent the decay of the genetic information of the species-repairers. In the absence of such large-scale perturbations, the genetic stability of species-repairers could be only supported by the mechanism of 'artificial selection' performed by other species of the community (see Section 5.8). For example, large animals could eat away decay individuals of species-repairers of plants. Finally, large animals can themselves play an important role in restoration of the perturbed environment, serving as a transport to bring nutrients to the perturbed areas in the form of excreta. Thus, large animals can be also classified as species-repairers. Finally, animals (especially birds) take an active part in distribution of seeds. (Note that we here consider as large all animals belonging to the lowest bar of the histogram of Figure 3.3, i.e. those with body size not smaller than 1 cm.)

Insects are able to perform the same perturbing function as large animals. Developing large population densities, insects completely destroy the biomass on large areas. The existence of large animals and insects that destroy the biomass, on the one hand, and the process of succession that restores the biomass and all environmental characteristics, on the other, makes it possible to envisage the process of substituting of a decay community by a normal one. Indeed, the competitive interaction of ecological communities is the cornerstone of the biotic regulation concept. Decay communities incapable of performing biotic regulation should be forced out and replaced by normal communities. However, unlike living beings, communities are not capable of reproduction, which makes the process of substituting a decay community by a new normal one not easy to envisage. The process can be organised as follows. A decay community that is not capable of the regulation of its own environment is identified by species-'destructors' such as large animals and insects. These organisms destroy the biomass of the community and initiate the process of succession on the perturbed area. When succession is finished and the environment restored, normal individuals of all climax species born in the adjacent normal climax communities colonise the restored area and combine into a new normal community. In such a way, the stability of organisation of climax communities is prevented from decay.

The role of insects as eliminators of decay communities is especially well known. In agriculture insects are called pests, as they completely destroy the biomass of artificially-planted species. Indeed, artificial communities are deprived of any regulatory capabilities, so that they are unambiguously identified by insects as decay communities. Figure 6.18 in Section 6.8 shows the increase in pest outbreaks with the growing degree of cultivation of territories previously occupied by natural ecological communities.

A very important point is that climax communities perform the biotic regulation of the environment. Modern species-repairers are, in principle, incapable of supporting a stable environment. Thus, were the whole of the biosphere perturbed to such a degree that only species-repairers were left, the mechanism of the biotic regulation of the environment would be completely violated. The global environment would then lose its stability.

The fact that species-repairers are deprived of the regulatory ability is clearly manifested by the following examples. At early stages of succession the biota colonising the perturbed area must form the biomass necessary for the further existence and replenish the soil stock of organic matter destroyed by the perturbation. This is inevitably accompanied by consumption of atmospheric CO_2 and its transformation into organic carbon. This process occurs independently of the excess or shortage of CO_2 in the atmosphere with respect to the biotic optimum. In other words, even if the CO_2 concentration is too low due to some reasons, the biota of the succession stage will in any case consume CO_2 from the atmosphere, enhancing the existing perturbation. In contrast, the climax community should react to the low CO_2 concentration in accordance with the biotic Le Chatelier principle, i.e. they should start to decompose organic matter, emitting CO_2 into the atmosphere and replenishing the shortage.

The analysis of reaction of climax and successional communities to anthropogenic atmospheric pollution also shows the difference in their reaction to perturbations (Gorshkov, 1994). All climax communities affected by atmospheric pollution display significant changes in population densities of some of their species and other measurable characteristics of the ecosystem, which can be interpreted as the negative feedback reaction to the perturbation. In contrast, successional communities show no reaction to the atmospheric pollution. Ironically, this fact is sometimes misinterpreted as the enhanced stability of successional communities with respect to atmospheric pollution.

If we continue the above analogy and compare succession with recovery from disease, the biotic regulation can be compared with work in a factory, while successional community can be compared with a patient who is on sick leave and is freed from his work. His organism is too weak to perform any work. All efforts of the organism are spent on the quickest possible recovery, after which the person again becomes capable of highly-qualified work.

It follows from the above discussion that effective biotic regulation of the environment on a large scale may be ensured only if non-perturbed climax communities constitute the overwhelming part of the biosphere. The share of successional communities incapable of biotic regulation should be low. This condition had been met for the preindustrial state of the biosphere where the perturbed biota was located along rivers and on those few territories affected by fires and windfalls (Section 6.8.3).

Homo sapiens is a typical example of an animal species-repairer that feels most comfortable in a successional community being surrounded by other species-repairers. Climax communities (e.g. dark spruce forest) seem gloomy and unhomely to people who usually prefer the bright green leafage and sunlight of successional communities. Note that most agricultural plants actively used by man are in fact species-repairers, i.e. their wild relatives are common in ecological communities on various stages of succession, but not in climax communities. Historically, people inhabited successional communities formed along rivers, lake shores and at the seaside, where perturbations are periodically introduced by the water fluxes.

With accumulation of scientific and technological knowledge, accompanied by increasing population numbers, people began to enlarge the occupied areas. They artificially destroyed the climax biota by the use of fires and cutting, which entailed formation of people-friendly successional communities on those areas. Arable lands and pastures appeared as an attempt of humans to stop the process of succession at a certain stage optimal for them. The resulting settled mode of agriculture is inevitably accompanied by adverse environmental changes as soil erosion, impaired water regime and general degradation of the environment finally leads to desertification of the inhabited area. All the modern agricultural techniques (those using crop rotation, fertilisation, melioration, etc.) slow down but do not stop the destructive processes (Doran *et al.*, 1998).

It is impossible to stop succession and stabilise the environment in a stationary state at some successional stage due to the inherent incapability of species-repairers to maintain a stable environment favourable for their own existence. The succession process can only be slowed down by introducing continuous perturbation of the environment. However, there is a threshold value of such additional perturbation beyond which the slow process of succession is replaced by the process of degradation of the environment. When the perturbation is stopped, succession resumes from the initial stage.

6.8 FOREST SUCCESSION: ANALYSIS OF EMPIRICAL EVIDENCE

In this section we discuss quantitative characteristics of the succession processes in forest communities on the basis of the available original data of the authors and review of relevant literature. Issues to be covered:

1. Formation of the environment by forest communities
2. Recovery dynamics
3. Fires, windfalls, insect invasions: natural periodicity
4. The climate issue
5. Current state of forest communities

6.8.1 Formation of the Environment by Forest Communities

Here we demonstrate that forest communities are able to form and maintain an environment drastically different from that on perturbed and waste lands deprived of natural vegetation cover.

Water regime Precipitation is one of the most important climate characteristics. The flux of precipitating water is significantly and non-randomly redistributed within the forest community. Up to 10–40% of the total amount of precipitation is intercepted by the tree canopy and used to moisten the plants themselves. The rest falls down to the forest floor (Kitredge, 1951; Siren, 1955; Molchanov, 1960; 1961; Jackson, 1971;

Rakhmanov, 1981; Dickinson, 1987, Whitmore, 1991). The community is able both to increase and decrease the amount of precipitation in different local areas. Such effect is due to the different structure of the tree canopy in different species of trees. The tree crown may work either to concentrate the rainfall near the tree trunk, as is the case for the aspen, or throw it off to the sides, as is the case for the fir. With the mean annual rainfall of 500–600 mm, precipitation in different local parts of the community varies from 100 to 1000 mm (Protopopov, 1975; Rakhmanov, 1981; Galenko, 1983).

Formation of the forest floor by communities significantly changes the amount of water accumulated by the upper soil horizons. In boreal forests, formation of forest floor enhances the moisture capacity of the upper soil horizons by approximately 2–5-fold (Molchanov, 1960; Gorshkov *et al.*, 1996; Gorshkov *et al.*, 1997). This phenomenon is responsible for considerable decrease of surface runoff. Forest communities diminish the coefficient of surface runoff by a factor of 3–11 compared to roads, ploughed fields and continuous burnt areas (Kitredge, 1951; Molchanov, 1960; Miller, 1976; Bormann and Likens, 1979; Paulyukyavichyus, 1989; Jarvis, 1989). Essentially this means that on a regional scale forests represent serious protection from large-scale floods.

Light and temperature regimes Vegetation cover transforms the physicochemical properties and spatial distribution of surfaces interacting with fluxes of solar radiation. In boreal forests the amount of solar energy incoming under the tree canopy is lowered by a factor of 2–50 as compared with open places (Karpov, 1969; Alexeyev, 1975) and by a factor of 30–100 in tropical forests (Chazdon and Field, 1984; Whitmore, 1991). Within a separate community, the amount of solar radiation also varies significantly changing by a factor of 3–30 (Alexeyev, 1975; Protopopov, 1975).

Forest communities significantly reduce diurnal variations in temperature and moisture (Jemisson, 1934; Siren, 1955; Molchanov, 1961; Goryshina, 1969; Spurr and Barnes, 1980; Jarvis, 1989). The maximal changes are observed in tropical forests where the diurnal temperature amplitude registered under the tree canopy is approximately 10 times lower than that above the tree canopy (Figure 6.8; Whitmore, 1991). In boreal forests, diurnal amplitudes of temperatures of the air and upper soil horizons may differ by 3–5 times (Figure 6.9).

However, as compared to the open place, changes in amplitudes of air temperature observed in forests of the temperate zone are not large and do not exceed 30–50% (Kitredge, 1951; Molchanov, 1961; Protopopov, 1975). The influence of boreal forest communities on the temperature regime is most conspicuous in smoothing out the temperature extremes (Figure 6.9). The absolute maximum of the forest floor temperature under the tree canopy in the southern part of the temperate zone never exceeds 30–35°C, whereas the temperature on bare ground may reach 50–70°C (Jemisson, 1934; see Table 6.2). In the boreal forest, the highest temperature of the forest floor never exceeds 26–30°C. Frequency of early autumn and late spring frosts is five times lower in the forest than in open space (Protopopov, 1975; Table 6.3).

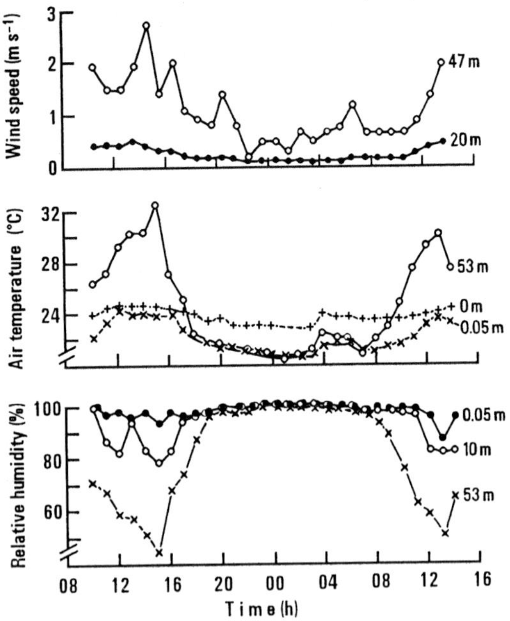

Figure 6.8. Daily march of wind speed, air temperature and relative humidity at various canopy levels in lowland rain forest at Pasoh, Malaya, 21–22 November 1973. Figures indicate height above ground level. Fluctuations of all the considered characteristics are substantially damped under the tree canopy. After (Aoki *et al.*, 1978) as cited in Whitmore (1991).

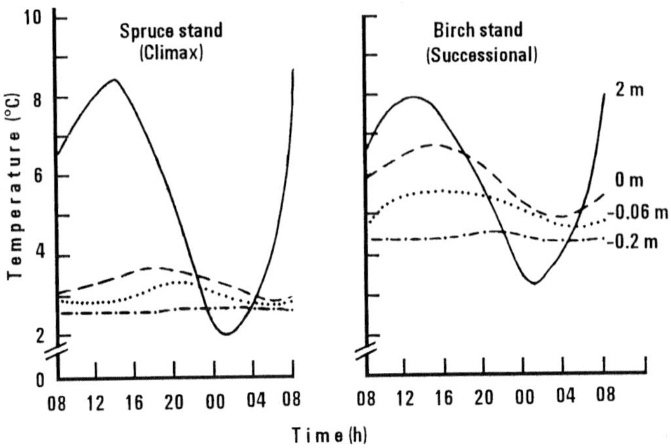

Figure 6.9. Daily variations in soil temperature observed in spruce stand (age of the last disturbance exceeds 150 years) and birch stand (age of the last disturbance is ~60 years), sunny day, 21 June 1951, Northern Finland. After (Siren, 1955). Positive and negative figures indicate height above and depth under the ground level respectively. Daily fluctuation of air temperature being about 6°C, soil temperature remains practically unchanged in the spruce forest and varies insignificantly (~1°C) in the birch forest.

Table 6.2. Maximum temperatures (°C) of the forest floor during the summer period in western white pine communities with different relative sum of basal area after cutting, Idaho, USA. Absolute maximum—the maximum temperature recorded during the whole period of observations. Average maximum—the maximum daily temperature averaged over the whole period of observations. Nonperturbed forest performs the most efficient regulation of the temperature regime preventing sharp temperature fluctuations. After Jemisson (1934).

Temperature characteristics	The degree of stand disturbance				
	Clear-cutting		Selective cutting (50%)		Without cutting
Absolute maximum, °C	70.0	>	50.0	>	34.4
Average maximum, °C	52.8	>	33.4	>	25.1

Table 6.3. Frequency of temperature extremes in the upper (0.2–0.3 cm) soil layer during vegetative season in middle taiga spruce forests with different age of perturbation. Note that the least perturbed forest communities are most efficient in damping temperature extremes. After Protopopov (1975).

Age of perturbation (clear cutting), years	Number of days in vegetation period with temperature (°C) more than						Number of days in vegetation period with the temperature (°C) less than						
	26	28	30	32	34	≥26	0	−2	−4	−6	−8	−10	≤0
2–6	62	44	28	22	6	162	16	10	6	2	4	2	42
18	26	6	–	–	–	32	6	2	2	1	–	–	15
35	19	2	–	–	–	21	5	1	5	–	–	–	11
>100	–	–	–	–	–	–	4	3	2	–	–	–	9
>120	–	–	–	–	–	–	3	3	2	–	–	–	8

The mat of mosses and lichens covering the forest floor essentially changes the dynamics of soil freezing (Kolomiets, 1961; Protopopov, 1975; Kershaw, 1975). According to the data of Kolomiets (1961) the temperature of the forest floor under the moss mat in non-perturbed fir forests never went below −18°C, while the air temperature dropped down to −48°C. Thus, natural forest communities maintain the temperature and water regimes within limits suitable for their existence prolonging the frost-free period and smoothing out the temperature extremes.

Physical and chemical characteristics of soil In the course of succession, forest communities essentially transform chemical characteristics of the soil substrate. They accumulate nitrogen, change concentrations and availability of other biogens, develop a special soil profile and form the forest floor.

Development of major components of the soil profile (excluding accumulation of humus in the lower soil horizons) takes place over the relatively short time of about

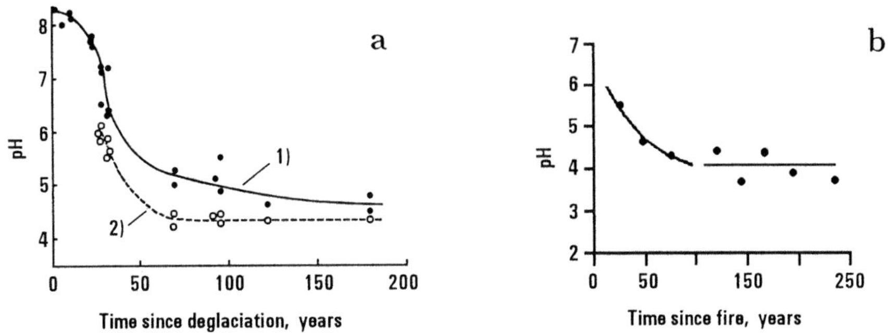

Figure 6.10. pH reaction in forest floor and surface horizons of mineral soil during succession. a) as a function of time since deglaciation, spruce forests, Glacier Bay, Alaska. 1)—mineral horizon (O-2″ horizon); 2)—forest floor. After Crocker and Major (1955); b) as a function of time since fire, boreal mix-wood (birch-spruce-fir) forests, northern taiga, South Quebec, Canada. After Brais *et al.* (1995).

100–300 years (Crocker and Major, 1955; Covington, 1976; Skvortsova *et al.*, 1983; Kovda and Rozanov, 1988; Bormann and Sidle, 1990). Crocker and Major (1955) reported drop in pH (over 2 units) during succession of spruce forests 100–150 years after deglaciation (Glacier Bay, Alaska) (Figure 6.10a). Analogues changes in pH of the upper soil horizons were registered during the recovery of mixed forests after fires in the western part of North America (Quebec) (Figure 6.10b). The amount of nitrogen in the upper soil horizons increases 50-fold and reaches its maximum over 100 years after formation of vegetation cover on deglaciated territories. Over 200 years after succession begins, the amount of nitrogen in the upper soil horizon may somewhat decrease as compared to its maximum value, being redistributed over other parts of the community. The total store of nitrogen in the community remains the same (Crocker and Major, 1955; Bormann and Sidle, 1990).

In boreal forests, concentrations of biogens in the forest floor (Ca, Mg, K, N) exceed by 10–100 times the respective concentrations in mineral abiotic horizons (Kovda and Rozanov, 1988). Formation or recovery of the forest litter during both primary and secondary successions takes on average 90–150 years (Crocker and Major, 1955; Covington, 1976; Sannikov and Sannikova, 1985; Bormann and Sidle, 1990; Gorshkov *et al.*, 1996; Figures 6.11 and 6.12).

Peculiarities of distribution of chemical elements along the soil profile and ratio between their soluble and insoluble forms at differing succession stages are naturally dependent on the community type. However, along with differences, there exist some fundamental features of succession that are common to all types of communities. One such feature is that, while at early succession stages biogens are dispersed along a deep soil profile, succession ends in a state when all biologically–important biogens are predominantly concentrated and biogenic cycles nearly closed within the upper soil horizon (Siren, 1955; Bormann and Likens, 1979; Kovda and Rozanov, 1988; Kellman and Ruolet, 1990). This can be monitored by the fact that plants of early successional stages are characterised by deep rootage reaching down to 1.5 m in depth, while in climax communities up to 90% of the rootage is concentrated within the upper 25 cm of the soil profile.

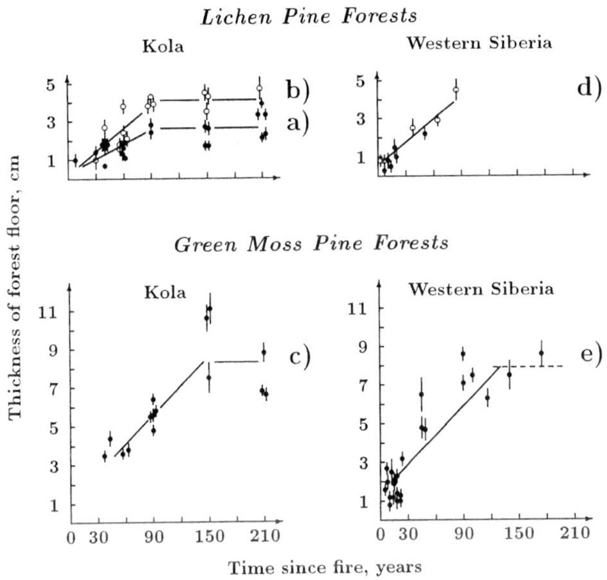

Figure 6.11. Thickness of forest litter in Scots pine forests as a function of time since fire. Kola peninsula: a) lichen site type (share of lichens in moss-lichen cover exceeds 70%), b) green moss-lichen site type (share of lichens in moss-lichen cover ranges between 30-70%), c) green moss site type (share of lichens in moss-lichen cover is under 30 %). South-western part of Western Siberia: d) lichen site type, e) green-moss site type. After (Gorshkov *et al.*, 1996).

Along with transformation of chemical properties of soil and formation of the forest floor during succession, the community changes (lowers) the density of upper mineral soil horizons. This occurs as a result of repeated processes of growth and death of roots and activity of animal species (Kitredge, 1951; Crocker and Major, 1955). In boreal forests, density of the upper soil horizons in both successional and climax communities averages $\sim 1\,g\,cm^{-3}$, which is at least two times lower than the density of the corresponding soil forming rocks estimated as $1.8\text{–}2.5\,g\,cm^{-1}$ (Kovda and Rozanov, 1988; Liski and Westman, 1995). As a result, soil moisture capacity and the rate of soil moisture and air exchange rise up.

Finally, communities at late stages of succession are capable of closing biogenic cycles to a high accuracy, whereas in perturbed communities the biogenic cycles are open which essentially entails environmental instability (Figure 6.13; Odum, 1971). The unique analysis of concentrations of biogens in brooks fed by water from strongly perturbed and non-perturbed areas was performed by Bormann and Likens (1979). They showed that water coming from a territory occupied by a 60-year-old forest practically does not contain any soil particles, nitrogen, calcium or potassium. Meanwhile in water coming from a clearly-cut territory concentrations of these elements sharply increase (see Figure 6.13).

Summing up, we have demonstrated that forest communities are able to form and maintain environmental characteristics that differ very significantly from initial characteristics of the occupied territory (ecotope).

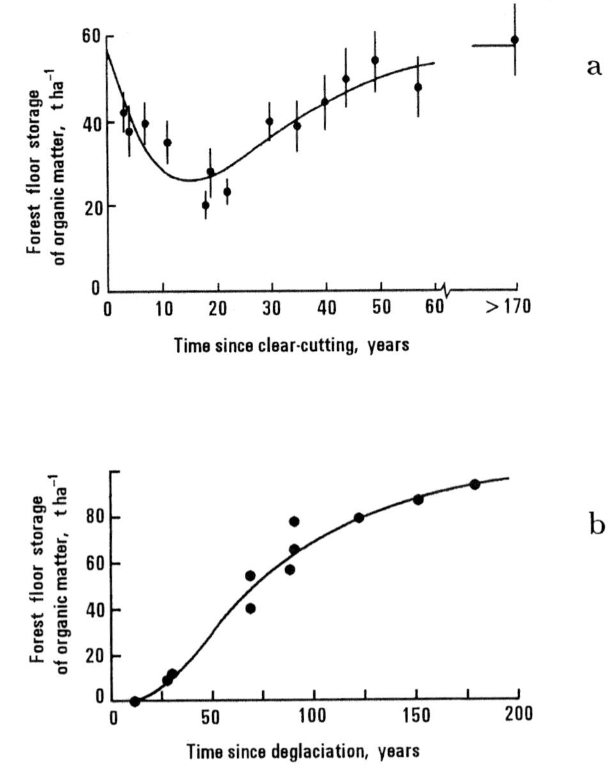

Figure 6.12. Dynamics of organic matter in forest floor (forest litter) during succession. a) as a function of the age of northern hardwood stands after clear-cutting. Points are means of 30 samples per stand with 95% confidence intervals. After Covington (1976). b) as a function of time since deglaciation, spruce forests, the Glacier Bay, Alaska. After Crocker and Major (1955).

6.8.2 Recovery Dynamics

Climax community type depends on characteristics of the ecotope which are not subject to biotic control, e.g. physicochemical characteristics of soil-forming rocks and local climatic characteristics. The community cannot change the total amount of solar radiation incident upon it, basic fractional composition of soil (i.e. size distribution of soil particles), characteristics of mezo-relief, etc.

Dominant species of climax communities are mostly determined by climatic characteristics and size distribution of particles of soil forming rocks (Whittaker, 1975; Liski and Westman, 1995; Begon *et al.*, 1996). In the boreal zone of European Russia, climax pine forests develop on sands, while spruce dominates on loamy sand, loam and clay (Siren, 1955; Lavrenko and Sochava, 1956; Karpov, 1969; Dyrenkov, 1984; Zjabchenko, 1984; Listov, 1986). Both spruce and pine forests develop on soil-forming rocks of varying chemical composition, basic as well as acid (Zaboyeva, 1975; Nikonov, 1987; Kovda and Rozanov, 1988; Gagarina, 1996).

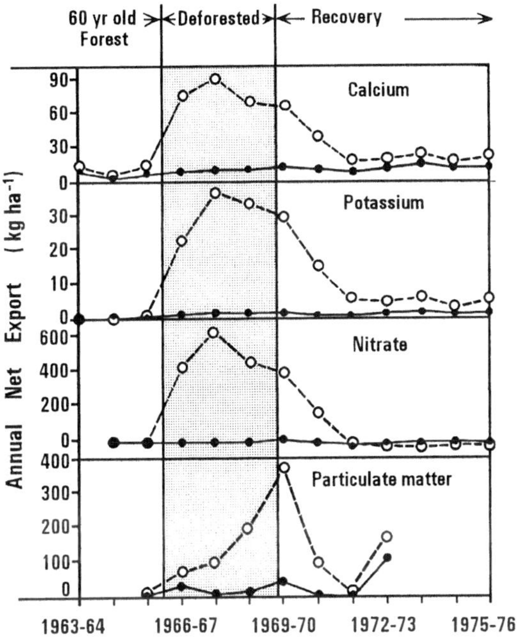

Figure 6.13. Export of chemical elements (calcium, potassium and nitrate) and particulate matter in stream water from an experimentally clear-cut watershed (white circles) and a forested reference watershed (black circles). Export of all elements increases hundreds of times after clear-cutting, which is a manifestation of rapid soil degradation. Modified data of (Likens *et al.*, 1978) compiled after Bormann and Likens (1979).

Following a perturbation comes the recovery period, during which both the forest community and its environment return to the initial non-perturbed state. Recovery of the climax community during succession is achieved due to work of different species-repairers that dominate the community at different stages of the successional process (Figure 6.14a,b).

The temporal order of dominant species-repairers and the rate of recovery of characteristics of forest communities of a definite type are the same in different parts of the natural area occupied by that type of community. For example, the rate of recovery of the forest floor after fire in the Kola peninsula and in the southern part of the Urals are nearly identical (see Figure 6.11). The chronosequence of lichen species change during postfire recovery of ground layer (in forests of the lichen site type) is uniform over the whole Northern hemisphere (see Figure 6.14b; Morneau and Payette, 1989; Ahti and Oksanen, 1990).

The major stages of the succession dynamics are summarised in Table 6.4. One can see that the time of community recovery is very short on a geological scale. In boreal forests, the majority of the community's characteristics is recovered in $\sim$150 years after the perturbation if no additional perturbations take place. The process of formation of vegetation cover on recently-deglaciated territories is characterised by the same time period. In the presence of periodical perturbations, e.g. on dunes or

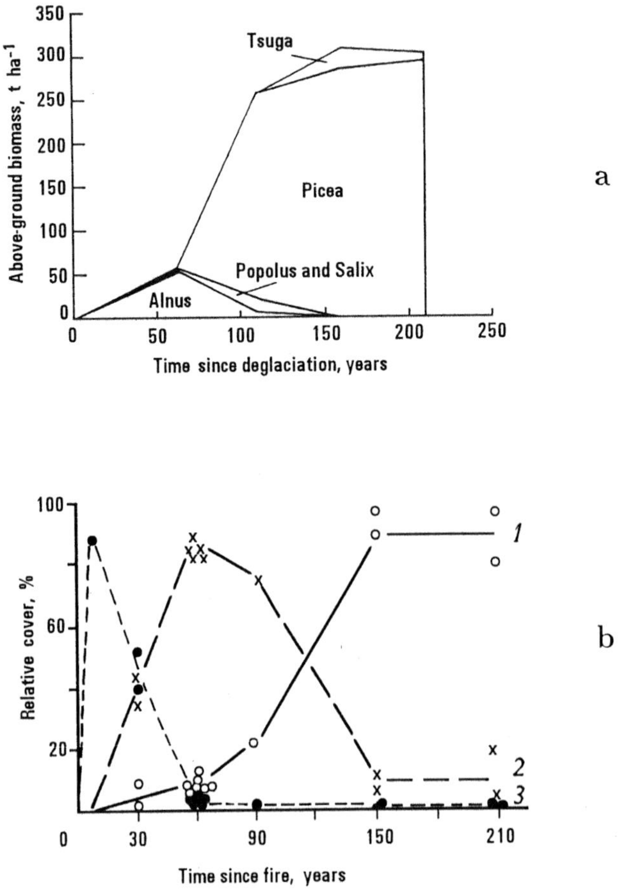

Figure 6.14. Change of dominant species during succession. a) distribution of the above-ground biomass by species along the Glacier Bay, Alaska, as a function of time since deglaciation (primary succession). After Bormann and Sidle (1990). b) changes in relative share of participation of selected lichen species and group of species in formation of lichen cover during postfire recovery in lichen pine forest in the Kola peninsula. *1—Cladina stellaris, 2—Cladina arbuscula, C. mitis, C. rangiferina, Cladonia uncialis, 3—Cladonia spp. (C. deformis, C. cornuta, C. crispata, C. gracilis)*. After (Gorshkov, 1993).

along rivers, the process of community formation takes longer (up to 2000 years), although the major successional stages and their order remain practically the same (Olson, 1958; Fonda, 1974).

An important feature of recovery dynamics is the fact that the species diversity (i.e. the total number of species that take part in the formation of a climax community) is among the first characteristics to stabilise (Figure 6.15). Its recovery takes no more than 30 years after perturbation. Analysis of the species diversity in the above defined sense is therefore of no help in differentiating climax communities capable of performing biotic regulation from those in later successional stages that are still deprived of regulatory abilities. On the other hand, when loss of species diversity does occur (as is the case, e.g. in all anthropogenically transformed

Table 6.4. Major stages of the recovery process.

Age of perturbation, years ago	Community characteristics	Sources of the data
30	Maximum chaos in distribution of the species cover (measured on the $1\,m^2$ scale)*. Recovery and stabilisation of the species composition (species diversity).	Bazzaz, 1975; Southwood *et al.*, 1978; Morneau and Payette, 1989; De Grandpré *et al.*, 1993; Gorshkov and Bakkal, 1996; Figure 6.15a,b.
50–100	Recovery and stabilisation of primary production, cumulative foliage area and basal area of the tree stand. Partial closure of biochemical cycles. Maximal concentrations of biogens in the upper soil horizons.	Whittaker and Woodwell, 1968; 1969; Molchanov, 1971; Whittaker, 1975; Zjabchenko, 1984; Bormann and Likens, 1979; Bormann and Sidle, 1990; Helli and Niemi, 1996; Figures 6.16 and 6.17.
~150	Recovery of the majority of community's characteristics: • Recovery and stabilisation of the natural distribution of productivity over different species in the community. • Stabilisation of the community biomass. • Stabilisation of the forest floor thickness and, consequently, of the store of biogens. • Stabilisation of the chemical composition of soil.	Crocker and Major, 1955; Siren, 1955; Whittaker, 1975; Covington, 1981; Sannikov and Sannikova, 1985; Morneau and Payette, 1989; Bormann and Sidle, 1990; Gorshkov, 1993; Gorshkov and Bakkal, 1996; Gorshkov *et al.* 1996. Figures 6.14a,b and 6.16.
~500	Complete recovery of natural community: • formation of the natural uneven-aged structure of the tree stand.	Odum, 1971; Whittaker, 1975; Dyrenkov, 1984; Volkov, 1998.

* This means that all plant species encountered in a $1\,m^2$ rectangle contribute equal shares into the vegetation cover.

areas), this not only means loss of biotic regulation on respective areas but is also an indication of complete disintegration of the natural organisation of biota on a local and regional scales.

One of the important features of the initial stages of the recovery process is a great increase in production of berries, mushrooms and other types of edible biomass,

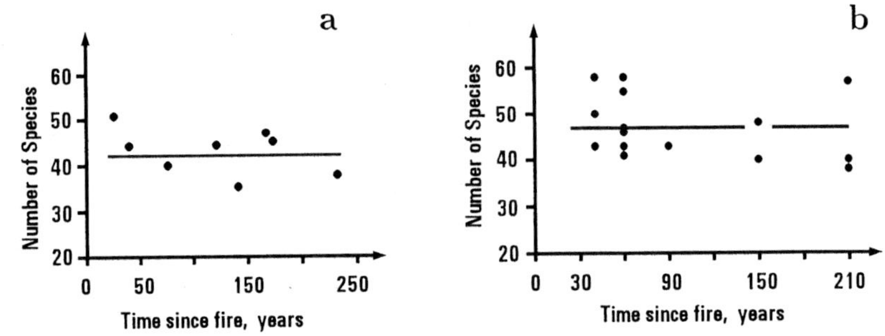

Figure 6.15. Species richness during postfire succession in different regions of boreal zone. a) vascular species richness, average values for $100\,m^2$ sample plots in relation to postfire age, boreal mix-wood (birch-spruce-fir) forests, northern taiga, South Québec, Canada. After De Grandpré *et al.*, 1993). b) total number of species of vascular plants, mosses and lichens in Scots pine forest of lichen site type in relation to postfire age estimated at $100\,m^2$ sample plots. Compiled after Gorshkov and Bakkal (1996).

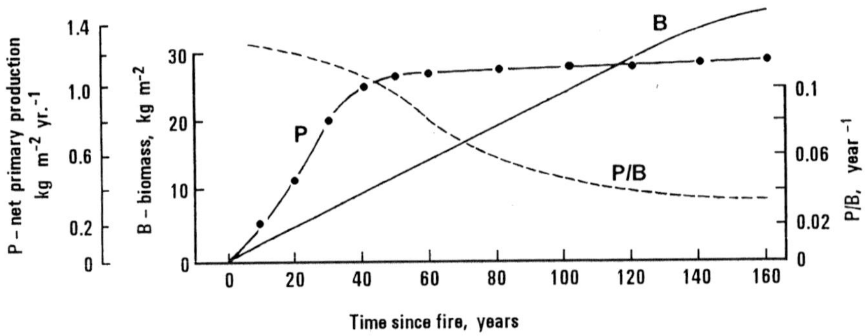

Figure 6.16. Annual above ground net primary production (P), plant biomass (B), and the ratio P/B during postfire recovery of oak forest at the Long Island, New York. After (Whittaker and Woodwell, 1968) as cited in (Begon *et al.*, 1996).

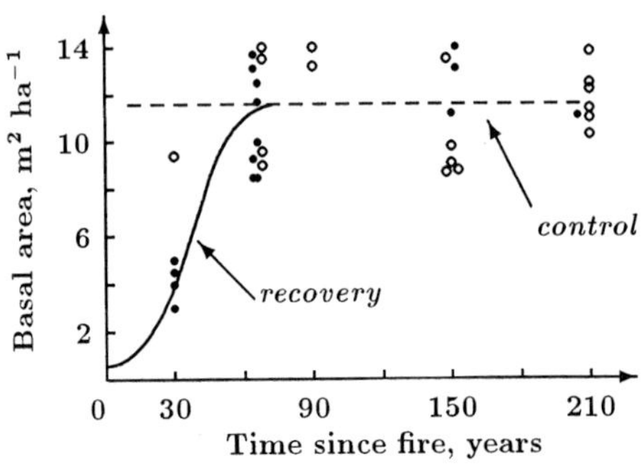

Figure 6.17. Relative basal area (i.e. sum of areas of cross-sections of tree stems (inclusive of bark) per unit area of the Earth's surface) measured at breast height (~1.3 m above the ground) (m^2ha^{-1}) during postfire recovery in Scots pine forests of lichen site type (Kola peninsula) by the degree of postfire tree stand damage and postfire age. Empty circles—communities with low degree of stand damage (less than 30%, control); filled circles—communities with high degree of stand damage (>70%, recovery). After (Gorshkov, 1998).

accompanied by a corresponding increase in population density of large animals (birds and mammals), as compared to their stationary population densities encountered in climax communities. The highest population density of animals is reached in communities 10–40 years after a perturbation. At this stage, the population density of animals is approximately 5 times greater than in communities perturbed 60–80 years ago and 10 times greater than in climax communities (Johnston and Odum, 1956; Kerzina, 1956; Reimers, 1972; Yurgenson, 1973; Ivanter, 1975; Kurkhinen, 1983; Vladyshevsky and Petrenko, 1987; Helli and Niemi, 1996). These data testify in favour of the important role of animals in the process of a community's recovery.

6.8.3 FIRES, WINDFALLS, INSECT INVASIONS: Natural Periodicity

Fires In the 1960s it became common in some scientific circles to speak about positive ecological functions of fires. Fires were even considered as an indispensable prerequisite of existence of some types of communities. Now this statement has turned to a dogma in the scientific ecological communities of the USA and Canada (Johnson, 1996).

However, within the biotic regulation concept, one inevitably comes to the conclusion that fire can by no means play any positive role in natural ecosystems. Indeed, biotic regulation is performed through very concrete, definite work of ecological communities. (Work is a use of free energy in a strictly specified direction, not chaotically.) Work requires energy, and all synthesised organic substances represent a source of energy that can be used for biotic regulation work. It is evident that in such a case those communities where the frequency of fires is minimised have the competitive advantage. This is because such communities have more organic matter, more energy to perform work on biotic regulation and, consequently, are able to keep their environment closer to the desired optimum. Such communities are naturally more competitive and force out communities that are adapted to frequent fires, i.e. communities where fires frequently eliminate the energy source. This logic is based on the fundamental principle of life organisation—the principle of stabilising selection and competitive interaction of living systems (see Chapter 2).

Thus it is impossible to speak about the positive ecological role of fires within the framework of the biotic regulation concept. The only thing that can be discussed is the inevitable minimum frequency of fires in nature, see below. Climax communities must have a mechanism that would allow them to recover after fires, because fires do occur in nature, though infrequently. In the boreal zone, postfire succession often goes through the stage of pine forests before the climax spruce is recovered. Pines require a lot of light and much drier soil than do spruces. As pines grow up and form a close canopy, it becomes difficult for young pines to grow owing to shadow, and they are gradually forced out by climax spruces, that are characterised by a stable mode of existence.

Continuing the analogy with recovery from disease, we can compare succession processes to the reaction of the immune system to infection. Successional pines, for example, can be compared to leukocytes that are especially abundant in the presence of an infectious agent, while keeping relatively low numbers in a healthy organism. But it would be absurd to say that leukocytes profit from the infection. Leukocytes are strongly correlated with the whole organism and their independent existence is impossible; leukocytes are 'happy' when there is no infection and they exist in small numbers, but the whole organism is healthy.

In very much the same manner, we can say that plants of various succession stages are 'happy' when there are no large-scale perturbations and the climax community maintains the optimal environment in a stable state, which cannot in principle be done by successional species. Using emotional terms, one can say that the climax community does 'care' about successional species. As discussed above, the climax community supports the existence of large animals and insects, whose ecological function is to introduce regular perturbations (e.g. make lawns in the forests, as do elephants) and, by doing so, form a favourable environment for successional species without threatening the environmental stability on a large scale and irrespective of the presence or absence of external perturbations. In other words, the climax community guarantees certain 'basic rights' for the successional species.

It is true that in climax forests successional species feature low population numbers. But why should we think that all species, like humans, care only about unlimited population growth? That huge population is all right, while normal population, compatible with long-term environmental stability, is too small? Unfortunately, such a way of thinking is very characteristic of the modern ecological mentality.

If we ignore the critical difference between the climax community that performs biotic regulation and successional community that is incapable of it, one can speak about the positive role of fires as promoting the inherently unstable successional communities.[11] However, bearing in mind that critical difference, one realises that promotion of successional communities weakens and may even completely destroy the regulatory potential of the biota, which finally leads to degradation of the environment over a range of scales. As we will now show, the natural frequency of fires is at least an order of magnitude lower than the modern one. In other words, the overwhelming majority of modern fires are of anthropogenic origin and represent but one of the many anthropogenic factors destabilising the contemporary biosphere and undermining the global environmental stability.

The present periodicity of forest fires averages between 40 and 100 years in well-drained sites and is estimated as 500 years for wetlands and mires (Rowe and Scotter, 1973; Zackrisson, 1977; Furyaev and Kireyev, 1979; Johnson, 1979; Heinselman, 1981; Listov, 1986; Engelmark, 1987; Gromtsev, 1993; Furyaev, 1996; Goldammer and Furyaev, 1996). So we can assume the modern periodicity of fires to be of the

[11] From the economical point of view, it is very convenient to make no difference between ecological functions of climax forests and successional forests that are maintained by fires. In climax communities where all characteristics are stable, the rate of biomass increment is equal to zero, while successional communities with the age of perturbation less than 60 years are characterised by the most rapid rate of biomass accumulation and are most profitable for use in the forest industry.

order of 100 years. According to the official statistics, 70–80% of forest fires are connected with human activity; the remaining 20–30% are treated as natural (Odum, 1971; Spurr and Barnes, 1980).

However, a special detailed investigation made in the continental regions of Canada showed that 100% of fires that occurred on the studied territory (400 km^2) during the four years of monitoring were caused by man (Penafiel and Terrado, 1977). Analogous data are available for fires in the pine forests in Japan, 99.9% of which were associated with human activity and only in 0.1% of cases the reason remained unknown (Nakagoshi *et al.*, 1987). It means that, at maximum, only one fire in a thousand is natural, 1 : 1000.

At the same time, due to the anthropogenic transformation of natural landscapes, the mean area affected by a single fire is essentially smaller nowadays than in the past. Among northern countries, the highest degree of the anthropogenic transformation of the landscape is registered in Finland, the lowest in Russia and Canada (World Resources, 1988; Syrjaenen *et al.*, 1994). According to the FAO statistics (FAO, 1995), from 1983–1994 the mean burnt area covers 0.8 ha in Finland, 50 ha in Russia and 180 ha in Canada, i.e. the mean area affected by a single fire in strongly transformed areas is about 100 times smaller than that in Russia and Canada. We can therefore estimate the relationship between the mean areas burnt in the past and nowadays as 100:1. It follows from the above two estimates that the ratio between modern and natural periodicity of fires on a given local area can be estimated as 1:10. In other words, the mean time between two successive fires on a given territory is now 10 times shorter than it is under natural environmental conditions. One can also put it a different way: the total area yearly affected by fires is nowadays 10 times larger than it was in the past when the most part of the biota remained in the non-perturbed state.

There is much speculation (Johnson, 1996) that the ratio between periodicity of natural and modern fires is equal to unity. Were that the case, one then could not explain the observed rise in fire frequency with increasing population density of the studied territories (Kurbatsky, 1964; Melekhov, 1971; Furyaev, 1996; Goldammer and Furyaev, 1996). Furthermore, the period of fire rotation during the last 3000 years was 5–7 times smaller compared with that during 3–6 thousand years ago (Tolonen, 1978; Bergeron *et al.*, 1996). As the periodicity of natural fires is unlikely to change, the observed increase in fire frequency following the growing population density of people can be due only to the increasing share of anthropogenically-caused fires. These data highlight the importance of the anthropogenic factor in fire periodicity and return us to the above estimate of 1:10 between the periods of rotation of modern and natural fires. Accordingly, taking into account that the modern periodicity of fires is about 100–500 years, the mean time period between two successive natural fires on a given territory should be of the order of ~1000–5000 years.

Windfalls In climax communities and those at late stages of succession, massive windfalls (i.e. uprooting of trees by wind on 3–30% of the considered territory) may be caused by winds at speeds higher than 30 m sec^{-1} (Belov, 1976; Turkov, 1979;

Skvortsova *et al.*, 1983; Quine *et al.*, 1995). In the majority of regions in the taiga zone of Siberia and the European part of Russia, the share of meteorological stations where such strong winds were registered at least once during the period 1936–1980, does not go beyond 6–17%. It is only in the Urals that this share exceeds 40%. Based on analysis of published data on 230 meteorological stations (*Scientific and applied* ..., 1988–1990), the annual frequency of such strong winds is estimated as 0.02–0.1 cases per station located on sea shores and in mountainous regions and 0.0005–0.002 cases per station located on plain inland territories. Accordingly, the mean time between two successive massive windfalls varies between 1400 and 5000 years.

Slower winds at speeds from 20 up to 30 m sec^{-1} are mainly responsible for local windfalls with the percentage of uprooted trees not exceeding one per cent (Belov, 1976; Turkov, 1979). During the same period of observations (1936–1980), such winds were registered by 83–100% of the meteorological stations.

In Europe, such wind speeds are most common in winter, but occur in Siberia in the spring. The probability of windfall increases with increasing degree of soil moistening (Turkov, 1979). When soil freezes over, the probability of windfalls approaches zero (Molchanov and Preobrazhensky, 1957; Repnevsky, 1961). At the southern border of the European boreal zone, frozen state of soil is observed from November to (and including) March, from October to March in Eastern Siberia.

Thus, given the average annual frequency of winds at speeds between 20 and 30 m sec^{-1} of about 0.17–0.35 cases per station, the recurrence of local windfall situations (i.e. when all the windfall-friendly environmental conditions coincide) can be estimated as 0.05–0.1 cases per station per year. Taking into account the relative area affected by local windfalls (less than 1%) as compared to massive windfalls, the rate of forest damaged by local windfalls is practically as low as that of rarer massive windfalls.

Note that as most meteorological stations are predominantly located in open country, the obtained values for recurrence of windfalls are sure to be overestimates of the real values characteristic to boreal forest communities.

Insect invasions and fungous diseases In modern literature on silviculture, much attention is paid to severe impacts imposed by insect outbreaks and fungi diseases on tree stands. It is assumed that such outbreaks may result in massive degradation of forests (Isayev *et al.*, 1984; Maslov *et al.*, 1988; Alfaro and Singh, 1997). According to model-based calculations, the recurrence of insects' invasions is estimated as 150–250 years (in communities of *Abies sibirica* as an example) (Isayev *et al.*, 1984).

In the meantime, analysis of geobotanical and historical literature devoted to description of large territories occupied by climax forests from the end of the 19th century onwards clearly indicates that, unlike fires and windfalls, neither insect invasions nor fungous diseases are ever cited as reasons for massive degradation of forest communities (Maak, 1859; Middendorf, 1867; Tkachenko, 1911; Borovikov, 1912; Kuznetsov, 1915; Krylov, 1919; Ivashkevich, 1929; Nikolayev, 1932; Turkov, 1979; Borisenkov and Pasetsky, 1988).

Massive outbreaks of xylophagous insects (those eating wood) are reported for territories affected by external perturbations like cutting, fires, windfalls or on

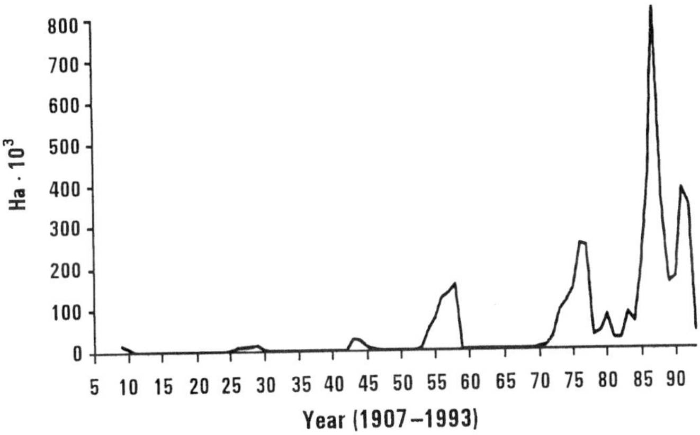

Figure 6.18. History of western spruce budworm (*Choristoneura occidentalis*) outbreaks in Douglas fir stand of the interior of British Columbia as measured by the extent of each outbreak (hectares of defoliation). Outbreaks have increased in frequency and intensity in recent decades following the enhanced degree of anthropogenic perturbation of the territory. After Alfaro and Singh (1997).

territories with intensive silviculture (Isayev and Utkin, 1963; Isayev and Petrenko, 1968; Galasyeva, 1977; Mozolevskaja *et al.*, 1979; Rozhkov and Massel, 1982; Isayev *et al.*, 1984; Vetrova, 1987).

In the majority of cases, massive outbreaks of phyllophagous insects (those eating leaves) are observed in successional forest communities (and adjacent areas) recovering after perturbations like clear-cutting, fire, windfall (with age of perturbation not more than 60–70 years) or in artificially managed forests (Vorontsov, 1963; Maslov *et al.*, 1988). The available data on massive outbreaks of these insects in old (80–240 years) tree stands in Siberia (Kolomiets, 1961; Kondakov, 1974; Isayev *et al.*, 1984; Ryapolov, 1987) do not contradict the above statement, as due to extensive anthropogenic transformation of Siberia at the end of the 19th century the average time of recurrence of fires in the studied areas is as high as 40–60 years (Furyaev and Kireyev, 1979; Furyaev, 1996), which implies a sufficient degree of perturbation of the territory.

Massive outbreaks of the western spruce budworm, *Choristoneura occidentalis*, in Canada and USA in the second half of the 20th century (Hardy *et al.*, 1986; Morin *et al.*, 1993; Bergeron *et al.*, 1996; Campbell and Liegel, 1996; Alfaro and Singh, 1997) apparently result from intensification of forest management, i.e. anthropogenic perturbation of forest communities, as is convincingly demonstrated by Figure 6.18 (Alfaro and Singh, 1997).

Summing up, all the so-called natural perturbations are in their essence very rare phenomena under natural environmental conditions in the absence of anthropogenic forcing. The natural periodicity of natural perturbations (1000–5000 years at minimum) exceeds the characteristic time of community's recovery by a factor of 10. That means that under natural conditions more than 90% of communities exist in a climax state and perform biotic regulation and only 10% are found at various stages of recovery process. Modern anthropogenic activities have more than halved the percentage of climax communities, which significantly weakens the cumulative regulatory potential of terrestrial biota.

6.8.4 The Climate Issue

Many researchers (see, e.g., Vasilevich, 1983; 1993; Rabotnov, 1983) assume that climatic trends represent an important factor governing the way of organisation of forest communities. According to this point of view, forest communities never reach a stable stationary state. Different species dominate the community depending on the current climatic situation. In other words, species that behave as successional under given climatic conditions may form a climax community when the climatic situation changes, and vice versa.

Such an attitude is in sharp contradiction with the biotic regulation concept and, consequently, with the total body of evidence testifying in favour of the latter. In particular, such an attitude appears inconsistent from the genetic and informational point of view. The ability of a climax community to maintain a stable environment requires correlated strictly specified work of all the species. Similarly, directional change of unfavourable environment towards the climax optimum, which is performed by successional species, is also a highly-ordered process. The information about optimal environmental characteristics and ways of their achievement, on which work of both climax and successional communities is based, can only be written in genomes of community species. As will be shown in Chapters 9–11, genetic composition of species remains unchanged during the whole span of species existence, which is of the order of a million years. Thus, successional and climax functions are fixed in particular species and cannot be interchanged.

However, besides conceptual objections, one can also name more concrete evidence that contradicts the above view. According to the available data, long-term climatic trends observed over time periods over 1000–10 000 year are insignificant. Temperature changes do not exceed ±1°C (see Figure 8.3), while precipitation varies by no more than 10%.

In contrast, the existing annual fluctuations of the average annual temperature constitute about 3°C, while annual precipitation varies by more than 30% (*Scientific and applied* ... , 1988–1990), which is several times greater than the long-term climatic trends. The observed annual fluctuations are known to affect the community only at early stages of recovery having no impact on climax communities. This is due to the fact that, as shown in Section 6.8.1, climax communities and those at late stages of recovery are able to smooth out physical fluctuations of the environment severalfold, so that the observed slight climatic trends are unlikely to change anything in the climax community organisation.

Furthermore, climax communities of one and the same type are known to occupy territories with very different climatic characteristics. For example, lichen pine forests formed on deep sands occupy territories with the annual mean temperature ranging between 6°C and −11°C. Comparison of communities from different regions show that they may differ in productivity and biomass distribution over the community's species, whereas the composition of dominant species (*Pinus silvestris*

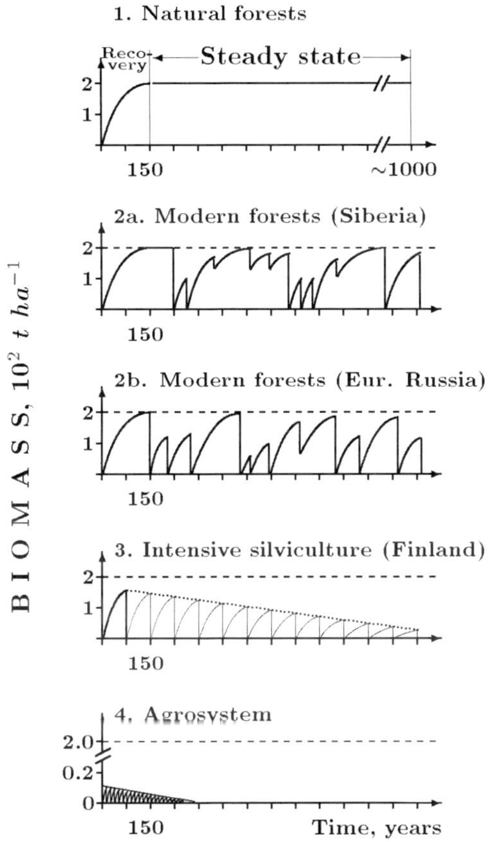

Figure 6.19. Schematic description of the state of a boreal forest community in different regimes of anthropogenic pressure. The monitored characteristic is the store of above-ground biomass. 1. Natural state of communities is shown based on the data discussed in this section. The recovery period corresponds to that in Table 6.4; store of biomass is cited after Whittaker (1975). 2a. European Russian North, Siberia and Far East; disturbances are predominantly caused by surface fires. 2b. European Russia; disturbances are predominantly caused by clear-cutting and, to a lesser degree, by fires. 3. Intensive silviculture (e.g. Finland). Note the declining biomass store. 4. Agrosystem (arable lands, pastures). Note the extreme unsustainability.

and *Cladina stellaris* (Opis.) Brodo) never changes (Listov, 1986; Sannikov, 1992; Gorshkov *et al.*, 1997; Ahti and Oksanen, 1990).

6.8.5 Current State of Forest Communities

Today only a small part of boreal forests exists under conditions of natural perturbations (fires, windfalls, insect invasions) occurring at their natural frequencies. For example, in Russia such forests are only found in Northern Europe (the Kola peninsula, the Northern and Middle Urals), certain regions of Northern and Middle Siberia and the Far East (Bryant *et al.*, 1997). The main type of perturbation in these forests is represented by natural fires, among which ground fires dominate (Korovin, 1996; Furyaev, 1996). The overwhelming majority of forest communities in these regions exist in a stationary stable state or close to that, Figures 6.19–1,2a.

In the rest of the European part of Russia, perturbations of forests are mainly due to clear-cuttings and, to a lesser degree, to fires (Forest Fund of Russia, 1995). Under such pressure forest communities may only occasionally reach the climax state (Figure 6.19–2b).

In the developed countries, territories previously covered by forests are represented now by artificially managed agricultural or silvicultural systems (Figure 6.19–3,4). In such a case, the natural stationary stable of community capable of biotic regulation is never reached. It should be noted that long-term land use is inherently unstable and leads to complete soil degradation in about two hundred years (Dokuchayev, 1954; Kovda and Rozanov, 1988; Zonn, 1991; Doran *et al.*, 1998). Intensive forest management can be also characterised as unstable due to substantial loss of biogens observed at early stages of recovery (Bormann and Likens, 1994). This loss has no time to be compensated over the too short process of recovery (50–60 years) during which the tree stand achieves the prescribed industrial standard and is cut down.

Summing up, the performed analysis of available experimental data on organisation and recovery dynamics of forest communities after strong perturbations show that the biota is able to completely compensate (up to a certain limit) even such strong perturbations of the environment as total fires, windfalls and others. This constitutes important empirical evidence in favour of the biotic regulation concept.

Anthropogenic activities seriously affect the functioning of natural ecological communities and, by doing so, undermine environmental stability on local and global scales.

7

Energy and Information

It is not heat that is sent by the Sun to the Earth, but highly ordered energy corresponding to a huge flux of information, which generates ordered physical processes and life of the biota on the Earth's surface. The natural biota absorbs solar radiation and processes it in molecular memory cells. Fluxes of information processed by the biota are used to support functioning of the mechanism of biotic regulation of the environment. Biotic information fluxes exceed by 20 orders of magnitude the information-processing capacity of modern civilisation. It follows that the existing natural biotic mechanism of maintenance of suitable-for-life environment is absolutely unique and cannot be replicated and replaced by technology however advanced the stage of technological progress.

7.1 ORDER AND DECAY

Ordered macroscopic processes surround us in our everyday life. Wind, formation of clouds, precipitation and flow of rivers are examples of ordered macroscopic processes of physical nature. The orderliness of these processes means that molecules of the involved substances move in a correlated fashion. For example, all molecules of water in a river have a downstream velocity component. During turbulent flow in whirlpools macroscopic groups of molecules feature identical angular velocities. The phenomenon of wind means that all the molecules of air have a common velocity component. Ordered motion of molecules can be opposed to their chaotic (non-correlated) thermal motion.

Molecules taking part in macroscopic motion interact with other molecules of the medium in which such motion occurs. Interaction of two molecules may produce a great number of final states. For example, during elastic collision a molecule may change the direction of its motion, while during inelastic collision it may transmit the whole of its energy to a molecule of the medium. Different final states of two interacting molecules are approximately equally probable. Thus, after interaction

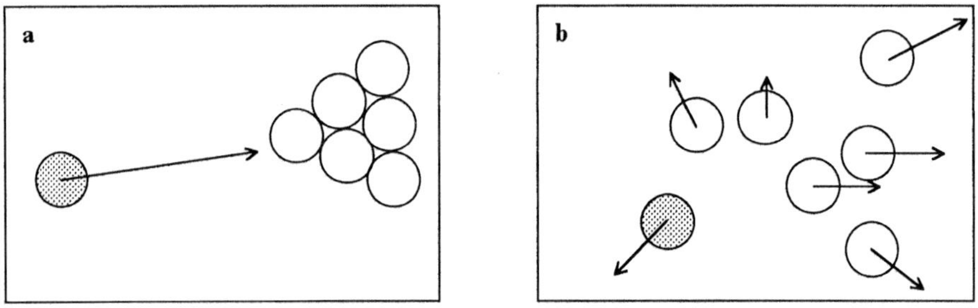

Figure 7.1. Highly-ordered energy of the rectilinear motion of the grey ball (a) converts (decays) to low-ordered energy of chaotically moving balls after collision (b). The energy (and impulse) quantity of the system remains the same; it is the energy (and impulse) quality that changes. Note that the reverse process (b → a) is improbable.

with the medium only an infinitely small fraction of molecules retain the characteristics of their initial ordered motion. Hence, an ordered correlated motion decays into disordered chaotic movement of molecules. It is the possibility of transition from a definite initial state to a large number of various final states that makes decay both possible, inevitable and irreversible.

Phenomena of this kind are well known from everyday experience. If a ball in billiards is given rectilinear motion in a definite direction (which is an ordered process, Figure 7.1) and collides with a group of motionless balls, the latter will move chaotically in different directions. As a result, the ordered motion of the first ball is terminated. The cumulative energy of chaotically moving balls after collision is equal to the energy of the initial ordered motion. Such a process is irreversible, because it is highly improbable that a few independent balls move in such a way that when they all meet, all but one of them stop and their cumulative energy is conveyed to a single ball that continues motion in a prescribed direction.

Biogeochemical cycles taking place in the Earth's environment present examples of ordered processes of chemical nature. Orderliness of these processes means that among all possible biochemical reactions, only a few strictly-defined reactions actually take place. This situation can be vividly illustrated when comparing a living and a dead organism. In a living organism all biochemical reactions are strictly defined and aimed at maintenance of biological homeostasis. Any deviation from the prescribed direction of any process in the living body leads to dysfunction of various parts of the organism, i.e. to disease. After death of the organism, its ordered biochemical reactions are stopped. The highly-ordered energy of the organic matter dissipates and converts to low-ordered thermal energy of the molecules. **According to the law of energy conservation, energies of the initial highly-ordered and final low-ordered states are equal. Thus, during ordered processes energy retains its quantity but loses its orderliness.**

All the diverse macroscopic processes observed on Earth are but various forms of decay of highly-ordered states of matter and of dissipation of highly-ordered energy they contain. The energy resources of Earth are formed by the totality of energy-rich states of matter on Earth. These resources exist in the form of kinetic or potential energy. Kinetic energy is the energy of correlated motion of particles, e.g. solar radiation, hydraulic energy of rivers and oceanic waves, energy of sea tide, energy of wind. Potential energy is the energy of correlated bonds between various parts of the matter, e.g. gravitational energy, chemical energy of organic matter, nuclear energy. Other forms of potential energy include the existing temperature gradient between the polar and equatorial regions of the Earth's surface and between surface and deep waters of the world ocean. The nuclear energy, the gravitational energy of redistribution of matter in the Earth's interior and the energy of sea tides, are all preserved since the time of the formation of the solar system. All the other forms of energy resources available on Earth are generated by solar radiation.

Decay processes are characterised by a certain direction and time of duration. The majority of phenomena either observed in everyday life or detected in the course of scientific experiments are irreversible processes of decay of certain initial state. In the absence of decay, no observable events develop in time. For example, the absolute geochronology of the Earth is only known because of the decay of unstable isotopes of chemical elements. **Thus, time only acquires a physical meaning when a process starts fading (decaying), i.e. when the considered process has a start and a finish.**

7.2 SOLAR ENERGY

Since decay and dissipation of energy go on permanently, any ordered process may be supported only if there exists an influx of ordered energy to it from another ordered process. On Earth it is solar energy that generates all natural ordered processes in the environment and dissipates finally converting to quantitatively equal energy of thermal radiation of Earth, which is emitted back to space.

Why is solar energy ordered compared with the chaotic thermal energy of Earth? Both solar radiation and thermal radiation of Earth consists of particles—photons. Absolute temperature measured in degrees Kelvin is proportional to the average energy of a single photon. Absolute temperature of the Sun is about $T_S = 6000\,\text{K}$ (Newkirk, 1980). Absolute global average temperature of Earth is about $T_E = 288\,\text{K}$ (i.e. about 15°C). Thus, average energy of one solar photon is about $T_S/T_E = 6000/288 \approx 20$ times larger than average energy of one thermal photon of Earth. According to the law of energy conservation, the cumulative energy of all solar photons coming to Earth is equal to the cumulative energy of thermal photons emitted by the Earth into space. It means that the number of thermal photons emitted by Earth into space is about 20 times larger than the number of solar photons reaching Earth's surface. Consequently, one solar photon decays on average into 20 thermal photons. Decay of solar photons gives rise to all ordered processes on Earth.

Imagine that we decrease gradually the temperature of the Sun but keep constant the total amount of solar energy reaching the Earth's surface. Temperature of the Earth's surface would then remain the same, because it is completely determined by the total amount of energy coming from the Sun. But cycles of all substances in Earth's environment would begin to slow down, because their rate is determined not by the total amount, but by the orderliness of the incoming energy. If the Sun's temperature were equal to that of Earth, all ordered processes on Earth would stop. Thus, it is not heat that is sent by Sun to Earth, as it is often stated, but energy, the primary characteristic of which is its high degree of orderliness as compared with the thermal radiation of Earth.

Since the middle of the 19th century the quantitative measure of disorderliness has been called entropy. In spite of its clear formal definition, this notion often remains vague for non-specialists. In the middle of the 20th century orderliness was quantitatively tied to a more transparent notion of information (see, for instance, Brillouin, 1956), which became especially popular when people throughout the world began to use personal computers widely. Information capacity of a system may be characterised by the available number of memory cells N. If all memory cells can be characterised by only two possible values of a certain variable, the total number of possible combinations of these values in all memory cells is 2^N. Such a memory system possesses the maximum possible amount of information equal to N bits when values of the measured variable are defined in all N memory cells. If states of N_1 cells remain unknown, the amount of information reduces to $N - N_1$. Finally, if the measured variable remains undefined in all memory cells, the information becomes zero while the entropy of the system reaches its maximum.

Solar photons interact with molecules of the Earth's surface that can be viewed as elementary memory cells of the environment. Solar photons can excite molecules, i.e. impart a certain amount of energy to molecules and increase their energy above the average thermal level. A good approximation is to assume that molecular memory cells are characterised by two states—excited (a definite state) and non-excited (indefinite chaotic state) compared with the average chaotic thermal level. During the process of decay, solar photons are able to excite molecules until their own energy becomes equal to the average energy of thermal photons of the Earth's surface. Each solar photon possesses an amount of energy equal to that of about 20 thermal photons of Earth. Consequently, one solar photon can excite about 20 molecules, i.e. impart information to about 20 molecular memory cells. Such consideration makes it possible to estimate the amount of information (in bit sec^{-1}) coming from the Sun to Earth. It is roughly equal to the number of thermal photons emitted from the Earth to space, because each thermal photon is emitted from an excited molecule, which represents a memory cell containing one bit of information. The number of Earth's thermal photons emitted to space in a unit of time is equal to the power Q ($Q \approx 2 \cdot 10^{17}$ W) of solar radiation reaching the Earth divided by the energy ε of one thermal photon, which is determined by the Earth's temperature T_E, $\varepsilon = k_B T_E$, where $T_E \approx 288$ K, k_B – Boltzmann constant, which is proportional to the reverse Avogadro number ($k_B = 1.4 \cdot 10^{-23}$ J K^{-1} molecule^{-1}). As far as one molecule represents a memory cell with two possible states, dimension molecule^{-1} in Boltzmann constant corresponds to bit^{-1}. Thus, for the estimate of information flux F coming from the Sun to Earth we obtain $F = Q/k_B T_E \approx 10^{38}$ bit sec^{-1} (Thribus and McIrvine, 1971; Nicolis, 1986). Note that the obtained expression for F is simply the traditional definition of the flux of entropy.

The obtained estimate of the amount of information sent by the Sun to Earth is based on the assumption that all thermal photons emitted by Earth come from excited molecules. In fact, if one solar photon possesses energy equal to that of $T_S/T_E \approx 20$ thermal photons, the maximum number of molecules it can excite is $T_S/T_E - 1 \approx 19$, because after 19 acts of excitation its energy becomes equal to that of one thermal photon. After that it cannot impart additional energy to molecules and, therefore, cannot excite them. So, only $(T_S/T_E - 1)/(T_S/T_E) \cdot 100\% = (T_S - T_E)/T_S \cdot 100\% \approx (19/20) \cdot 100\% = 95\%$ of Earth's thermal photons come from excited molecules and characterise information flux coming from the Sun to Earth. Note that the ratio $(T_S - T_E)/T_S$ characterises the well-known Carnot efficiency of the solar radiation on Earth. If the Sun's temperature were equal to that of Earth, solar photons would have the same energy as thermal photons of Earth and could not excite any molecules on the Earth's surface. Evidently, in such a case the information flux from Sun to Earth would be equal to zero. Although the total number of Earth's thermal photons would then remain the same, none of them would come from excited molecules.

Distribution of available energy over different kinds of ordered processes on Earth is given in Table 7.1. Note that only a negligible part of ordered processes is generated by geothermal power and power of tides. Tidal power is determined by movement of a stationary wave of water masses across oceanic surface. It is mostly caused by the gradient of gravitational field of the Moon at the distance equal to Earth's radius. The corresponding impact of the Sun is considerably smaller owing to a considerably larger distance between Earth and the Sun as compared to that between Earth and Moon. Tidal power represents ordered power of mechanic movement and can be converted into any other ordered processes. Geothermal power is mainly represented by heat flux coming from the Earth's interior. Being lost in the total thermal radiation of Earth, it cannot generate ordered processes. Only a small part of the geothermal power exists in a highly-ordered state, e.g. in volcanoes, geysers and in the vicinity of oceanic rifts. The ordered part of geothermal power constitutes about one millionth of the solar radiation (Table 7.1). Thus, life that uses ordered geothermal power in oceanic depths and is based on chemosynthesis is characterised by cumulative production which is about one million times lower than that of life based on sunlight and photosynthesis. On a global scale, chemosynthesising life is negligibly small and taken altogether corresponds to photosynthesising life on an island with an area of about $10\,\text{km} \times 10\,\text{km}$.

7.3 STORES AND FLUXES OF INFORMATION IN NATURAL BIOTA AND CIVILISATION

Today there is little doubt that the global environment as well as the natural biota of Earth are degrading under the increasing pressure of anthropogenic activities. Until very lately, however, the observed adverse changes of the environment and human-induced degradation of the natural biota have been envisaged as two independent processes, although having a common cause. A common view has been that people should fight on the two fronts simultaneously. People should both restore the

Table 7.1. Energy fluxes at the Earth's surface.

Power source/sink	Power		References
	10^{12} W	Relative to the solar power	
Solar Power and Processes of its Dissipation			
Total solar power coming from Sun to Earth	$1.7 \cdot 10^5$	1.0	[1]
Solar power absorbed by the Earth's surfaces[a]	$8 \cdot 10^4$	0.47	[2], [3]
Evaporation from the total Earth's surface	$4 \cdot 10^4$	0.24	[2], [3]
Evaporation from land (evapotranspiration)	$5 \cdot 10^3$	$3 \cdot 10^{-2}$	[4], [5]
Heat fluxes from equator to the poles:			[6], [7], [8]
atmospheric	$3 \cdot 10^3$	$2 \cdot 10^{-2}$	
oceanic	$2 \cdot 10^3$	10^{-2}	
Wind power	$2 \cdot 10^3$	10^{-2}	[5], [9]
Oceanic waves	10^3	$6 \cdot 10^{-3}$	[5]
Maximum hydraulic power of rivers	3	$6 \cdot 10^{-5}$	[4], [10]
Available for humans			
wind and rivers' hydraulic power	1	$6 \cdot 10^{-6}$	[5], [9], [10]
Moonlight	0.5	$3 \cdot 10^{-6}$	[15]
Biota			
Transpiration	$3 \cdot 10^3$	$2 \cdot 10^{-2}$	[5]
Photosynthesis	10^2	$6 \cdot 10^{-4}$	[11]
Non-solar Sources of Power			
Total flux of geothermal heat	30	$2 \cdot 10^{-4}$	[10], [12], [13]
volcanoes and geysers	0.3	$2 \cdot 10^{-6}$	
chemosynthesising life	10^{-4}	$6 \cdot 10^{-10}$	
Tidal power	1	$6 \cdot 10^{-6}$	[14]
Humankind in the end of the 20th century			
Energy consumption (fossil fuel combustion mostly)	10	$6 \cdot 10^{-5}$	[10], [16]
Consumption of the net primary production of the biosphere	9	$6 \cdot 10^{-5}$	[17]

[a] Solar power absorbed by the Earth's surface is equal to the total solar power coming from the Sun to Earth minus solar power reflected by Earth back to space minus solar power absorbed by the atmosphere.

References: [1] Willson, 1984; [2] Ramanathan, 1987; [3] Schneider, 1989; [4] Lvovitch, 1974; [5] Brutsaert, 1982; [6] Kellogg and Schneider, 1974; [7] Peixoto and Oort, 1984; [8] Chahine, 1992; [9] Gustavson, 1979; [10] Skinner, 1986; [11] Whittaker and Likens, 1975; [12] Berman, 1975; [13] Starr, 1971; [14] Hubbert, 1971; [15] Allen, 1955; [16] Starke, 1987, 1990; [17] Gorshkov, 1995; Figure 1.1.

environment eliminating industrial wastes and help the natural biota preventing loss of biodiversity.

Such setting of the problem has resulted in endless discussions about how to feed the growing population, stop the global changes and prevent extinction of endangered species. Those who think that all these issues can be (easily) settled, are called *optimists*. People who question the solvability of this task are labelled *pessimists* and *alarmists*. As we demonstrate here, the very setting of the problem is incorrect. In reality—fortunately—we do not need to save the environment and biota each taken separately, which is indeed impossible. The only thing to be done is to lessen the burden of anthropogenic impact currently imposed on the natural biota. After that the latter will be able to restore and further maintain an environment suitable for humans.

In order to show that this is the only possibility to ensure environmental stability, let us compare quantitative information characteristics of the natural biota and modern civilisation.

The maximum rate of information processing by the human brain does not exceed 100 bit sec^{-1} (short-term memory).[1] Rate of information acquisition in the long-term memory is lower and does not presumably exceed 10 bit sec^{-1} (Ninio, 1998). Information is acquired most actively during the first 20 years of life of the individual, i.e. during about $6 \cdot 10^8$ sec. It can be safely assumed that the amount of information acquired later in life does not change the order of magnitude of the total store. Thus, the amount of information stored in human memory can be estimated as $6 \cdot 10^8$ sec $\cdot$ 10 bit $sec^{-1} = 6 \cdot 10^9$ bit.

Processing and storage of information are inevitably linked to certain material carriers. Information about the surrounding world acquired by our brain is read from macroscopic natural memory cells, that are defined by the brain itself. When we look at a rose, for example, we can perceive and remember it as an integral unit (flower) or, at a closer look, as an assemblage of petals. But we cannot see and remember a rose as an assemblage of atoms, because our organs of sense (and, therefore, our brain) do not give us such a possibility. In the brains of animals (including humans) storage and processing of information that is read from macroscopic memory cells is performed in microscopic memory cells of molecular size. In the brain, therefore, macroscopic processes that take place in the environment and organism itself are imitated on a microscopic scale and are detached from their macroscopic natural material carriers. Such a situation may give an impression of independence of the world that is created by our brain from the real world. As a result, the idea of *soul* which is independent of the body, emerges. Thus, soul can be defined as the cumulative non-genetic information acquired during one's life-span and stored in the individual memory.

The unique ability of *Homo sapiens* to transmit the information acquired in memory to further generations forms the basis for such notions as *immortal soul* and *the future life*. Memory information (i.e. soul in the above definition) of people who contributed considerably to the cultural heritage continue to exist in memories of people of further generations. Memory information of people who did not make

[1] The minimum time interval that can be discerned by human organs of sense is not less than 10^{-2} sec. Among organs of sense, vision is characterised by the largest rate of information processing. Perception of visual information in humans is based on innate and acquired symbols stored in the long-term memory cells. Acquisition of visual information is based on successive perception of symbols of different level of specificity (Loftus and Bell, 1975), e.g. forest → tree → leaf. Thus, it is a good approximation to assume that at every moment only one symbol (bit) is perceived. Therefore, the maximum rate of acquisition of information does not exceed 100 bit sec^{-1}.

such contributions and were forgotten by their offspring dies (vanishes) as soon as the last person who remembers those people dies. Note that 'soul' in the above definition is a property of all locomotive animals that have memory and the ability to learn. However, the soul of all animals except humans inevitably decays with death of the organism. Only humans are able to transmit part of their memory information to further generations. Thus, *immortal soul* is a unique property of people.

Cultural information of civilisation is, for the most part, represented by information about an artificial human-created environment. Information stored in books and computers is lost unless it is contained in the memory of certain living members of the society. Thus, the amount of cultural information stored by civilisation can be estimated as the sum of non-overlapping elements of memory of all living people. An upper estimate of the amount of cultural information of modern civilisation can be obtained by multiplying the current population of Earth ($\sim 6 \cdot 10^9$ people) by the average individual memory store of information ($\sim 6 \cdot 10^9$ bit), which gives a value of about 10^{19} bit. This figure is evidently a gross overestimate of the real value, because the majority of memory information is the same in all contemporary people. The unique non-overlapping information of civilisation is stored in memories of *specialists*—scientists, craftsmen, writers, musicians, artists. Working specialists constitute not more than about 10% of the whole population (multiplier 10^{-1}). Each field of knowledge can exist sustainably with no less than 100 specialists working in this field and sharing the same memory information (multiplier 10^{-2}). Thus, the real value of information store of modern civilisation can be obtained by multiplying the upper estimate by 10^{-3}, which gives about 10^{16} bit.

Note that the amount of individual memory information of animals coincides by the order of magnitude with that of humans. Large animals live longer but perceive information more slowly due to lower rates of metabolism (Section 3.2). Small animals are characterised by shorter lifespans, but higher metabolic rates and, consequently, higher rates of information acquisition. Thus, human society differs from a herd of animals not by the absolute amount of individual memory of its members, but by the presence of differentiation of information over memories of different members, that is, by the presence of specialists (professionals). It may be argued that social insects, for example ants, are also characterised by a high degree of specialisation. The critical difference between specialisation in humans and other animals lies in the fact that humans continuously increase the amount of cultural information in each field of specialisation, while in animals the amount of information does not change from generation to generation. If we compare, for example, architectural knowledge of humans and ants, we realise that ants of a certain species always make the same type of anthills. A particular anthill construction is one of the genetically encoded species-specific characteristics. Meanwhile the modern architecture of humans differs greatly from what our ancestors used to build. Modern architecture comprises both traditional and newly-acquired knowledge.

Let us now estimate the store of genetic information in the biota. Genetic information of most species of the biosphere is written in polymer double-strand molecules of DNA, which represent various sequences of the existing four different monomer units – nucleotide base pairs (bp). That is, the genetic alphabet consists of four different letters. Human genome contains $G = 3 \cdot 10^9$ bp (Lewin, 1987), i.e. $3 \cdot 10^9$ memory cells each of which can be characterised by one of the four different values. The store of genetic information in the human genome is equal to $\log_2 4^G = 2G = 6 \cdot 10^9$ bit. Note that the stores of genetic and non-genetic (memory) information in humans are of the same order of magnitude.

To quantify the store of the genetic information of the natural biota as a whole it is necessary to multiply the information content of an average genome by the total number of species in the biosphere, which is equal to about 10^7 species (Thomas, 1990). The average genome size can be taken equal to 10^9 bp, which is the average genome size of insects that constitute the majority of species in the biosphere. Intraspecific genetic variability is for the most part represented by decay changes of the genome (i.e. loss of information, Chapters 9 and 10) and cannot influence the obtained estimate. **Thus, the total amount of genetic information stored in the natural biota is of the order of 10^{16} bit and coincides in the order of magnitude with the information store of the modern civilisation.**

The amount of information that can be stored in a modern PC is of the order of 10^9 bit, i.e. it coincides by the order of magnitude with the amounts of genetic and individual memory information of one person. The cumulative memory capacity of modern computer technologies is large enough to store both the cultural information of modern civilisation and the genetic information of the natural biota.

Thus, when it comes to *stores of information* and memory capacity, modern civilisation is not inferior to the natural biota (Figure 7.2). However, life is a process. It is characterised by *fluxes of information* and the amount of work that can be performed in a unit of time by living organisms interacting with the environment. As far as the temporal characteristics of information are concerned, there exists an insurmountable quantitative gap between the natural biota and civilisation. We show this below.

Let us now estimate the major information fluxes. The species composition of the natural biota changes with evolutionary transitions of old species to new related ones. The genetic information of the biosphere changes in the course of evolution. Closely related species differ from each other in about 1% of their genetic information (Lewin, 1987). The average species life-span is about three million years (Chapter 11). Thus, 1% of the genetic information changes in a single act of speciation every $3 \cdot 10^6$ years. Consequently, a complete turnover of the genetic information of the natural biota (which we estimate as 10^{16} bit, Figure 7.2) takes about $3 \cdot 10^8$ years, i.e. about 10^{16} sec. Thus, the rate of evolution (i.e. the rate of change of genetic information of the natural biota in the course of evolution) is approximately equal to 10^{16} bit/10^{16} sec $= 1$ bit sec^{-1} (Figure 7.2). This extremely low rate of information change has been sufficient to ensure sustainable development of the biosphere, that is, to support evolution of the natural biota in such a manner that the latter has been able to compensate directional adverse environmental changes of cosmic and geophysical nature during the whole period of life existence, i.e. during nearly four billion years (Hayes, 1996).

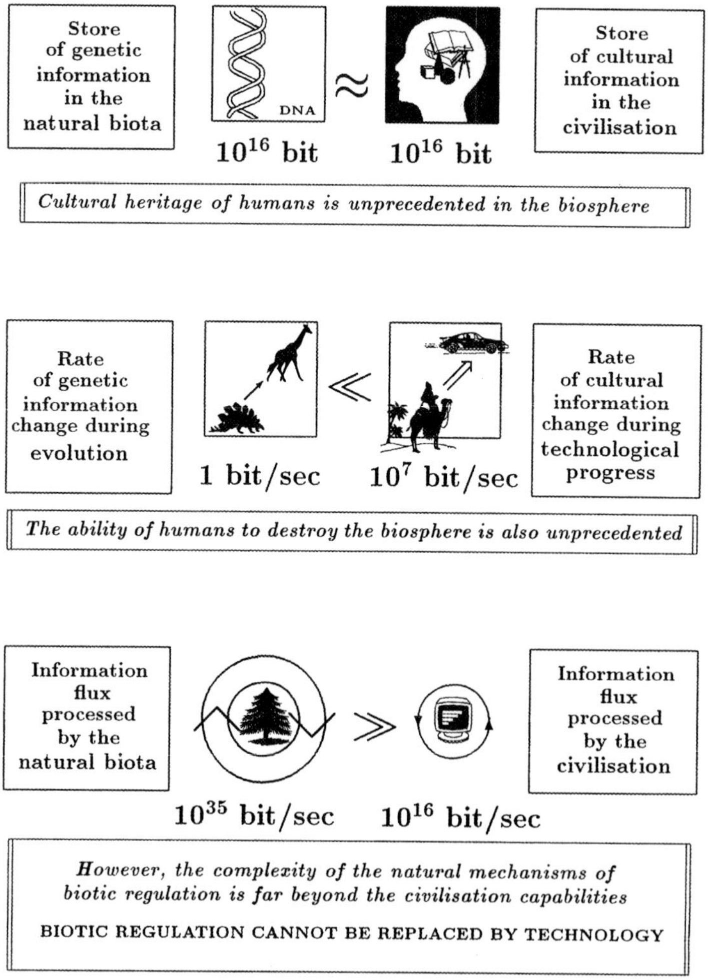

Figure 7.2. Major informational characteristics of the natural biota and modern civilisation.

The rate of information change during the technological progress of civilisation is determined by the ability of people to generate and assimilate new cultural information. According to the above estimates, the present-day population of Earth can assimilate no more than $6 \cdot 10^9 \cdot 10\,\text{bit sec}^{-1} = 6 \cdot 10^{10}\,\text{bit sec}^{-1}$. The present-day rate of technological progress depends on the average time of renewal of modern technological systems, which is of the order of 10 years, i.e. about $3 \cdot 10^8$ sec. Given that the store of information of modern civilisation is of the order of 10^{16} bit, the modern rate of civilisation progress is about $10^{16}\,\text{bit}/3 \cdot 10^8\,\text{sec} \approx 3 \cdot 10^7\,\text{bit sec}^{-1}$. (This estimate is obtained under the reasonable assumption that the majority of cultural information of modern civilisation is represented by information stored in the memory of specialists dealing with modern technologies.) Thus, the ratio of the amount of newly-generated information to the amount of assimilated information for modern people does not exceed $3 \cdot 10^7/6 \cdot 10^{10} \approx 10^{-3}$. Note that the same value characterises the average ratio of the amount of unique non-overlapping cultural information to the whole amount of the cultural information stored in the individual memory (see above).

The obtained information rate of the progress of modern civilisation, $3 \cdot 10^7$ bit sec^{-1}, exceeds the information rate of evolution, 1 bit sec^{-1}, by more than seven orders of magnitude (Figure 7.2). This provides an explanation for the unprecedented (as compared to all other extant and extinct species) 'competitiveness' of *Homo sapiens*, as far as its potential to destroy the environment is concerned.

An estimate of the total information flux going through all modern computers can be obtained by multiplying the average information flux in one PC, $\sim 10^8$ bit sec^{-1}, by the total number of people owning computers. Assuming that at present there is one PC for every hundred people we obtain that the total flux of information in computers of modern civilisation is of the order of 10^{16} bit sec^{-1}. This figure will hardly ever increase by more than six orders of magnitude (by providing computers for all people on the planet and ensuring a four orders of magnitude's increase of information flux in each PC), i.e. up to 10^{22} bit sec^{-1}.

Note, however, that even the present-day global computer information flux, 10^{16} bit sec^{-1}, exceeds the assimilation capacity of the brain of modern people, 10^{10} bit sec^{-1}, by a factor of a million. Computers work on the basis of programs designed by people and significantly speed up processing of information. But this only makes sense while people are still able to check and control the outgoing flux of information. All the information that is generated by computers and other mass-media devices (TV, cinema, video, theatre, music, etc.) above that threshold is nothing else but informational pollution of the environment (Gore, 1992). Informational pollution affects all the five organs of sense and, of the various types of pollution, presents the most dangerous threat to the mental health of humans.

Let us now estimate fluxes of information processed by living objects. In humans, metabolic power of existence in adults is equal to $q = 140$ W (Section 3.1). The body temperature of humans is approximately equal to $T_b \approx 37°\text{C} = 310$ K. The average thermal energy of molecules in the human body is equal to $k_B T_b$, where k_B is the Boltzmann constant ($k_B = 1.4 \cdot 10^{-23}$ J K^{-1}). Thus, $k_B T_b$ gives the order of the average amount of energy necessary to excite a molecule, i.e. the additional energy committed to a molecule as compared with the average thermal level. Assuming as above that one molecule corresponds to one memory cell with two possible states (excited and non-excited), i.e. 1 molecule $\approx$ 1 bit, we obtain that the information flux going through all living cells of the human body is equal to $q/k_B T_b \approx 3 \cdot 10^{22}$ bit sec^{-1}. This value exceeds the asymptotic information power of possible future computers and by more than 12 orders of magnitude exceeds the assimilation capacity of modern humanity.

The human body contains about 10^{14} living cells. Thus, every living cell processes on average about $3 \cdot 10^{22}/10^{14} \approx 10^8$ bit sec^{-1}, which is equal to information flux realised in a modern PC. The biosphere contains about 10^{28} living cells (Section 3.8).

Thus, the natural biota of Earth as a whole processes about $10^8 \cdot 10^{27} = 10^{35}$ bit sec^{-1}, which is about 20 orders of magnitude more than the information flux of all computers of modern civilisation (Figure 7.2). Unlike computers, molecular memory cells of living cells are directly coupled with the environment. Thus, the whole flux of information processed by biota is immediately used for correct interaction with the environment.

Energy consumption of modern humankind is equal to 10^{13} W, which is only an order of magnitude less than the photosynthetic power of the natural biota (Table 7.1). It may seem that people only need to increase their energy consumption tenfold to be able to perform the same amount of work on regulation of the environment as the natural biota does. In reality, however, due to the huge difference in the rates of information processing between the natural biota and civilisation, any kind of anthropogenic energy use is inevitably characterised by remarkably low efficiency. By *low efficiency* we do not mean here a high percentage of energy loss. Modern technologies permit to minimise energy loss and achieve high efficiency coefficients when transporting energy or converting it to kinetic energy of working devices. Rather, *low efficiency* refers to the low information content of most processes generated by humankind with help of energy use.

Indeed, humankind spends a large portion of its energy on transport, i.e. moving macroscopic objects—cars, trains, people, etc. Motion of a macroscopic object is totally determined by only four figures—its mass and three co-ordinates of the velocity vector. Thus, motion of macroscopic objects can be fully described by a very small amount of information coded in corresponding macroscopic memory cells. It is in principle possible to convert efficiently the kinetic energy of moving macroscopic objects to gravitational or electric energy that could be further used for generation of complex correlated molecular processes similar to those taking place in a living cell. In fact, however, the kinetic energy of transported objects does not generate any ordered, information-rich processes. It is spent on friction, and finally dissipates, converting to heat.

Note that macroscopic motion can be found in the natural biota as well (e.g. locomotive animals). However, the amount of energy allocated by the natural biota to this low-efficient channel of energy use does not exceed 1% of the total biotic energy consumption (Section 3.7, Figure 3.3). Meanwhile people spend on transport more than one-third of the consumed energy. The remaining part of anthropogenically-utilised energy is spent even more wastefully with respect to the information content of the generated processes (e.g. heating of buildings).

A similar wasteful character of energy expenditure is observed in such phenomena as hurricanes, tornadoes, earthquakes, falls of large meteorites, etc. In all these phenomena huge amounts of energy are released and dissipate in a non-controllable fashion, breaking down ordered structures in the natural biota, civilisation and environment. For this reason, all these phenomena present a serious hazard to people and are called *natural calamities*. **For the same reason, the non-controllable dissipation of anthropogenically-utilised huge fluxes of energy is becoming an *anthropogenic calamity* for the natural biota, global environment of Earth and, ultimately, for people themselves.**

Thus, the *total amount of energy* consumed by humankind does not characterise the amount of work that can be done by people in order to stabilise the environment. We have seen that of critical importance here is the *total flux of information* that can be processed by modern civilisation. And, as far as information fluxes are concerned, modern civilisation is hopelessly inferior to the natural biota (Figure 7.2).

As noted above, the information flux gap between the natural biota and civilisation will hardly ever be reduced by more than six orders of magnitude and become less than 10^{14} bit sec^{-1}. But even if we imagine that the gap is surmounted, and people have somehow been able to process as much information as the natural biota, it is easy to show that it is impossible to create a technological operating system equivalent to that of the biotic regulation of the environment. This becomes evident from the analysis of the well-known problem of automatic and manual control.

Manual control of technological processes is performed on the basis of genetic and cultural information of people. It consists in creation of feedback loops connecting the human brain and external fluxes of information. The rate of manual control is limited by the maximum rate of acquisition of information by the human brain, which is determined genetically as well as by all other major properties of the human organism. Automatic control functions on the basis of computer programs. Rate of automatic control is limited by technological properties of computers and can exceed the rate of manual control by millions of times.

However, the automatic control can be safely used in only those cases when the basic computer program leans upon well-developed theories. Such theories have been tested in so many independent experiments that people are absolutely sure in their predictions and do not need to check the results of automatic control. So, for example, the known exact physical laws allow the automatic control of space flights. But any unexpected situation not accounted for in the computer program inevitably forces the astronaut to switch the systems from automatic to manual control, the rate of all operations being immediately reduced by millions of times. Creation of any programme of automatic control is a very time-consuming process. Before it can be put into operation, all aspects of the work of all its elements need to be carefully tested by people, which is performed slowly by manual control.

Biotic regulation of the environment is equivalent to an operating system where the characteristic rate of information processing exceeds the maximum possible rate of manual control by 34 orders of magnitude and by 20 orders of magnitude—the rate of automatic control provided by all computers of modern civilisation. Biotic regulation is based on genetic programmes of biological species of the biosphere. Thus, from an anthropocentric standpoint, biotic regulation can be viewed as an automatically-controlled operating system where the programme of automatic control has been tested for reliability in an experiment lasting for several billion years (during the whole period of life existence). Humankind is unable to create adequate mathematical models of the natural biota. It is impossible to create a technological system equivalent to the natural biota, where each micron of the Earth's surface is controlled by dozens of independently functioning unicellular and multicellular organisms, each living cell processing an information flux similar to that of a modern PC. Finally, the genetic programme of the natural biota cannot be substituted by any technological programme of automatic control (even if this technological programme is characterised by fluxes of energy and information similar to those in the natural biota), because the search for appropriate technological decisions and their testing is performed by people and can take billions of years.

It follows unambiguously from the above that people are in principle unable to perform management of the natural biota, to improve or intensify its regulatory mechanism. Such attempts could be compared to the attempts of a domesticated animal (e.g. a cow) to improve functioning of a computer that governs its food supply. In very much the same manner as non-human animals are incapable of acquiring cultural knowledge of civilisation, humans are unable to manage fluxes of information similar to those processed by the natural biota. It is impossible to manage natural forests and wetlands to make them regulate the environment 'better' than they do. Humans are only able to manage artificial biological systems that are completely deprived of the regulatory ability, e.g. agricultural fields, where the only characteristic to be controlled is the productivity of the planted culture. Note also that even such primitive management always occurs at the expense of environmental stability, i.e. all other environmental characteristics (except for the productivity) change in a direction unfavourable for humans.

Genetic information of the natural biota totally changes every $3 \cdot 10^8$ years (see above). Thus, during the whole period of life existence ($3.8 \cdot 10^9$ years) there were no more than a dozen completely different programmes of biotic regulation of the environment. A working programme of biotic regulation is presumably unique for each particular epoch. Evolution of the biotic regulatory programme is possible due to acting geophysical and cosmic processes. That is, directional changes in parameters that cannot in principle be controlled by biota (e.g. solar activity) may lead to a situation when the old regulatory programme is no longer the most effective one. As a result, there opens a possibility for a new, more effective regulatory programme of the biota to establish as a result of genetic modifications (i.e. appearance of new species) in the old programme. New regulatory programmes appearing in the course of evolution are exposed to a thorough experimental testing.

The relative degree of *unsteadiness* in the work of a computer can be defined as the ratio of the rate of human-induced changes in the computer program to the total flux of information processed by the computer. The relative unsteadiness of the regulatory programme of the natural biota, similarly defined, is fantastically low, $1\ \text{bit sec}^{-1}/10^{35}\ \text{bit sec}^{-1} = 10^{-35}$. (Rate of programme change corresponds to the rate of information change in the course of evolution, $1\ \text{bit sec}^{-1}$. The total information flux processed by the natural biota is equal to $10^{35}\ \text{bit sec}^{-1}$, Figure 7.2.) It means that each working regulatory programme is maintained by the natural biota in a steady state for the maximum possible periods of time.

Summing up, we have seen that while the store of cultural information of modern civilisation is of the same order of magnitude as the store of genetic information in the natural biota, the natural biota is able to process 20 orders of magnitude more information in a time unit than modern civilisation (Figure 7.2). There is no sense in attempting to substitute the reliable working mechanism of the biotic regulation of the environment by a technological system. All such attempts are doomed to fail, as shown above and, which is more, the very setting of such a goal is dangerous. The only promising strategy for modern humankind is the strategy of preservation of the remaining natural biota and restoration of its global regulatory potential.

7.4 ECOLOGICAL INFORMATION OF LARGE ANIMALS

In unicellular organisms, nearly all cellular processes have an immediate impact on the external environment. Some multicellular organisms (plants mostly) have also preserved this quality thanks to a very large effective surface area of their organs—leaves, branches, roots. In large locomotive animals, on the contrary, most metabolic processes take place in the internal organs that do not have direct contact with the external environment. The regulatory biotic impact on the environment is evidently proportional to the surface area of metabolically active (living) biomass. The ratio of the surface area of living biomass to the territory occupied by the organism is proportional to the projection area index (leaf area index)[2] (Section 3.8). This index is of the order of 10 for immobile organisms (plants, bacteria, fungi) (Section 3.3) and of the order of 10^{-7} for large locomotive animals (Sections 3.7 and 3.8). Thus, the regulatory potential of large animals is a hundred million times weaker than that of plants, bacteria and fungi. With increasing body size the variety of the organism's possible interactions with the environment becomes poorer. Complex metabolic processes taking place in the internal milieu of a large organism are mostly aimed at converting metabolic power of the organism to the mechanical power of locomotion, feeding (consuming biomass) and work of excretory organs. These relatively simple (information-poor) actions do not require the high complexity of biochemical organisation of cellular processes that was inherited by multicellular organisms from their unicellular predecessors. The same work could be easily performed by artificial devices, far more primitive than organisms of large animals. Meanwhile small animals specialise in very fine and complex (information-rich) interactions with vegetation, that cannot be performed by any technological means of civilisation (e.g. pollination of plants by insects). Thus, contrary to common opinion, large animals represent the most primitive part of the natural biota with respect to environmental control.

Natural biota is from time to time exposed to various catastrophes of abiotic as well as of biotic nature (e.g. fires, windfalls, volcanic eruptions, occasional invasions of alien species, etc.). Thus, natural biota must include as its integral part a certain set of species that could ensure recovery of the ecosystem after such disturbances. It

[2] For trees the projection area index is roughly equal to the average number of times the tree's leaves overlap each other along a vertical axis (leaf area index), i.e. how thick (in terms of leaves) would be the tree's foliage when all the leaves are put on the projection area of the tree. Note that for plants the ratio of the surface area of metabolically active biomass to the territory occupied by the organism is roughly equal to leaf area index multiplied by a factor of two (taking into account the upper and lower surfaces of a leaf). For locomotive animals the projection area index is equal to the ratio of the area of the organism's projection on the Earth's surface to the feeding territory of the organism.

is natural to call such species *species-repairers* (see Section 6.7). Evidently, species-repairers should achieve high population numbers and high population density immediately after disturbances and exist in lower numbers in the absence of disturbances. External disturbances do not occur at regular intervals. If a time interval between two consecutive disturbances (e.g. fires) becomes too long, the 'unemployed' species-repairers may lose their genetic stability, as soon as the process of stabilising selection is most efficient in dense populations. Thus, in order to 'train' species-repairers, natural biota has to support *species-disturbers* that disturb the ecosystem at regular intervals and, by doing so, periodically activate species-repairers, preventing them from genetic decay. Primitive man was among such species-disturbers together with many other large animals (e.g. elephants making large lawns amidst tropical forests). This issue has been previously discussed in Sections 4.7 and 6.7.

Homo sapiens has preserved the strategy of behaviour based on the genetic programme of positive and negative emotions inherited from large animals. Natural ecological role of man in the biosphere was to use fire and other means (e.g. stubbing up trees to get arable land) in order to introduce periodical disturbances of local ecosystems and activate species-repairers. Thus, the genetically-based programme of environmental impact of humans is aimed towards transformation of the natural environmental conditions (e.g. forest) into more favourable artificial ones (e.g. field), i.e. creation of an artificial 'internal' environment forcing out a less favourable external environment. This can be regarded as a further advancement of the principle of large body size (i.e. when more and more space is occupied by the internal milieu of the organism), which, as explained above, reduces the ability of *Homo sapiens* as a species to regulate the external environment. Cultural information accumulated by humankind in the course of historical development of civilisation has enabled people to impose a huge impact on the biosphere. But, as ever, the strategy of this impact is confined to the genetic programme of behaviour of *Homo sapiens*. Thus, most human–environment interactions result in large-scale perturbations of the natural biota and the forcing out of the natural 'wild' environment. It is this unprecedented scale of these processes that is now threatening the future existence of humanity.

8

Unique Nature of Climate Stability on Earth

In this chapter evidence is presented in favour of the statement that Earth's modern climate is physically unstable. In the absence of natural biota it should change spontaneously to one of the two possible physically stable states—that of complete glaciation of the Earth's surface and that of complete evaporation of the Earth's hydrosphere, both of which are life-incompatible. The existing climate has remained suitable for life during the last four billion years. It means that natural biota of Earth is the only mechanism that keeps the Earth's climate in a state suitable for life.

8.1 MAJOR CLIMATIC CHARACTERISTICS OF EARTH

All biogeochemical cycles in the environment represent steady ordered processes maintained at the expense of solar energy. A steady state of all substances cycling in the environment means that masses and concentrations of all substances remain, on average, constant in all local areas of the environment despite continuous exchange of substances between different areas. It means that for all substances sources are equal to sinks in all areas of the environment. If this equality is broken, masses and concentrations of corresponding substances begin to change. Rates of such changes are determined by properties of the corresponding sources and sinks.

Atmospheric oxygen and carbon dioxide are in physical thermodynamic equilibrium with oxygen and carbon dioxide dissolved in rivers, lakes, seas and oceans. On average, water solutions of these two substances are saturated. Solubility of these two gases increases with decreasing temperature. Thus, in polar regions there is a net flux of these gases from the atmosphere to the ocean, while in equatorial regions there is a reverse flux of gases from the ocean to the atmosphere. Physical cycles of oxygen and carbon dioxide depend mainly on the average temperature of the Earth's surface and on the difference between average temperatures of polar and equatorial regions. These two parameters are determined by the total amount of energy received from the Sun and in practice do not depend on informational characteristics of solar energy that are determined by the difference between temperatures of Earth and Sun (see Chapter 7). Thus, on a global scale, physical cycling of oxygen and carbon dioxide can be described by traditional thermodynamics.

Water is present in the biosphere in all the three possible phases—solid, liquid and gaseous. Conditions for thermodynamic equilibrium between different water phases are met only in very small local areas. On a global scale, the three phases of water are not in physical thermodynamic equilibrium. If this were the case, water vapour would be present in the atmosphere in a saturated concentration. In reality, the observed global mean value of relative humidity, which is equal to the relative difference between the saturated and real concentration of water vapour, is less than 100%.

As noted in Section 6.6, atmospheric water vapour concentration changes hundreds of times in different regions of the planet. The observed non-random unevenness of distribution of atmospheric water vapour over the Earth's surface represents a departure from the chaotic state of thermodynamic equilibrium. Its ultimate cause is the orderliness of the solar radiation. Unevenness of water vapour distribution is a prerequisite for the alternating processes of evaporation and precipitation that are vitally important for the terrestrial biota.

Thus, for some substances (e.g. oxygen and CO_2) the steady state of cycling is close to physical thermodynamic equilibrium and is totally determined by Earth's average temperature and atmospheric pressure. For other substances (e.g. water) the steady state of cycling is far from physical thermodynamic equilibrium. Cycling of such substances represents a highly-ordered process and depends on the degree of orderliness of the solar energy, that is, on the difference between the Sun's and Earth's temperatures (Chapter 7).

Some of the most vital questions of modern fundamental ecology are as follows. What factors determine the average temperature of the Earth's surface? Are concentrations and masses of globally cycling substances constant and to what accuracy? Is the observed state of global environment a stable one? Are there any other stable states of Earth's environment? Are they suitable for life? Can the present-day global environment of Earth transform into one of the other possible stable states and how much time will it take? We address below some of these issues.

The orbital position occupied by the Earth in the solar system determines the flux of the solar energy incident upon the planet outside the atmosphere. It is called the solar constant I (Table 8.1). About one third of this flux is reflected back into space, predominantly by clouds in the atmosphere and by ice cover at the Earth's surface. The relative part of solar radiation reflected back into space is called the planetary albedo (Table 8.1). Due to non-zero values of albedo, Earth and other planets of the solar system are visible in space. Solar radiation absorbed by the Earth's surface generates cycling processes in the biosphere and finally converts into heat, that is, thermal radiation. In the atmosphere thermal radiation of the Earth's surface is trapped by the so-called greenhouse gases. Molecules of the greenhouse gases emit thermal photons in all possible directions, so that about half of this radiation returns to the Earth's surface where it is re-emitted to the atmosphere and so forth. This phenomenon is known as the greenhouse effect and can be quantified as the relative part of thermal radiation of the Earth's surface that is effectively

Table 8.1. Planetary energy and temperature characteristics (after Mitchell, 1989).

Planet	Solar constant	$A = B = 0$ (Orbital temperature	$A > 0,\ B = 0$ (Effective planetary temperature)		$A > 0,\ B > 0$ (Average values at planetary surface)	
	I, W/m^2	t_o, °C	A, %	t_e, °C	B, %	t, °C
Mars	589	−48	15	−56	7	−53
Venus	2613	+58	75	−41	99	+460
Modern Earth (2)	1367	+5	30	−18	40	+15
Earth, if total ice cover (1)	1367	+5	80	−90	7	−85
Earth, if total evaporation of oceans (3)	1367	+5	75	−80	99	+400

I is the solar constant, i.e. the power of solar radiation incident upon unit of perpendicular area outside the planet's atmosphere. $I/4$ is the average power of solar radiation incident upon unit of Earth's surface area, see Section 8.5. Orbital temperature t_o (degrees Celsius) is uniquely determined by I, i.e. by the orbital position occupied by the planet in the solar system. Under the approximation of the black body radiation, I and t_o are connected by the expression $T_o = (I/4\sigma)^{1/4}$, where $T_o = t_o + 273$ is the absolute orbital temperature in degrees Kelvin, $\sigma = 5.67 \cdot 10^{-8}\ \mathrm{W\,m^{-2}\,K^{-4}}$ is the Stephan–Boltzmann constant.
A is the planetary albedo, i.e. fraction of solar radiation reflected by the Earth back to space. The remaining fraction of solar radiation, $(1 - A)I/4$, is absorbed by the planet and is equal to the planet's heat emission to space. This flux can be directly measured from outer space by artificial satellites. It determines the effective planetary temperature t_e as $T_e = [(1 - A)I/4\sigma]^{1/4} = T_o(1 + A)^{1/4}$, where $T_e = t_e + 273$.
B is the relative greenhouse effect, i.e. the fraction of the planet's heat radiation that is reflected by the atmosphere back to the planet's surface. The heat flux at the Earth surface q is equal to $q = (1 - A)I/4 + Bq$, that is $q = (1 - A)I/4(1 - B)$. Accordingly, temperature at the Earth's surface t (degrees Celsius) is determined by values of planetary albedo and greenhouse effect: $T = (q/4\sigma)^{1/4} = T_o[(1 - A)/(1 - B)]^{-1/4}$, where $T = t + 273$.
In the state of total ice cover of Earth, the value of planetary albedo is taken as equal to the albedo of ice and snow sheets, the value of greenhouse effect was taken as on Mars. In the state of total evaporation of oceans on Earth, the values of albedo and greenhouse effect are taken as on Venus. Note that changes of A and B lead to drastic changes of temperature on the planet's surface.

reflected back to the Earth's surface by the atmosphere. Modern atmosphere of Earth effectively reflects about 40 percent of the thermal radiation, while on Venus this effect amounts to 99 percent (Table 8.1).

Temperature of the Earth's surface is uniquely determined by the cumulative flux of thermal radiation from the Earth's surface, which is proportional to the fourth power of absolute surface temperature. Irrespective of what orbital position is occupied by the planet, this flux is totally determined by the planetary albedo and greenhouse effect (see notes to Table 8.1). Thus, the planet's surface temperature can assume almost arbitrary values depending on values of albedo and greenhouse effect, the latter being completely determined by inherent environmental characteristics of the planet. In other words, the orbital position does not guarantee a suitable-for-life environment on the planet.

Physical density and gaseous composition of the atmosphere determine atmospheric absorption of less than one-third of the incoming solar radiation. Thus, the majority of solar radiation reaches the Earth's surface making possible photosynthesis and, therefore, life of the whole biota of Earth. If atmospheric density were several times higher than at present, sunlight could not practically reach the Earth's surface which would remain dark, similar to the present situation at oceanic depths. Under such conditions, no modern life would then be possible on the Earth's surface. If, on the contrary, atmospheric density were several times lower than it is now, the atmosphere would feature some life-incompatible characteristics. In particular, concentration of oxygen would be insufficient for breathing, while concentration of CO_2 would be insufficient for photosynthesis (Brody, 1945; Larcher, 1980).

The gaseous composition of the atmosphere is no less important for life than the atmospheric density. Modern atmosphere is relatively transparent to the incident sunlight, but opaque with regard to thermal radiation of the Earth's surface. Absorption of thermal radiation is performed by greenhouse gases. Gases absorb and emit radiation at frequencies that correspond to transitions between discrete energy levels of their molecules. Transitions between some energy levels of molecules of greenhouse gases are in resonance with characteristic frequencies of the thermal radiation of Earth. The cumulative mass of all greenhouse gases is less than two thousandths of the total mass of the atmosphere. The major greenhouse gas is water vapour. Its volume content in the atmosphere is less than 0.3 percent. The second important greenhouse gas is carbon dioxide with a volume content of about 0.03 percent. The remaining part of the greenhouse warming is determined by methane, nitrous oxide and ozone. Their cumulative volume content does not exceed 0.0003 percent (Mitchell, 1989). Note also, that ozone which is present in the atmosphere in negligibly small concentrations, ensures an almost complete absorption of ultraviolet radiation harmful to terrestrial life.

Thus, we have seen that the properties of the modern atmosphere are in a unique concert with requirements of life. Minor changes in atmospheric properties can be fatal to any life on Earth.

In a stationary state—that is, when the average energy content of the planet remains constant over time—the power of the thermal radiation released by Earth into space is naturally equal to the power of the solar radiation absorbed by the planet. The latter is totally determined by the solar constant and the planetary albedo (see notes to Table 8.1). Thus, thermal radiation released by Earth into space does not depend on the greenhouse effect and is characterised by the so-called effective planetary temperature (Table 8.1). This temperature characterises the upper radiative layers of the atmosphere from which the thermal radiation is emitted directly into space. Thus, the effective planetary temperature can be measured from outer space. One cannot judge about the temperature of the planet's surface from the effective planetary temperature.

The greenhouse effect increases temperature of the planet's surface as compared with its effective planetary temperature. It leads to formation of a temperature gradient between the planet's surface and the upper radiative layers of the atmosphere. A well-known manifestation of this gradient is the existence of eternal snow caps on tops of high mountains. This gradient, so-called lapse rate, constitutes 5.5°C km^{-1} and remains nearly constant up to about 6 km above the sea level. At that altitude the temperature difference between the Earth's surface and the atmosphere equals 33°C, which gives the magnitude of the greenhouse effect measured on a temperature scale. The atmospheric layer at height of about 6 km represents an effective radiating surface that determines thermal radiation of the planet emitted directly into space.

In the stratosphere the density of atmospheric gases rapidly decreases with height, while the temperature of these rarefied gases increases and exceeds not only the temperature of the effective radiating tropospheric layer, but even that of the Earth's surface. As soon as the thermal radiation of Earth measured from outer space is actually emitted by molecules of the upper tropospheric radioactive layers, the increased temperature of the stratosphere means that molecules of the rarefied stratospheric gases are not in thermal equilibrium with long-wave (thermal) radiation of the planet. This means that the stratosphere and processes taking place there impose no considerable impact on the heat balance of the planet and cannot influence the climate of Earth. The fact that the thin ozone layer absorbs ultraviolet radiation is essential for life on the planet, but does not seriously influence the heat balance of the planet.

In the absence of greenhouse gases the average temperature of the Earth's surface would be equal to the effective planetary temperature of the planet, $t_e = -18°C$ (Table 8.1). There would be no temperature gradient present in the atmosphere. The snow cover of high-latitude territories would gradually spread down over the whole Earth's surface. Thus, the planetary albedo would increase up to the value characteristic of snow and ice, that reflect about 80% of the solar radiation (Mitchell, 1989). As a result, the average global temperature would drop to about −90°C (Table 8.1), a state where no life is possible.

Hydrosphere of Earth is sufficient for total glaciation of the Earth's surface. At the same time, Earth's hydrosphere is the main source of atmospheric water vapour, the major greenhouse gas. Absolute concentration of water vapour in the atmosphere, as well as the relative humidity, vary greatly both in space and time, changing several times as compared with their average values. However, the average concentration of the water vapour changes proportionally to its saturated concentration (the maximum possible concentration under given conditions). Saturated concentration of water vapour grows exponentially with increasing temperature in accordance with Boltzmann distribution and the Clausius–Clapeyron equation (Landau *et al.*, 1965). The concentration of water vapour is negligibly small at low temperatures. With increasing temperature, concentration of water vapour grows exponentially and causes an increase of the greenhouse effect that leads to further increase of temperature and so forth. At 100°C water vapour pressure becomes equal to the atmospheric pressure and water begins to boil. At higher temperatures the atmospheric pressure is totally determined by the pressure of water vapour, which continues exponential growth with increasing temperature (Landau *et al.*, 1965). The process ends in a state of complete evaporation of the hydrosphere, when processes of evaporation and condensation of clouds occur in the atmosphere and do not pertain to the Earth's surface. As a result, the equilibrium temperature of the Earth's surface jumps up to several hundred degrees Celsius due to the catastrophic increase of greenhouse warming, as on Venus (Table 8.1). In this state life is not possible either.

The existing positive feedback between the surface temperature and atmospheric water vapour concentration points to the physical instability of the current climate of Earth, where hydrosphere is maintained predominantly in the liquid phase (water of oceans, rivers, lakes). The solid phase (ice of glaciers) constitutes less than 14% and the gaseous phase (atmospheric water vapour) less than two millionth parts of the total amount of liquid water which is mostly concentrated in the oceans (Lvovitch, 1974), see Table 6.1.

When values of albedo and greenhouse effect are temperature-independent, the surface temperature of the planet $T = T_s$ is stable. It is determined by the balance of the absorbed solar radiation and the emitted thermal radiation. The stability of stationary surface temperature T_s is ensured by the following negative feedback. When the surface temperature accidentally increases, $T > T_s$, the emitted flux of thermal radiation increases as well, according to the Stephan–Boltzmann law. As a result, the planet loses more energy than it receives and cools down back to T_s. When the surface temperature decreases, $T < T_s$, the planet loses less energy than it receives. As a result, temperature increases back to T_s.

There are two physically stable states where values of albedo and greenhouse effect remain constant in a wide temperature interval. These are the state of total glaciation of the Earth's surface at temperatures about −90°C and the state of total evaporation of the Earth's oceans at temperatures close to 400°C (Table 8.1). In both states life is impossible. Constancy of albedo and greenhouse effect in these states is determined by the fact that in both states water exists predominantly in only one phase—solid at low and gaseous at high temperatures.

Under modern climatic conditions water exists in all three phases. Values of albedo and greenhouse effect depend on temperature. For example, with increasing temperature the ice shields melt and planetary albedo decreases, while atmospheric water vapour concentration grows and the greenhouse effect increases. Stable existence of life during the last four billion years gives unambiguous evidence that the modern, suitable-for-life state of the Earth's climate is stable and spontaneous transitions to both lifeless states are forbidden. The degree of stability of the modern climate depends on the rates of changes of albedo and greenhouse effect with temperature. This is discussed in sections to follow.

8.2 SPECTRAL CHARACTERISTICS OF THERMAL RADIATION

Spectral distribution of energy over radiation wave frequencies (or inverse wave numbers) is the main characteristic of any flux of radiation. Those parts of the Earth's surface that are not covered by ice or snow absorb almost all incident radiation. It means that in a stationary state, when the surface temperature remains constant, the energy spectre of radiation emitted by the Earth's surface does not depend on properties of its molecules, but is determined by temperature alone. Such radiation is known as the black-body radiation and is described by the well-known Planck function.

Unlike the Earth's surface, atmosphere only absorbs thermal radiation at certain spectral intervals defined by the presence of absorption bands of the greenhouse gases. Thus, thermal radiation of the atmosphere cannot be reduced to black-body radiation depending on the gas temperature alone. Rather, thermal radiation of the atmosphere depends on concentrations of greenhouse gases and their height distribution. In order to evaluate temperature dependence of the greenhouse effect, it is necessary to understand how radiating properties of the atmosphere depend on concentrations of the greenhouse gases.

Energy spectre of thermal radiation of the Earth's surface $q(T)$ over radiation frequencies ω can be written as follows:

$$q(T) = \sum_i \Delta\omega_i I(\omega_i, T) = \Delta\Omega \bar{I}(\bar{\omega}, T) = \sigma T^4 \tag{8.2.1}$$

Here $I(\omega, T)$ is the observed spectral density of thermal radiation emitted from the Earth's surface, which is close to the Planck function; T is the observed global mean absolute temperature of the Earth's surface. According to the properties of the Planck function, the effective width of the spectre $\Delta\Omega$ grows linearly with increasing T, while the maximum and the mean values of $I(\omega, T)$ increase proportionally to the third power of T. Thus, $q(T)$ grows proportionally to T^4. Proportionality coefficient σ is close to the Stephan–Boltzmann constant characterising black-body radiation. The average frequency $\bar{\omega}$ grows linearly with T.

Unlike the Earth's surface which absorbs the most part of the incoming radiation, atmospheric gases absorb radiation at some frequencies but allow radiation at other frequencies to pass through unimpeded. Greenhouse gases absorb thermal radiation in those parts of the spectre (8.2.1) that correspond to their molecular absorption bands. The latter are determined by molecular properties of the gases and only weakly depend on atmospheric temperature and density. In some parts of the spectre (8.2.1) there are no absorbers at all or their concentration is very low, so that the absorption of thermal radiation in the atmosphere in such parts of the spectre is not complete. Such parts of the spectre are called spectral windows. Thermal radiation of the Earth's surface emitting into space through spectral windows is characterised by spectral intensity $I(\omega, T)$ determined by the surface temperature T. Thermal radiation of the planet measured in the outer space, q_e, can be thus represented as a sum of the spectral window radiation and radiation emitted from the upper radiating layer of the atmosphere with temperature T_e:

$$q_e(T_e) = \sum_{i \neq k} \Delta\omega_i I_e(\omega_i, T_e) + \sum_k \Delta\omega_k I(\omega_k, T) \approx \sigma_e T_e^4, \tag{8.2.2}$$

where frequency intervals $\Delta\omega_k$ stand for spectral windows. Here $I_e(\omega, T_e)$ is the observed spectral intensity of the upper radiating layer of the atmosphere. Note that $I_e(\omega, T_e)$ can be substantially different from the Planck function $I(\omega, T)$, because the atmosphere is not a black-body radiator. In the modern atmosphere only a minor part of the terrestrial radiation escapes directly into space through the spectral windows (Barry and Chorley, 1987). This makes it possible to consider q_e approximately as a function of T_e only, neglecting the minor contribution of the second sum in (8.2.2). The constant σ_e in (8.2.2) may differ from the Stephan–Boltzmann constant by several tens of percent.

8.3 TRADITIONAL ESTIMATES OF THE CONTRIBUTIONS FROM DIFFERENT GREENHOUSE GASES TO THE GREENHOUSE EFFECT

The observed global mean fluxes of terrestrial radiation $q(T)$, atmospheric radiation $q_e(T_e)$ and their difference $f \equiv q_e - q$ are as follows:

$$q = \sigma T^4 = 390\,\mathrm{W\,m^{-2}}, \qquad T = 288\,\mathrm{K}\ (15^\circ\mathrm{C}),$$

$$q_e = 240\,\mathrm{W\,m^{-2}} \approx \sigma_e T_e^4, \qquad T_e \approx 255\,\mathrm{K}\ (-18^\circ\mathrm{C}),$$

$$f = q - q_e = 150\,\mathrm{W\,m^{-2}}. \tag{8.3.1}$$

Note that f has a meaning of absolute greenhouse effect in energy units (greenhouse forcing).

In a stationary state the energy lost by the Earth back into space, q_e, is balanced by the energy received by the Earth from the Sun, i.e. q_e is completely determined by values of I and A, $q_e = (I/4)(1 - A)$. The atmosphere is relatively transparent to the solar radiation, which is transformed into heat predominantly at the Earth's surface. It is therefore q_e that serves as the primer radiation flux. It interacts with the greenhouse gases in the atmosphere and initialises a cascade of absorption-emission processes, the result being the observed greenhouse effect. Introducing relative greenhouse effect $B \equiv f/q$ (Raval and Ramanathan, 1989) we may write:

$$q = q_e + Bq \qquad \text{or} \qquad q = q_e/(1 - B) \tag{8.3.2}$$

The relative greenhouse effect B can be interpreted as the fraction of terrestrial radiation that is reflected by the atmosphere back to the planet's surface. Note that both absolute, f, and relative, B, values of the greenhouse effect depend on two temperatures, T and T_e.

The absolute greenhouse effect f can be represented as a sum of spectral components, similar to q (8.2.1) and q_e (8.2.2). Using (8.2.1), (8.2.2) and the definition of f (8.3.1) we have:

$$f(T, T_e) = \sum_{i \neq k} \Delta\omega_i [I(\omega_i, T) - I(\omega_i, T_e)] \tag{8.3.3}$$

Here spectral intervals $\Delta\omega_i$ are the same as in (8.2.2). Naturally, spectral windows $\Delta\omega_k$ (8.2.2) do not make any contribution to (8.3.3).

Modern greenhouse effect on Earth is by more than 90 percent determined by atmospheric water vapour and carbon dioxide (Mitchell, 1989). These two gases absorb thermal radiation over a wide range of frequencies spanning almost the whole spectre of terrestrial radiation. Thus, terms $\Delta\omega_{H_2O}$ and $\Delta\omega_{CO_2}$ are the major contributors to (8.3.3):

$$f(T, T_e) \approx \Delta\omega_{H_2O}\,\Delta I(\bar{\omega}_{H_2O}) + \Delta\omega_{CO_2}\,\Delta I(\bar{\omega}_{CO_2}),$$
$$\Delta I(\omega) \equiv I(\omega, T) - I(\omega, T_e) \tag{8.3.4}$$

Using the Planck function $I(\omega, T)$ and corresponding spectral widths $\Delta\omega_{H_2O}$ and $\Delta\omega_{CO_2}$ known from spectroscopic measurements and taking into account a few minor correction coefficients we obtain the following estimates for the greenhouse contributions of water vapour and carbon dioxide (Mitchell, 1989):

$$f_{\Delta\omega_{H_2O}} = \Delta\omega_{H_2O}\,\Delta I(\bar{\omega}_{H_2O}) \approx 100\,\mathrm{W\,m^{-2}},$$
$$f_{\Delta\omega_{CO_2}} = \Delta\omega_{CO_2}\,\Delta I(\bar{\omega}_{CO_2}) \approx 50\,\mathrm{W\,m^{-2}}. \tag{8.3.5}$$

The cumulative contribution of the two gases is approximately equal to the total value of f (8.3.1), which justifies use of the Planck function in (8.3.5) instead of the empirically observed values of spectral intensity $I(\omega)$. According to (8.3.5) the contribution of carbon dioxide is two times less than that of water vapour.

Note that values of concentrations of the two gases did not enter the calculations of the result (8.3.5). The only fact used implicitly was that these gases are present in the atmosphere in sufficiently large (saturated) concentrations to ensure almost complete absorption of terrestrial radiation in the respective parts of the spectre, which makes it possible to use the Planck function. It was assumed that after absorption has been saturated, further increase of the gas concentration does not impose any major impact on the greenhouse effect.

Concentration dependence of the greenhouse effect is only observed for the minor greenhouse gases which are present in very small quantities and have absorption bands in spectral windows where there are no other absorbers. Greenhouse contributions of such gases (chlorofluorocarbons mostly) grow linearly with concentration until the latter is high enough to ensure almost complete absorption in the middle of the absorption bands. Further increase of concentration will result in a very slow (square root or logarithmic) broadening of the absorption interval $\Delta\omega$ and a correspondingly slow increase in the greenhouse effect due to saturation of the wings of the absorption lines, provided that there are no other saturated gases absorbing at the same frequencies (Mitchell, 1989).

Such a situation is realised for the atmospheric CO_2. Thus, the widely discussed potential global warming is traditionally calculated on the basis of logarithmic growth of the CO_2 greenhouse contribution (8.3.5) with concentration (Ramanathan *et al.*, 1987; Mitchell, 1989; IPCC, 1994).

However, the results (8.3.5) obtained with use of (8.3.3) and (8.3.4) do not actually represent contributions of real *concentrations* of the greenhouse gases to the greenhouse effect but, rather, contributions of those spectral *intervals* that correspond to absorption bands of CO_2 and H_2O. The calculations were based on the known values of temperature T_e of the upper radiating layer of the atmosphere and temperature T of the Earth's surface. Values of T_e and T entered the calculations as empirical parameters. In reality, however, the difference $T_e - T$ is completely determined by and strongly dependent upon atmospheric concentrations of the greenhouse gases.

True greenhouse contribution of a gas can be measured as the flux of energy re-emitted by this gas back to the Earth's surface. True greenhouse contributions of gases are not additive. It means that a mixture of greenhouse gases does not produce a greenhouse effect equal to the sum of greenhouse effects caused by each gas alone. Energy absorbed by certain gas A *in a given spectral interval* is partly re-emitted back to the Earth's surface heating it. The Earth's surface emits then more radiation *in all spectral intervals* due to re-distribution of the absorbed energy over the whole black-body spectre. In the presence of gas A other greenhouse gases have therefore an opportunity of absorbing and, consequently, re-emitting more thermal radiation in their own spectral intervals.

As far as energy absorbed by one gas in a given spectral interval can be transmitted to another gas and thus re-distributed over other spectral intervals, greenhouse contributions from different spectral intervals calculated as in (8.3.5) do not give information about true greenhouse contributions of gases. The real dependence of the greenhouse effect on concentrations of the greenhouse gases, which is critical to evaluating stability of the Earth's climatic system, in such consideration appears to be hidden.

8.4 DEPENDENCE OF THE GREENHOUSE EFFECT ON CONCENTRATIONS OF THE GREENHOUSE GASES

Let us consider a one-dimensional atmosphere consisting of N greenhouse gases with non-overlapping absorption intervals $\Delta\omega_l (l = 1, 2, \ldots, N)$ covering the whole range $\Delta\Omega$ of terrestrial radiation frequencies, $\sum_{l=1}^{N} \Delta\omega_l = \Delta\Omega$ (8.2.1). Let δ_l be the relative portion of energy of thermal radiation of the Earth's surface corresponding to the absorption interval of the l-th greenhouse gas. We obtain from (8.2.1):

$$\delta_l = \frac{\Delta\omega_l I(\omega_l, T)}{\sigma T^4}; \qquad \sum_{l=1}^{N} \delta_l = 1. \tag{8.4.1}$$

Let n_l be the number of optically dense layers of the l-th gas present in the atmosphere, i.e. the number of such layers that with sufficient accuracy ensure complete absorption of the thermal radiation in the respective part of the spectre. The number of optically dense layers is proportional to the so-called optical thickness. Thickness of a single optically dense layer can be approximated as the average length of free path of thermal photons between two successive collisions with molecules of the gas. The total number of optically dense layers is thus proportional to the gas concentration given the constant height of the atmosphere. Each optically dense layer absorbs thermal radiation in the corresponding spectral interval and reradiates in all possible directions, i.e. up and down in the one-dimensional atmosphere.

Let $x_{l,k}$ be the flux of thermal radiation emitted by the k-th layer of the l-th gas in either upward or downward direction (we consider these two fluxes equal). $x_{l,1}$ stands for radiation of the upper radiative layer emitted directly into space.

Energy balance equations for Earth as a whole, (8.4.2), each layer of each greenhouse gas, (8.4.3), and the Earth's surface, (8.4.4), can be written in the following form, see Figure 8.1:

$$q_e = \sum_{l=1}^{N} x_{l,1}, \tag{8.4.2}$$

$$2x_{l,1} = x_{l,2}, 2x_{l,2} = x_{l,1} + x_{l,3,\dots,}$$

$$2x_{l,k} = x_{l,k-1} + x_{l,k+1,\dots,} 2x_{l,n_l} = x_{l,n_l-1} + q\delta_l, \tag{8.4.3}$$

$$q = q_e + \sum_{l=1}^{N} x_{l,n_l}. \tag{8.4.4}$$

Here q and q_e are fluxes of radiation emitted from the Earth's surface and the upper radiating layers, respectively. Note that q_e is equal to the flux of solar radiation absorbed by the Earth's surface. Equation (8.4.4) can be deduced from all the other equations. Figure 8.1 gives an idea of the described situation for $N = 1$ and $\delta_l = 1$ (i.e. when only one gas is present).

The recurrent equations (8.4.3) reflect the fact that any layer absorbs radiation emitted from the two neighbouring layers only. Radiation that is emitted from more distant layers does not reach this particular layer being completely absorbed by the intermediate layers. Solving the system (8.4.3) we obtain that $x_{l,k} = kx_{l,1}$, $(k = 1, 2, \dots, n_l)$, which can be easily tested. Using this expression in the last equation of (8.4.3) and solving the latter together with (8.4.2), we arrive at the following expression for q:

$$q = \frac{q_e}{\dfrac{\delta_l}{n_1+1} + \dfrac{\delta_2}{n_2+1} + \dots + \dfrac{\delta_N}{n_N+1}} \equiv \frac{q_e}{b}, \quad \text{where} \quad b = \sum_{l=1}^{N} \frac{\delta_l}{n_l+1}$$

$$\text{and} \quad \sum_{l=1}^{N} \delta_l = 1 \tag{8.4.5}$$

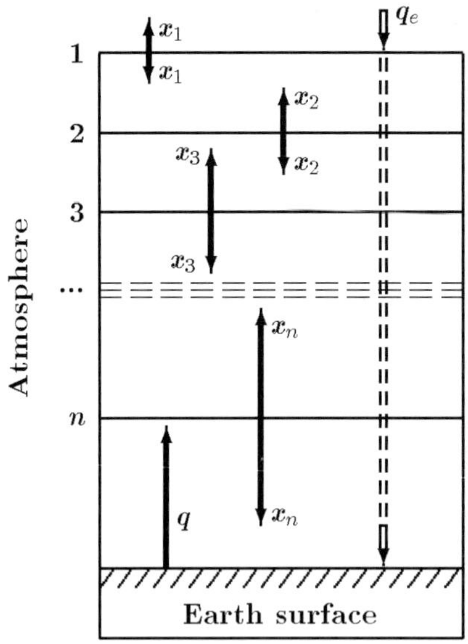

$$q_e = x_1 \quad (8.4.2)$$

$$\left.\begin{aligned} 2x_2 &= x_1 + x_3 \\ &\cdots\cdots \\ 2x_k &= x_{k-1} + x_{k+1} \\ &\cdots\cdots \\ 2x_n &= q + x_{n-1} \end{aligned}\right\} \quad (8.4.3)$$

$$q = x_n + q_e \quad (8.4.4)$$

Figure 8.1. Dependence of the greenhouse effect on the number of optically dense layers n. $x_k (k = 1, 2, 3, \ldots, n)$ is the intensity of thermal radiation emitted by the k-th layer upwards and downwards; q is the intensity of thermal emission by the Earth's surface; q_e is the intensity of solar radiation absorbed by the Earth's surface, which is equal to the intensity of radiation of the upper radiative layer, x_1. The atmosphere is considered to be completely transparent to solar radiation. It is assumed that the considered gas ensures complete absorption of thermal radiation over the whole terrestrial spectre, $\delta_1 = 1$.

Comparing (8.4.5) and the definition of the relative greenhouse effect B (8.3.2) we obtain for the relative and absolute values of the greenhouse effect B and f:

$$B = 1 - b = 1 - \sum_{l=1}^{N} \frac{\delta_l}{n_l + 1}; \qquad f = Bq \tag{8.4.6}$$

If the number of optically dense layers n_l is of the same order of magnitude for all gases, the flux of terrestrial radiation q and the absolute greenhouse effect f increase infinitely with growing n_l, while b tends monotonously to zero and B tends to unity. This is especially clear in the case of one greenhouse gas ($N = 1$) with n layers. Then $B = 1 - 1/(n+1)$ and $f = q_e n$. Expressions (8.4.5) and (8.4.6) make it clear that contributions of different greenhouse gases into the greenhouse effect are not additive, as opposed to the results of (8.3.5).

Let us now consider a situation when one of the N gases is absent, $n_k = 0$. Then there is a spectral window in the frequency interval $\Delta\omega_k$. Terrestrial radiation in this part of the spectre passes unimpeded through the atmosphere directly into space and is equal to $\delta_k q$. In such a case an increase in the number of optically dense layers n_l and, consequently, in concentrations of the remaining greenhouse gases cannot lead to an infinite increase of the terrestrial radiation q. Increase in q is saturated when all the $(N-1)$ items with non-zero n_l in the denominator of fraction (8.4.5) become infinitely small as compared to the finite k-th item corresponding to the spectral window (at $n_k = 0$ it is simply equal to δ_k). The saturated values of q, B and f are then as follows:

$$q = q_e/\delta_k, \qquad B = 1 - \delta_k, \qquad f = q_e(1-\delta_k)/\delta_k \tag{8.4.7}$$

The effect of saturation has a very clear explanation. Higher concentrations enable the greenhouse gases to re-emit more radiation, heating the Earth's surface and increasing the surface temperature and terrestrial radiation q. As a result, more and more radiation is released into space through the spectral window while less and less radiation is emitted from the upper radiating atmospheric layers ($x_{l,1} \to 0$), the total amount of the released radiation being limited by the absorbed solar energy q_e. Saturation corresponds to the case when practically all the radiation passes through the spectral window, $q_e = \delta_k q$.

The role of the spectral windows in the modern atmosphere can be estimated comparing the share of window radiation with the amount of absorbed solar energy q_e. Observations show that radiation escaping the Earth's surface through the spectral windows constitutes less than 10% of the total released radiation (Barry and Chorley, 1987), $\delta_k q \leq 0.1 q_e$. Thus, the situation is that in the modern atmosphere there is very far from saturation of the greenhouse effect owing to the spectral windows.

In the absence of clouds the limiting contribution to the greenhouse effect comes from the l-th gas that is characterised by the maximum value of $\delta_l/(n_l+1)$ in (8.4.5). If the concentration of that gas remains constant, changes in concentrations of the other gases do not have any considerable effect on q and f. When the concentration of this gas and, hence, n_l, increases, the term $\delta_l/(n_l+1)$ decreases, and values of q and f grow practically linearly with n_l until the diminishing value of $\delta_l/(n_l+1)$ becomes equal to that of some other gas, $\delta_m/(n_m+1)$. After that the concentration of the m-th gas becomes the limiting factor for q and f and so on.

Clouds absorb terrestrial radiation rather evenly over the whole thermal spectre contributing to all spectral intervals $\Delta\omega_l$. Thus, the number of optically dense layers of clouds n_0 should be added to the number of layers of every greenhouse gas giving the following expression for b:

$$b = \sum_{l=1}^{N} \frac{\delta_l}{n_l + n_0 + 1} \tag{8.4.8}$$

In a case when the major contribution to the greenhouse effect comes from clouds $(n_0 + 1 > n_l)$, b can be expanded into a Taylor series in terms of small quantities $n_l/(n_0+1)$. Then we arrive at the following expression for the absolute greenhouse effect $f \equiv q_e B/b$:

$$f = q_e \left\{ n_0 + \sum_{l=1}^{N} \delta_l n_l \right\}, \qquad n_l < n_0 + 1 \tag{8.4.9}$$

Note that contributions from different greenhouse gases are additive only in such limiting case.

For the modern atmosphere where the greenhouse gases (CO_2 and H_2O mostly) absorb thermal radiation practically over almost the whole terrestrial spectre, see (8.3.5), expression (8.4.9) can be written as follows:

$$f \equiv q_e B/b = q_e\{n_0 + \delta_{H_2O}\, n_{H_2O} + \delta_{CO_2}\, n_{CO_2}\} \qquad (8.4.10)$$

The available estimates of the contribution of clouds to the thermal radiation of the modern atmosphere (Raval and Ramanathan, 1989; Kondratyev, 1999) coincides by the order of magnitude with contributions from CO_2 and water vapour. Thus, for the modern atmosphere the expression (8.4.10) is true to the accuracy of the order of magnitude only. However, with increasing cloudiness its contribution into greenhouse heating becomes dominant and the accuracy of (8.4.10) should increase. The absolute greenhouse effect f grows then proportionally to the mass of clouds, while contributions from concentrations of CO_2 and water vapour become less and less important.

In the Sections that follow we discuss the influence of the obtained temperature dependence of the greenhouse effect on the stability of the Earth's climate.

Note that the existence of the observed vertical temperature gradient in the atmosphere means that the energy of excitation of greenhouse gases' molecules owing to absorption of thermal radiation is fairly evenly distributed over all energetic degrees of freedom of air molecules, including degrees of freedom of chaotic (thermal) movement. This is achieved via rapid energy exchange during molecular collisions. Were there no such even distribution (e.g. if the lifetime of the excited states of greenhouse molecules were much shorter than the time interval between two successive molecular collisions), no vertical temperature gradient could form in the atmosphere. In such a case the air temperature in the upper troposphere would coincide with the temperature of the Earth's surface, although the greenhouse effect could remain absolutely the same. That is, the upper troposphere would be as warm as the Earth's surface is now, i.e. with a temperature much higher than the effective planetary temperature T_e (Table 8.1). (To stress the difference: in the absence of the greenhouse effect the Earth's surface would be as *cold* as the upper troposphere is now.)

This suggests that the greenhouse effect (i.e. heating of the Earth's surface above the effective planetary temperature) is not necessarily coupled to the presence of the vertical temperature gradient. Hence, the dependence of the greenhouse effect on the concentrations of the greenhouse gases cannot be calculated on the basis of the observed temperature gradient alone.

With account made for distribution of the excitation energy of greenhouse molecules over all molecular energetic degrees of freedom, the observed constant vertical temperature gradient, and, consequently, the linear dependence of the air temperature on height, follows easily from the obtained solution of Eqs. (8.4.2–3), $x_{l,k} = kx_{l,1}$, if one recalls the small ratio of the difference between the surface and upper tropospheric temperature, $T - T_e$, to the absolute effective surface temperature T.

8.5 POSSIBLE CLIMATES ON EARTH AND THEIR STABILITY

Energy balance for a unit of the Earth's surface area consists in the fact the rate of energy content change per unit area is equal to the difference between the average flux of solar radiation absorbed by the Earth and the average flux of thermal radiation emitted by the Earth to space.

Energy content is equal to cT, where c is the average heat capacity per unit area of the Earth's surface and T is its absolute temperature. Due to rotation of the Earth, the solar flux I incident upon the Earth's cross section area πr_E^2, where r_E is the Earth's radius, is distributed over the whole planet's surface area, $4\pi r_E^2$. As a result, the average flux of solar radiation per unit of the Earth's surface area is equal to $I/4$. The absorbed solar radiation is equal to $aI/4$, where $a \equiv 1 - A$, A is the planetary albedo, that is, the fraction of solar radiation reflected by the planet back to space. The net flux of heat from the Earth to space is equal to the flux of heat from the Earth's surface σT^4 (Stephan–Boltzmann law) multiplied by the coefficient $b \equiv 1 - B$, where B is the relative greenhouse effect and describes the part of terrestrial radiation effectively re-emitted by the atmosphere back to the Earth's surface.

The energy balance equation for a unit area of the Earth's surface can be written as follows:

$$c\frac{dT}{dt} = \frac{I}{4}a - \sigma T^4 b \equiv -\frac{dU}{dT} \tag{8.5.1}$$

We introduced in (8.5.1) the potential Lyapunov function U characterising stability of the energy balance equation. The only variable in (8.5.1) is the temperature T. Coefficients a and b are also temperature-dependent. The potential function U is also temperature-dependent. It is chosen so that the negative value of its first temperature derivative is equal to the rate of energy content increment. Equation (8.5.1) is based on the law of energy conservation. Its accuracy is determined by the accuracy of characterisation of the whole Earth's surface by a global mean temperature. Minor deviations of the thermal radiation of Earth from blackbody radiation in some parts of thermal spectre are taken into account in coefficient b. Processes of convection, water evaporation and condensation are ordered processes that are generated by highly-ordered solar energy (see Chapter 7). These processes, as well as absorption of solar radiation by the lower atmosphere, represent intermediate stages of dissipation of the solar energy into the thermal radiation of the Earth's surface.

In a stationary state, when the energy content does not change, $c\dfrac{dT}{dt} = 0$, the derivative of U turns to zero, and, consequently, U has an extreme—maximum or minimum. The right-hand part of the equality in (8.5.1) also turns to zero, and this equality determines a stationary temperature $T = T_S$:

$$\frac{I}{4}a - \sigma T_S^4 b = 0, \qquad \text{or} \qquad T_S = T_o\left(\frac{a}{b}\right)^{1/4},$$

$$\text{where} \qquad T_o \equiv \left(\frac{I}{4\sigma}\right)^{1/4} = 278\,\text{K} \tag{8.5.2}$$

(See notes to Table 8.1 for numerical values of σ and I.) If $a = b = 1$, which means that both the planetary albedo A and greenhouse effect B are equal to zero, the stationary temperature of the Earth's surface would be equal to $T_o = 278\,\mathrm{K}$ (5°C). This temperature, which is totally determined by I, that is, by the planet's location in the solar system, can be called the planet's orbital temperature (Table 8.1).

Note that here and below we imply that the notion of stationary state describes a state where the *average* energy content cT does not change with time. Here all oscillatory processes after averaging over time periods longer than periods of oscillation are included, as well as chaotic fluctuations that do not change the average energy content.

When the first derivative of U is equal to zero, the sign of its second derivative $U'' \equiv \dfrac{d^2U}{dT^2}$ determines the character of the extreme. It is a minimum, when U'' is positive, and maximum, when U'' is negative. The sign of U'' allows one to judge about stability of stationary solutions of (8.5.1). This can be shown as follows.

In the neighbourhood of the stationary point $T = T_S$ the first derivative $U' \equiv \dfrac{dU}{dT}$ can be expressed as a Taylor power series in terms of a small deviation $x \equiv T - T_S$:

$$U'(T) = U'|_{T=T_S} + U''|_{T=T_S}(T - T_S) = U''|_{T=T_S}(T - T_S) = U''|_{T=T_S}x$$

Equation (8.5.1) can be then written as follows:

$$\frac{dx}{dt} = -kx, \qquad \text{where} \qquad k \equiv \frac{1}{c}U''|_{T=T_S}$$

Heat capacity c is positive, so that the sign of coefficient k coincides with that of U''. Solution of the above equation looks like $x = x_0 e^{-kt}$, where x_0 is an arbitrary constant, t stands for time.[1] Thus, when $k > 0$ (U has a minimum), any initial deviation of temperature T from the stationary value T_S exponentially damps out, which means that the stationary state is stable. In such case k can be interpreted as coefficient of relaxation, while the reciprocal value k^{-1} characterises the time of recovery of the stationary state after a perturbation. On the contrary, when $k < 0$ and U has a maximum at $T = T_S$, any deviation exponentially grows with time. In such case the stationary state $T = T_S$ is unstable.

We have seen therefore that stability of stationary states of (8.5.1) can be readily illustrated by the character of function U in very much the same manner as the gravitational potential of Earth can be described by relief of the surface. Stable stationary states correspond to minima (pits) of function U. The degree of stability depends on a pit's depth. Unstable states correspond to maxima (hills) of function U. Any deviation from the stationary state leads to sliding down the hill to one of the two nearest pits located to the left and to the right of the hill.

[1] The value $\lambda \equiv U''|_{T=T_S} = kc$ has the dimension of the flux of entropy, $\mathrm{W\,m^{-2}\,K^{-1}}$, and is usually called climate sensitivity (Kondratyev, 1999).

Values and signs of U'' and relaxation coefficient k are unambiguously determined by temperature dependence of functions a and b in (8.5.1) and, consequently, temperature dependencies of albedo A and greenhouse effect B. Expanding functions a and b in Taylor power series with respect to small deviations $x \equiv T - T_S$ and using relation (8.5.2) we obtain the following expression for U'' in the stationary point T_S using temperature derivatives of a and b, a' and b':

$$U''|_{T=T_S} = \frac{I}{4}\frac{a}{T_S}(4+\beta-\alpha), \qquad \text{where} \qquad \alpha \equiv \left(\frac{a'T}{a}\right)_{T=T_S},$$

$$\beta \equiv \left(\frac{b'T}{b}\right)_{T=T_S}. \tag{8.5.3}$$

From (3) it is evident that a stationary state is stable when $\alpha - \beta < 4$ and unstable when $\alpha - \beta > 4$.

It is convenient to seek solution of (8.5.2) in a graphical form drawing the curve $y_1(T) = T_o\left(\frac{a(T)}{b(T)}\right)^{1/4}$ and looking for intersections of this curve with the line $y_2(T) = T$.

8.6 PHYSICAL STABILITY OF THE EARTH'S CLIMATE

Let us now consider the particular physical behaviour of the greenhouse effect and albedo and functions $a(T)$ and $b(T)$ when the temperature T changes from the state of total glaciation **1** to the state of complete evaporation of the hydrosphere **3**, Table 8.1.

The stationary steady state of an ice-covered Earth lies in the interval of temperatures lower than −80°C. In this state all the major components of the environment, including atmospheric CO_2, are present in the solid phase. As soon as the solid phases of most components persist over a broad interval of low temperatures, one can assume that in the vicinity of the stationary state **1** neither the greenhouse effect nor albedo depend on temperature. Albedo of the ice-covered Earth should be equal to that of snow cover, i.e. to 80% (North *et al.*, 1981; Mitchell, 1989) making the value of $a(T)$ equal to 0.2. The relative greenhouse effect on Mars where the mean surface temperature is higher than −80°C and gaseous carbon dioxide is relatively abundant, is about 7% (Pollack, 1979; Kasting *et al.*, 1988; Mitchell, 1989). Thus, for the ice-covered Earth the relative greenhouse effect $B(T)$ does not presumably exceed 1–5%, and the value of $b(T)$ is not more than 0.95. We use below $b(T) = 0.95$ for the state of total glaciation.

The stationary steady state of complete evaporation of the hydrosphere corresponds to global mean surface temperature higher than 400°C (Table 8.1) and atmospheric pressure exceeding the present one by a factor of several hundred. As soon as all the hydrosphere is evaporated, the atmospheric concentration of water does not further change with temperature. Thus, it is reasonable to assume that in state **3** neither albedo nor greenhouse effect change considerably with temperature, similarly to the situation in the stationary state **1**. On Venus the surface temperature is about 460°C, while the planetary albedo (which is completely due to the cloudiness) is equal to that of the ice-covered Earth (Table 8.1). Thus we may take the value of albedo in state **3** equal to 80% and value of $a(T) = 0.2$, as in state **1**. The relative greenhouse effect on Venus can be calculated from the difference between the temperature T on the planet's surface and the effective temperature T_e of thermal radiation of the planet measured from the outer space. The effective thermal radiation of Venus is equal to $q_e = \sigma T_e^4 = Ia/4 = 163\,\mathrm{W\,m^{-2}}$, $T = 232\,\mathrm{K}$. Thermal radiation of the planet's surface is calculated as $q = \sigma T^4 = (Ia/b)/4 = 16\,000\,\mathrm{W\,m^{-2}}$, $T = 730\,\mathrm{K}$ (Table 8.1). From these values and (8.3.2) we obtain

$$b \equiv 1 - B = (T_e/T_s)^4 = 163\,\mathrm{W\,m^{-2}}/16\,000\,\mathrm{W\,m^{-2}} = 0.010$$

It is reasonable to accept the obtained value of $b(T) = 0.010$ for the stationary state of the totally evaporated hydrosphere on Earth. The atmosphere of Venus consists of carbon dioxide to the extent of 96% and has a pressure of about 93 bars[2] (Kasting *et al.*, 1988). On Earth the cumulative mass of the oceans exceeds mass of the modern atmosphere by 300 times (Allen, 1955). In the state of total evaporation of the hydrosphere the atmospheric pressure would be about 300 bars. In such a case, the terrestrial H_2O will find itself, similarly to CO_2 on Venus, above the critical point where the differences between gases and liquids vanishes (Landau *et al.*, 1965). Thus it is natural to assume, with allowance made for the cloudiness, see Section 8.4, that in the state of total evaporation of the hydrosphere the relative greenhouse effect B on Earth would be at least not less than it is on Venus, making the value of $b \equiv 1 - B$ not greater than 0.010.

The chosen values of $a(T)$ and $b(T)$ in states **1** and **3** completely determine the mean global temperature of the Earth's surface in these two states, see (8.5.2) and Table 8.1.

Let us now consider possible physical mechanisms of transition from the stable state of the ice-covered Earth **1** to the stable state of total evaporation of the hydrosphere **3**.

The major part of the modern greenhouse effect on Earth is due to the atmospheric water vapour. Atmospheric concentration of water vapour varies greatly in space and time. However, the average concentration of the water vapour changes proportionally to its saturated concentration (Ramanathan *et al.*, 1987; Raval and Ramanathan, 1989). Saturated concentration of water vapour, as well as the saturated partial pressure, p_{H_2O}, grows exponentially with increasing temperature in accordance with the Clausius–Clapeyron equation (Landau *et al.*, 1965):

$$p_{H_2O}(T) = Ce^{-\frac{T_{H_2O}}{T}} \equiv e^{-\frac{T_{H_2O}}{T}+\varepsilon}, \qquad T_{H_2O} \equiv \frac{Q_{H_2O}}{R} = 4871\,\mathrm{K} \tag{8.6.2}$$

Here Q_{H_2O} is the latent heat of evaporation of 1 mole of water vapour, R is the gas

[2] The global mean atmospheric pressure on Earth is equal to 1.013 bar, 1 bar $= 10\,\mathrm{N\,m^{-2}}$ (Allen, 1955).

constant, T_{H_2O} is the effective temperature characterising the energetic of evaporation process, $C \equiv e^{\varepsilon}$ is a temperature-independent constant.

As is shown in Section 8.4, the greenhouse effect is predominantly determined by the greenhouse gas that leads to formation of clouds and creates an absorption interval comparable with the total width of the thermal spectre. The absolute greenhouse effect grows linearly with concentration of clouds, (8.4.9). On Earth the most important greenhouse component responsible for cloud formation is the water vapour. Atmospheric concentration of water vapour changes with temperature according to (8.6.1).

In a stationary state we may write for the thermal radiation of the Earth's surface $q(T)$:

$$q(T) = \frac{I}{4}a + Bq(T) \tag{8.6.2}$$

Here I is the solar constant, $(I/4)a$ defines the amount of the absorbed solar energy ($a \equiv 1 - A$, where A is the planetary albedo), and is equal to the amount of thermal energy released by the planet into space, $(I/4)a = q_e$. Term $Bq(T) \equiv (1 - b)q(T) = f$ describes the additional radiation of the Earth's surface due to the greenhouse effect.

Using (8.6.1) and the results of Section 8.4, according to which the absolute greenhouse effect should grow proportionally to the concentration of atmospheric water, we may write

$$f \equiv Bq(T) = \frac{I}{4}ae^{-\frac{T_{H_2O}}{T}+\varepsilon} \tag{8.6.3}$$

Using (8.6.2) and (8.6.3) we obtain the following expression for $b(T) \equiv 1 - B$:

$$b(T) = \frac{1}{1 + \exp\left(-\frac{T_{H_2O}}{T} + \varepsilon\right)} \tag{8.6.4}$$

Expression (8.6.4) does not take into account the fact that the Earth's hydrosphere has a finite mass. At sufficiently large values of ε the value of b (8.6.4) diminishes almost infinitely. In reality, however, the decrease of b is stopped when all the hydrosphere is evaporated and there is no further increase in water vapour concentration and change of function $b(T)$ with increasing temperature. Thus, it is reasonable to specify the limiting value of b as $b_{min} = 0.010$, making it equal to the value of b observed on Venus, as discussed above. This can be done by adding the term 0.010 to the whole fraction (8.6.4). Parameter ε can be determined from the condition that at the modern mean global temperature $T = 288$ K the value of $b(T)$ given by (8.6.4) is equal to the observed value, $b(288) = 0.60$. The obtained value of ε is equal to 16.5. Noting that $T_{H_2O} = 4871$ K we finally obtain for $b(T)$:

$$b(T) = \frac{1}{1 + \exp\left(-\frac{4871}{T} + 16.5\right)} + 0.01 \tag{8.6.5}$$

The character of dependence of the planetary albedo on temperature remains to a large extent unknown. Its basic features, however, can be taken into account if the unknown function $a(T)$ is approximated by a Gaussian curve. The limiting value of $a_{\min} = 0.2$ at high and low temperatures corresponds to the states of the total glaciation and total evaporation of the hydrosphere, where the planetary albedo $A \equiv 1 - a$ is at its maximum due to the high reflectivity of ice cover and cloudss",4>, respectively. The observed modern value of a is 0.7 ($A = 0.3$). We assume that the maximum value $a_{\max} = 0.8$. It corresponds to the state of already totally melted ice-cover but still predominantly liquid hydrosphere, which is chosen at a global mean temperature $T_m = 298$ K (25°C). We thus obtain for $a(T)$:

$$a(T) = 0.20 + 0.54 \exp\left[-\frac{(T - T_m)^2}{(\Delta T)^2}\right]; \qquad T_m = 298\,\text{K}\ (25^\circ\text{C});$$
$$\Delta T = 40\,\text{K}\ (40^\circ\text{C}) \tag{8.6.6}$$

Proportionality coefficient corresponding to the exponential term and the effective width of the curve ΔT are retrieved from the condition that at the modern mean global temperature $T = 288$ K the value of $a(T)$ given by the formula (8.6.6) is equal to the observed value, $a(288) = 0.70$, and the condition that Earth is totally covered with ice at -15°C. Variations of all the chosen parameters within physically reasonable limits do not change the results to be obtained.

Note that following temperature changes accompanying transition from the state of complete glaciation of the Earth's surface to the state of complete evaporation of the hydrosphere, the absolute greenhouse effect changes a hundredfold as compared to the two- to threefold change in the planetary albedo at maximum. Hence, the temperature dependence of the albedo has practically no impact on the solutions of (8.5.2) and their stability.

Figure 8.2-Ia shows the graphical solution of equation (8.5.2) with $a(T)$ and $b(T)$ specified by (8.6.6) and (8.6.5). Potential function U obtained by integrating (8.5.1) gives information on stability of the obtained three stationary states (Figure 8.2-Ib, c) (the integration constant was chosen such that the potential function U turns to zero in the stationary state of total evaporation of the hydrosphere). Stationary states of total glaciation and total evaporation of the hydrosphere are stable, as could be expected, while the stationary state corresponding to the modern climate proves to be unstable.

The obtained results suggest that the physical mechanisms alone cannot account for the observed substantial stability of the Earth's climate with regard to the global mean temperature during the four billion years of life existence. In the following section we address the problem of the biotic nature of climate stability on Earth.

8.7 BIOTIC STABILITY OF THE MODERN CLIMATE OF EARTH

According to the available paleodata global mean temperature of the Earth's surface never went beyond the interval from 5°C to 25°C (Savin, 1977; Watts, 1982;

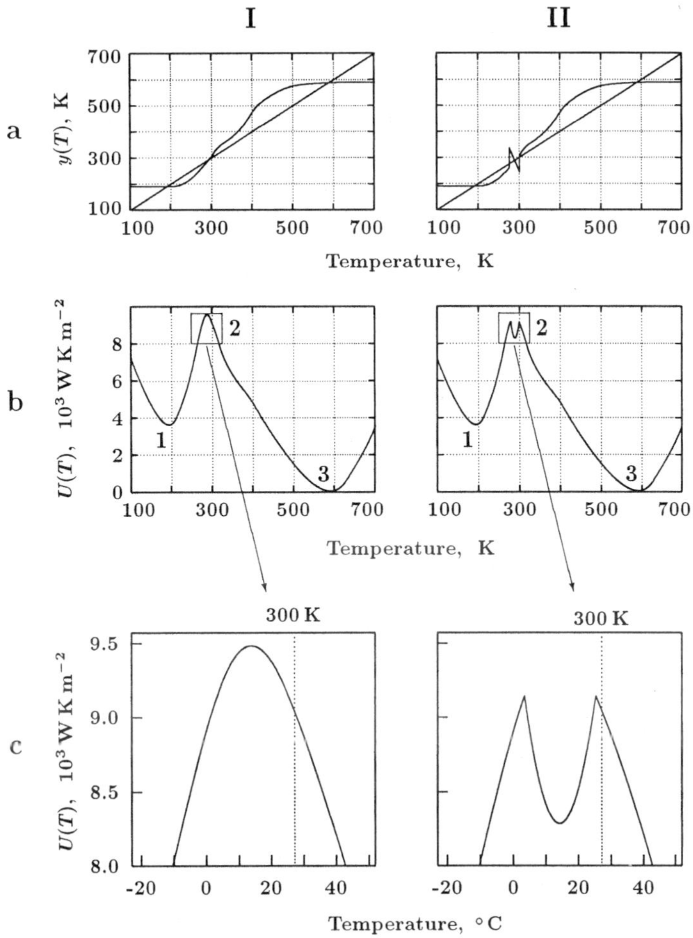

Figure 8.2. Physical (I) and biotic (II) stability of the global mean surface temperature on Earth. **a** – graphical solution for determination of stationary states **1**, **2**, **3** (8.5.2). Intercepts of the two lines are the stationary states corresponding to the extreme of potential function U (8.5.1). **b, c** – Potential (Lyapunov) function describing stability of the Earth's climate. The minima correspond to stable states. The maxima correspond to unstable states.

Berggren and Van Couvering, 1984; see also Figure 8.3), fluctuating around a stable mean value of about 15°C. It means that there exists a sufficiently strong negative feedback between changes in temperature and albedo and greenhouse effect, e.g. when an increase in temperature causes a reduced greenhouse effect that lowers the temperature back to the initial value.

The maximum possible stability of the stationary state **2** can be retrieved from the condition of the maximum physically possible negative feedback in the temperature interval 5°C $< T <$ 25°C. The maximum physically possible feedback can be illustrated by a hypothetical situation when the relative greenhouse effect B increases with decreasing temperature up to the maximum possible value, which is

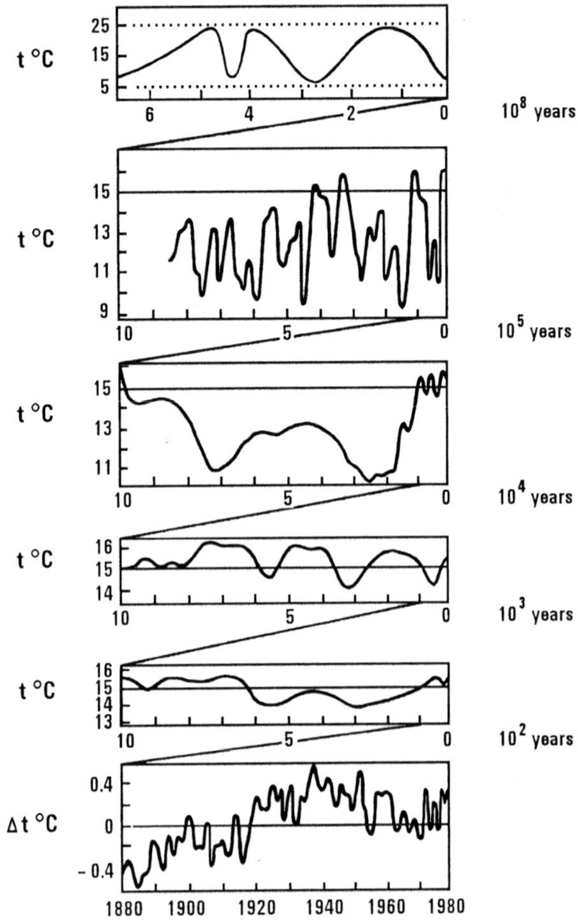

Figure 8.3. Time variations of the mean global surface temperature on Earth. (After Savin, 1977; Watts, 1982; Berggren and Van Couvering, 1984.)

realised at the lower boundary of the considered temperature interval, while the albedo A decreases with decreasing temperature and attains there the minimum possible value, $B(5°\text{C}) = 1$, $A(5°\text{C}) = 0$. The same situation is realised when the limiting values of albedo and greenhouse effect correspond to the upper boundary of the considered temperature interval, $B(25°\text{C}) = 0$, $A(25°\text{C}) = 1$.

Figures 8.2-IIa, b, c show the graphical solution of equation (8.5.2) and the potential function U (8.5.1) for a model example of weaker negative feedback:

$$A(T) = 0.3 + (t - t_S)\,0.02; \qquad B(T) = 0.4 - (t - t_S)\,0.02;$$
$$5°\text{C} < t < 25°\text{C}; \qquad t_S = 15°\text{C}$$

In this example the boundary values of greenhouse effect and albedo are not the maximum (minimum) possible ones: $A(5°C) = 0.1$, $B(5°C) = 0.6$, $A(25°C) = 0.5$, $B(25°C) = 0.2$. We assumed here that the global mean surface temperature in the stable stationary state chosen by life coincides with temperature of unstable physical stationary state. In this case the potential biotic pit (Figure 8.2-IIb, c) is symmetrical.

The real behaviour of albedo and greenhouse effect in the temperature interval $5°C < T < 25°C$ can be presumably deduced only empirically on the basis of detailed studies of paleodata, which is not our goal here. The natural biota of Earth uses highly ordered solar energy to ensure negative feedback of climatic perturbations in accordance with the genetic information of biological species. This is performed on the basis of non-equilibrium processes that cannot be accounted for in models based on physicochemical properties of the atmosphere alone. If the biosphere is introduced into the model as a physicochemical system, this critical property of the biota appears to be lost irrespective of whatever large number of empirical parameters that we are able to process in the model calculations. In other words, if we compare the biota to a programmed machine which is able to use the external flux of ordered energy in different regimes, we cannot predict this machine's functioning under different conditions without knowing the machine's program and judging by the characteristics of its current regime. As shown in Chapter 7, functioning of the biota is characterised by a degree of complexity that cannot in principle be modelled, while the biotic climatic impact (manifested in biotic power of control of the water cycle, carbon cycle, etc.) is huge. This calls for caution when interpreting predictions of mathematical climate models.

We have shown that the observed stability of the Earth's climate with regard to the mean global surface temperature points to the complex nature of the temperature dependencies of albedo and greenhouse effect within the life-compatible temperature interval. We have shown that, on the basis of the known physical atmospheric properties alone, there are no grounds to expect such anomalies (Figure 8.2-IIa) in behaviour of either albedo or greenhouse effect. This, in its turn, points to existence of a biotic mechanism of climate control within the interval of life-compatible temperatures. Such control can be based only on highly-ordered processes that are generated by solar radiation due to large differences between temperatures of solar radiation (short-wave radiation of the Sun) and terrestrial thermal radiation (long-wave radiation of the Earth).

The main quantitative characteristic of orderliness of processes generated in the course of any type of energy transformation is the difference between temperatures of the initial and final states, see Section 7.2. The final state of all processes taking place in the environment corresponds to chaotic thermal energy of the Earth's surface, which is characterised by global mean surface temperature $T_E \sim 300\,K$. Solar energy is characterised by temperature $T_S \sim 6000\,K$. The relative difference between these temperatures constitutes

$$\eta_S = \frac{T_S - T_E}{T_S} \sim 0.95 \qquad (8.7.1)$$

In thermodynamics the variable η_S has the meaning of the efficiency of energy transformation (Landau *et al.*, 1965). This is the maximum possible work efficiency of a reversible heat machine based on a succession of equilibrium thermodynamic processes, see also Section 7.2. The same variable characterises the efficiency of highly-ordered non-equilibrium processes generated by solar energy in the Earth's environment.

All processes of global circulation in the ocean and atmosphere are due to the temperature difference between the equator and the poles, which is of the order of $\Delta T_E \sim 30°C$. The efficiency of ordered processes generated by global circulation are characterised by

$$\eta_E = \frac{\Delta T_E}{T_E} \sim 0.1, \tag{8.7.2}$$

which is an order of magnitude lower than the value of η_S (8.7.1). It means that a unit energy of global circulation generates in the Earth's environment 10 times less energy of ordered processes than the same unit of solar energy. The amounts of solar energy consumed by the global biota and by the global circulation processes are of the same order of magnitude (see Section 7.3, Table 7.1).

Hence, stability of the modern climate can be explained taking into account highly-ordered biotic processes generated by solar energy. Ordered processes that are due to global circulation make an order of magnitude lower contribution and may be neglected. There are no grounds to expect that detailed elaboration of global circulation models yields reliable predictions of climate change.

Until now, studies devoted to the climate stability problem have been based on quantitative characteristics of energy fluxes, while the degree of orderliness of corresponding processes has never been taken into account (North *et al.*, 1981; Ramanathan *et al.*, 1987; Tang and Weaver, 1995; Li *et al.*, 1997). This is predominantly due to the fact that most processes generated at the Earth's surface by solar energy go through stages that are very far from thermodynamic equilibrium and cannot be formally described within traditional thermodynamics. However, the parameter η, (8.7.1) and (8.7.2), is valid for all processes, including non-equilibrium ones, and can be used when estimating their orderliness. Note also that all models concerned about climate stability are based on empirical characteristics of the modern climate that enter the models as parameters. It is not surprising that in all such models the modern climate proves to be stable. The question *why* the observed characteristics of climate ensure its stability is rarely addressed.

Highly-ordered processes generated by solar energy with efficiency close to (8.7.1) are of biological nature and take place only in the biota. That means that the climate stability on Earth can be due only to the regulatory function of the natural biota. Ordered processes aimed at maintenance of climate stability should be necessarily characterised by strictly defined succession, direction and duration, i.e. fluxes and stores of information. Such information is contained in the genetic programme of biological species combined into ecological communities of the global biota. Natural ecological communities are characterised by strictly specified population densities of all species that undergo directional changes in response to environmental perturbations. Anthropogenic perturbation of natural ecological communities is manifested in artificial change of the natural population density distribution of species and violation of genetic programmes of natural species due to creating new sorts of plants and breeds of animals. This leads to degradation of the stabilising environmental potential of the biota.

During large-scale natural perturbations of the environment of the past (e.g. glaciations, volcanic eruptions, meteorite fall, etc.) the regulatory potential of the biota decreased in proportion to reduction of areas occupied by the natural biota, but remained substantial. For example, were the area occupied by the global biota to reduce by half, the global regulatory potential would halve as well. That means that the rate of biotic compensation of external perturbations would be reduced twofold. Estimates of relaxation coefficients of the nonperturbed biota presented in Section 6.4 show that the characteristic time of complete recovery of the initial stable state of the environment after natural external perturbations is of the order of 10 years for the nonperturbed biota. Were the global biota reduced by half, that time would only double, which may not have any drastic consequences to the environmental stability.

In contrast, anthropogenic cultivation of territories occupied by natural biota has a catastrophic impact on environmental stability. The perturbed biota loses information necessary for correct regulation of the environment, but retains the absolute huge value of environmental impact. The destabilising impact of the perturbed biota may be larger than stabilising impact of the remaining intact territories, see Figure 6.6a. Hence, we again come to the conclusion that in order to ensure long-term stability of climate on Earth natural biota should be restored and protected on the majority of the continental surface. As one of the immediate measures, one can name considerable reduction and ultimate abandonment of forest use, see Section 6.5.

9

Genetic Bases of Biotic Regulation and Life Stability: Theoretical Consideration

Biotic regulation of the environment is performed on the basis of complex processes pertaining to huge fluxes and stores of information (Chapter 7). The information needed to ensure a negative feedback response to environmental perturbations may only be stored in DNA molecules of species genomes. Mechanisms preventing the genetic information of species from spontaneous decay are considered in this chapter.

9.1 ORGANISATION OF GENETIC INFORMATION OF SPECIES

Genetic information of most species of the biosphere is written in polymer double-strand molecules of DNA, which consist of the two mutually-complementary spirals. There are four different monomer units composing the double-helix DNA molecule. They are known as nucleotide pairs or base pairs (bp). In other words, the genetic text of the species-specific properties is written with the help of the four-letter alphabet. The complete genetic text of all the properties of the individual is called its genome. Genomes of higher organisms are very large and consist of separate chromosomes (separate books, to continue the analogy).

Replication of DNA is performed by unwinding the double helix and building a complementary daughter strand of nucleotide pairs at each of the single parent strand, so that two copies of the parent DNA appear. Random errors that arise in the course of such copying and other cell activities are known as mutations. They include random change, insertion or deletion of nucleotide pairs (these mutations are referred to as *point mutations*) and deletion, insertion or transposition of whole fragments of DNA (these mutations are known as chromosome mutations or *macromutations*) (Ayala and Kiger, 1984; Lewin, 1987).

The probability of random mutation depends on the quantum characteristics of the polymer molecule and does not correlate with information written into that molecule (the number of misprints in the book does not depend on the book's content). That probability is reduced by approximately a millionfold by a

DNA-synthesis catalytic system such as proofreading and mismatch repair (Loeb and Kunkel, 1982; Goodman and Fygenson, 1998), which is itself encoded into the DNA macromolecule. The proofreading and other repairing catalysts operate in one and the same way upon any section of the DNA molecule, independent of information written into them. One may reasonably assume that the probability of random mutation, ν, per nucleotide site per division is a universal value typical for the whole genome.[1]

Synthesis of proteins begins with transcription of the genetic information of the DNA to single-chain polymer molecules of the RNA, which is followed by translation of the RNA information to proteins, which are polymer molecules built of 20 aminoacids (protein letters). Genomes of certain viruses are built directly of RNA molecules. Replication of RNA molecules is not controlled by the proofreading and mismatch repair systems. Thus the probability of random mutations in RNA viruses is on average a million times higher than that in organisms having their genetic information written into the DNA molecules (Holland *et al.*, 1982; Domingo *et al.*, 1996).

Rates of all processes, including those of mutagenesis and cell division, are proportional to the metabolic rate of the organism. If we assume that the system of catalytic reparation is also universal for any DNA genome (Radman and Wagner, 1988), one may expect that the probability of random mutation, ν, per base pair (bp) per division (d) is also a universal value for most DNA-based organisms. Such an assumption agrees satisfactorily with the experimental data available (Drake, 1969, 1974; Radman and Wagner, 1988; Orr, 1995), according to which ν is of the order of

$$\nu \sim 10^{-10}\,(\mathrm{bp})^{-1}\,\mathrm{d}^{-1} \tag{9.1.1}$$

Deviations from (9.1.1) in either direction, observed in various species, do not usually exceed an order of magnitude (Drake, 1974).

Genes consist on average of about 10^3 nucleotide pairs. Thus the probability ν_g of a random mutation per gene per division is of the order of

$$\nu_g \sim 10^{-7}\,(\mathrm{gene})^{-1}\,\mathrm{d}^{-1} \tag{9.1.2}$$

Probabilities of mutual mutational substitutions between the four nucleotide pairs differ slightly from pair to pair (Kimura, 1983; Jukes, 1987). The cumulative probability of various types of macromutations pertaining to a given site does not exceed the average probability of point mutation, (9.1.1), by more than a factor of five (Ayala and Kiger, 1984). Thus, recalling that the value of (9.1.1) is found to the accuracy of an order of magnitude only, we may further use that value as a representative estimate of the probability of a single random mutation of any type in a given site.

[1] The probability of random mutation in certain small parts of the genome may increase in cases where special 'mutator' genes are present, that is, coding enzymes that identify these particular parts of the genome and increase the probability of mutations in them, for example blocking the system of proofreading (Ayala and Kiger, 1984). Decay of mutator genes results in the failure of synthesis of the respective enzymes and in the evening out of the rate of mutations in the genome as a whole.

A most important characteristic of genome stability is the average number of mutations per birth of an individual, or, which is essentially the same, per genome per generation. Below this characteristic is called the genome decay rate, μ. Let the size of the genome (that is the total number of nucleotide pairs in the genome) be G. The number of cell divisions in the germ line is denoted k. Then the dimensionless value of the rate of genome decay μ is:

$$\mu = G\nu k \tag{9.1.3}$$

Size of the protein-coding region G_g of the genome varies from 100% of the G value in bacteria to about 10% in mammals and about 1% in certain fishes and salamanders (see below Table 10.1 in Section 10.3). The decay rate of the coding region μ_g is equal to

$$\mu_g = G_g \nu k, \tag{9.1.4}$$

where G_g is measured in nucleotide pairs. If G_g is measured in genes instead of nucleotide pairs, formula (9.1.4) assumes the form $\mu_g = G_g \nu_g k$.

For bacteria we have $G = G_g \sim 10^6$ bp, $k = 1$ division per generation, $\mu = \mu_g = 10^{-4}$ (Drake, 1991, 1993). For mammals we have $G \sim 3 \cdot 10^9$ bp, $G_g \sim 0.1G \sim 3 \cdot 10^8$ bp, $k \sim (20\text{–}40)$ divisions per generation,[2] $\mu \sim 10$ and $\mu_g \sim 1$.

When $\mu > 1$, one should account for the probability of the appearance of multiple mutations. If multiple mutations remain completely uncorrelated, the probability $p(n)$ of the appearance of n mutations is known to be described by the Poisson distribution:

$$p(n) = \frac{\mu^n}{n!} e^{-\mu} \tag{9.1.5}$$

The Poisson distribution characterises a multiplicity of mutually independent events. We have for the average number of the observed events $\bar{n}$, for variance $\overline{(n - \bar{n})^2}$, and standard deviation $\sqrt{\overline{(n - \bar{n})^2}}$:

$$\bar{n} = \overline{(n - \bar{n})^2} = \mu, \qquad \frac{\sqrt{\overline{(n - \bar{n})^2}}}{\bar{n}} = \frac{1}{\sqrt{\bar{n}}} \tag{9.1.6}$$

Poisson distribution is characterised by the equality between the variance and the mean.

Relations (9.1.6) may serve as criteria for testing the independence of multiple processes and whether we stay within a Poisson distribution or not. When several individuals are born, the average number of mutations is the same for each of them, the difference lying in random localisation of those mutations in the genome.

An essential property of mutations is their irreversibility, that is inherent to all decay processes (see Section 7.1). It means that mutations result in an increase in the overall number of possible states of the genome. Mutations increase the genetic

[2] Calculated as the number of doublings of the zygote needed to obtain from $10^6 \approx 2^{20}$ to $10^{12} \approx 2^{40}$ germ cells (Raven and Johnson, 1988; Vogel and Rathenberg, 1975; Vogel and Motulsky, 1979, 1982, 1986).

chaos and decrease orderliness. Probabilities of the direct and reverse mutations in any fixed nucleotide site are approximately the same. However, a direct mutation may occur in any site of the genome, whereas a reverse mutation has to occur in a particular site (namely where the direct mutation did). If the size of the genome is $G \sim 10^{10}$ bp, and $\mu \sim 1$, one mutation appears on average in any individual born. Meanwhile a particular reverse mutation appears only in one individual in every 10^{10} individuals. Thus one may always neglect inverse mutations as compared with direct ones.

Generally speaking, besides the decay (deleterious) mutations lowering the competitive capacity of individuals, neutral mutations (those not affecting the competitive capacity of an individual) and progressive ones (those increasing that capacity) are also possible. By definition, neutral mutations pertain to those parts of the genome where no meaningful information useful for the species is stored. Progressive mutations are able to spread quickly over the whole population. The relative number of progressive as well as neutral mutations is much less than the total number of decay mutations (see Section 11.3). Thus the value of μ (9.1.1) specifically controls the rate of genome decay. When μ is small ($\mu \ll 1$), it is equal to the ratio between the number of offspring carrying hereditary mistakes to the number of offspring free of such mistakes in the genome, as compared with the parent genome.

9.2 POPULATION IN THE ABSENCE OF STABILISING SELECTION

In this section we estimate the characteristic time of genome decay in the absence of stabilising selection.

Consider a clonal population which at the initial time moment $t = 0$ consists of genetically identical individuals. The clone genome may be considered a reference of genetic information, so that we calculate all the mutations taking place in that population with respect to that reference. Let us also assume that such a reference genome is a normal one, i.e. it contains all the information necessary for the survival and correct actions of individuals aimed at stabilisation of the environment within the ecological community.

The notion of the normal genome calls for a constructive definition, which would make it possible to differentiate, using some empirical procedure, the normal genome from decay ones. That definition, as well as the problem of degeneration of the normal genome, is discussed in Section 9.5 below. To simplify our treatment, we consider the normal genome to be a unique one here, and all the deviations from it to be of a decay (deleterious) nature.

Due to spontaneous decay mutations, new decay individuals featuring deviations from the normal genome appear in the population with every new generation. Every mutation results in a nucleotide substitution. We denote the number of decay substitutions in the genome as n. The number of individuals with a genome carrying n substitutions is denoted as N_n. Then, assuming that mutations are random and follow the Poisson distribution (9.1.5), we obtain the following equations to describe the change of the number $N_n(t)$ with time t:

$$\begin{aligned}
\dot{N}_0 &= (B_0 - d_0)\, N_0, \\
\dot{N}_1 &= (B_1 - d_1)\, N_1 + \mu B_0 N_0, \\
\dot{N}_2 &= (B_2 - d_2)\, N_2 + \mu B_1 N_1 + \frac{\mu^2}{2!} B_0 N_0, \\
&\vdots \\
\dot{N}_n &= (B_n - d_n)\, N_n + \mu B_{n-1}\, N_{n-1} + \cdots + \frac{\mu^n}{n!} B_0 N_0, \\
\dot{N}_n &\equiv dN_n/dt, \quad B_n \equiv b_n e^{-\mu}, \quad N_0(0) = N_0 = N(0), \quad N_n(0) = 0 \quad \text{for} \quad n \geqslant 0
\end{aligned} \tag{9.2.1}$$

Here N_0 is the number of normal individuals (those without deviations from the reference genome), b_n and d_n are the average birth rate (fertility) and death rate (mortality), respectively.

We have $d_n = \tau_n^{-1}$, where τ_n is the average lifetime of an individual with n decay changes in its genome. In a stationary state of constant population number τ_n coincides with the lifespan of a single generation.

The coefficient b_n (fertility) is equal to the average number of offspring generated by a single individual per unit time. B_n is the share of these offspring containing no changes in their genome, as compared to the parent genome.

Eqs. (9.2.1) are organised along the following scheme. Normal individuals may only be born to normal individuals (we ignore the reverse mutations). Thus, the rate of change of the number of normal individuals $\dot{N}_0$ depends only on the number of normal individuals itself. Note that only a certain share of offspring remain genetically identical to their parents, so that among all offspring born to N_0 normal individuals, only $N_0 e^{-\mu}$ are normal (i.e. they carry no deviations from the normal genome). Individuals with one mutational substitution (N_1) may be born to normal individuals (in such a case one mutation has to occur in the germ line) as well as to individuals that already have one genetic substitution (in such a case no additional mutations occur in the germ line). Thus the rate $\dot{N}_1$ depends on both N_0 and N_1. Similarly, $\dot{N}_2$ depends on N_0, N_1 and N_2 and so on. Coefficients governing that dependence come from the Poisson distribution (9.1.5).

Accumulation of deleterious decay substitutions in the genome results in gradual erosion of normal genetic information necessary for existence of individuals. We may thus assume that accumulation of deleterious substitutions goes as far as the lethal threshold $n_L : n < n_L$. At $n > n_L$ the number of lethal cases at early stages of development sharply rises, eliminating individuals with $n > n_L$ from the population.

A certain part of the genome of any organism, from 0.01% to 1% of the total size of the genome, remains conservative and does not change from generation to generation, whatever the external conditions (Lewin, 1987). That means that any changes in that part of the genome, including any arbitrary point mutation, bring about a lethal outcome. One may say that individuals with such changes are immediately eliminated from the population. The rest of the genome is capable of accumulating decay substitutions[3] up to the lethal threshold n_L. All the observed phenotypic differences between individuals are due to differences in the non-conservative part of their genomes.

In the absence of competitive interaction within the population (when it is exponentially expanding or exists under artificial laboratory conditions), both the birth and death rates of all viable individuals may be assumed equal to each other, i.e. $b_0 = b_n \equiv b$ and $d_0 = d_n \equiv d$ when $n < n_L$.

Summing up separately the right-hand and the left-hand parts of Eq. (9.2.1) we arrive at the following relation, which holds true to the accuracy of negligibly small terms of the order of $e^{-\mu}\mu^{n_L}/n_L!$:

$$\dot{N} = (b - d)\,N, \qquad N = \sum_{n=0}^{n_L} N_n, \qquad \mu/n_L \ll 1 \tag{9.2.2}$$

Eq. (9.2.2) demonstrates that coefficients b and d have a meaning of birth rate and death rate of individuals independent of the number of decay genome substitutions that they contain.

As follows from (9.2.2), at $b = d$ the population number N remains constant, while the number of normal individuals continuously decreases due to spontaneous decay process. The number of individuals with n decay substitutions in their genomes is equal to zero at the initial moment, increases with time up to a certain maximum and then falls off again. Evidently, with an increasing number of decay substitutions n the time when the maximum is reached, increases as well.

The exact solution of Eqs. (9.2.1) has the form:

$$\begin{aligned}
N_0 &= N_0(0)\,e^{-(d-B)t}, \qquad B \equiv be^{-\mu},\\
N_1 &= N_0(0)\,e^{-(d-B)t}\,\mu Bt,\\
N_2 &= N_0(0)\,e^{-(d-B)t}\,\frac{\mu^2}{2!}\{(Bt)^2 + Bt\},\\
&\vdots\\
N_n &= N_0(0)\,e^{-(d-B)t}\,\frac{\mu^n}{k!}\left\{(Bt)^n + \sum_{l=1}^{n-1}(Bt)^{n-l}\,c_{ln}\right\}
\end{aligned} \tag{9.2.3}$$

Here $N_0(0)$ is the total population number at the initial moment of time $t = 0$, when it coincides with the number of normal individuals. We do not write out in full the constant coefficients c_{ln}, that are obtained when sequentially integrating Eqs. (9.2.1). When the total population number N remains constant, i.e. when $b = d$ and $N = N_0(0)$, the expressions following $N_0(0)$ in the right-hand part of (9.2.3) are in fact population frequencies $p_n \equiv N_n/N$ of individuals with n decay substitutions.

Braces in (9.2.3) comprise a series of powers of (Bt), beginning from the n-th and down to the zero power for each variable N_n. At $Bt \gg 1$ the first term $(Bt)^n$ makes

[3] *Mutation* is a process which results in a (mutational) *substitution*.

the most significant contribution into the sum in braces. When the number of normal individuals remains constant, i.e. when $B = d$ and $N_0 = N_0(0)$, an approximate solution of (9.2.1) assumes the form $N_n = N_0(\mu Bt)^n/n!$, if in (9.2.3) we neglect all the power terms except $(Bt)^n$. Using this relation,[4] when summing up all individuals in the population we arrive at the following expression for the total population number N:

$$N = N_0 \exp(\mu Bt) \tag{9.2.4}$$

Note that the condition of stationarity of the *number of normal individuals*, $B = d$, corresponds to a e^μ-fold increase in fertility as compared with the condition of stationary *total population number* N, $b \equiv Be^\mu = d$.

For the population frequency p_n of individuals with n decay substitutions we have from (9.2.4):

$$p_n \equiv \frac{N_n}{N} = \frac{(\bar{n})^n}{n!}\exp(-\bar{n}), \qquad \bar{n} = \mu Bt = \mu dt \ll n_L \tag{9.2.5}$$

Here μ is the rate of genome decay (9.1.3) and $Bt = dt$ is the number of generations changed in the population during the time t. As is clear from (9.2.5), the frequency p_n of occurrence of decay individuals with n decay substitutions has the form of a Poisson distribution, its multiplicity $\bar{n} = \mu Bt$ growing linearly with time.

We have seen, therefore, that in the absence of stabilising selection any information encoded into the species genome inevitably decays. Normal individuals that contain genetic information necessary for correct actions on stabilisation of the environment, disappear from the population. Their place is occupied by decay individuals with substantially degraded genetic information. Such individuals are unable to perform any meaningful work on stabilisation of the environment, but remain viable and fertile. Decay individuals differ from normal ones in their morphological and behavioural properties. Some of these differences may prove to be useful to people. Decay individuals of natural biological species often form the basis for new breeds of domesticated animals and sorts of plants.

Relations (9.2.5) hold until the Poisson multiplicity $\bar{n}$ does not approach the lethal threshold n_L. When $\bar{n}$ comes close to n_L, the Poisson peak shrinks to a vertical line at n_L and all individuals in the population acquire a number of decay substitutions approximately equal to n_L. As a result, the population becomes inviable.

The condition of inviability of the population may be approximately written as the equality of the average Poisson multiplicity $\bar{n} = \mu dt$ and n_L. This condition determines the number dt of generations and the absolute time t during which the

[4] Note that summing introduces additional inaccuracy into the obtained result. When Bt is sufficiently large, we are able to neglect all terms $(Bt)^l$ with power l less than n for each N_n in (9.2.3), because it is $(Bt)^n$ that makes the largest contribution to N_n. When we then sum all N_n to get the total population number N, linear terms (Bt) omitted in all the expressions for N_n with $n > 1$, appear important for N_1 and change the constant multiplier of the latter. Similarly, terms $(Bt)^2$ omitted in all N_n with $n > 2$, change the constant multiplier of N_2 in the resulting sum, and so on. However, such inaccuracy does not significantly change the results obtained.

population becomes extinct in the absence of competitive interaction of its individuals:

$$\bar{n} = \mu dt = n_L, \qquad t = \tau \frac{n_L}{\mu}, \tag{9.2.6}$$

where $\tau \equiv d^{-1}$ is the average lifetime of the individual. Obviously, the lethal number of decay substitutions n_L cannot exceed the total genome size G. Recalling that $\mu = \nu kG$ (9.1.3) we may thus obtain an upper estimate of the time t using G instead of n_L in (9.2.6):

$$t = \tau \frac{n_L}{\mu} < \tau \frac{G}{\nu kG} = \frac{\tau}{\nu k} \approx 10^{10} \frac{\tau}{k} \tag{9.2.7}$$

For bacteria the average lifetime τ is about 10^{-3} years, $k = 1$ and, consequently, $t < 10^7$ years. For most mammals $\tau \sim 1$ year, k (the number of divisions in the germ line) is about 30, and $t < 3 \cdot 10^8$. In both cases the time of complete decay of the genetic information of a species exceeds significantly the average time of species existence, which is of the order of several million years for most species of the biosphere (Chapter 11).

The real value of n_L is obviously less than G, making the time of population extinction t shorter than the obtained upper estimate (9.2.7). The lower limit for the time t may be obtained using the observed values of genetic polymorphism in populations of different biological species. For example, if genomes of two humans are compared, on average one comparison in every 500 nucleotide sites will discover different nucleotide pairs (Hofker *et al.*, 1986). Thus, the number of genetic substitutions tolerated by the genome without the individual's viability being affected is about $0.002G$. The number of genetic substitutions causing death of the individual should be substantially larger than that. Assuming that n_L is of the order of $0.01G$ we obtain from (9.2.7) that the time of population extinction t for mammals is of the order of $t \sim 3 \cdot 10^6$, which is again not less than the average time of a species existence (Avise *et al.*, 1998).

Thus we come to the conclusion that even in a situation of complete absence of stabilising selection, genetic information of species decays very slowly. It means that species may exist for a long period of time under distorted environmental conditions outside their natural ecological niche, i.e. in a situation when the competitive capacity of normal individuals is reduced and competitive interaction weakened. In such a situation, genetic information of species undergoes continuous decay, the viability of individuals not being threatened, however. When the normal environment is restored, individuals with the least number of deviations from the normal genome regain their maximum competitiveness (see Section 9.8).

The slow mode of genome decay is ultimately due to the high fidelity of the process of replication of DNA molecules, i.e. the small probability of mutations per site per cell division ν, see (9.2.7). Slow rate of decay of genome organisation guarantees for persistence of biological species under any natural fluctuations of environment. This can be easily explained in the framework of the biotic regulation concept. Each species within the community performs certain unique work on environmental stabilisation, so that genetic decay and ultimate extinction of any species would adversely affect the wellbeing of the community. Thus the most competitive communities are those where all species have the maximum possible fidelity of the DNA copying process. As a result, the high-fidelity system of DNA copying is used by the overwhelming majority of living beings in the biosphere.

If there were no biotic regulation and species were continuously adapting to changing environment instead of stabilising it, species with low fidelity of DNA copying would be more advantageous, because low-fidelity replication produces more genetic variants, which form the basis for genetic adaptation. In such a case the pronounced large-scale evolutionary tendency towards increasing fidelity of DNA synthesis would be difficult to explain.

9.3 STABILISATION OF GENETIC INFORMATION OF SPECIES

According to the available paleodata, morphological properties of most species are conserved during the whole period of species existence (Gould and Eldridge, 1993). Correlation between the morphological and genetic constancy of the existing species enables one to assert that the genetic information of species is also conserved during the majority of the period of species existence (Jackson, 1990).

It means that the existing species are not in the process of continuous decay of their genetic programme. There should be a stabilising mechanism present in the population that would limit accumulation of decay substitutions in the genome well before the lethal threshold n_L is reached. Such a mechanism would prevent erosion of the genetic information that is necessary for correct actions of individuals on stabilisation of the environment, but not indispensable for their viability. The fact that such information does exist, is demonstrated by domesticated animals and plants appearing as the result of artificial selection and gene engineering. Being perfectly viable and fertile, these organisms are unable to persist sustainably in natural ecological communities. In the course of artificial genetic modifications their genome has lost the information necessary for correct interaction with other species of the community.

Let us now consider possible ways of genetic stabilisation of a population in terms of Eqs. (9.2.1). The most efficient functioning of a population of individuals within a community is ensured when normal individuals comprise the overwhelming majority in the population. As is clear from Eqs. (9.2.1), numbers N_n of individuals with n deviations from the normal genome is stationary when either the birth rate b_n is decreased, or the death rate d_n is increased for decay individuals, i.e. for those with $n \geq 1$. Such an effect may be reached when competitive interaction of individuals is switched on in the population and the normal genome is associated with the maximum competitive capacity.

For convenience we introduce relative coefficients of birth and death rate for decay individuals:

$$\beta_n \equiv B_n/B_0 = b_n/b_0, \qquad \delta_n \equiv d_n/d_0 \tag{9.3.1}$$

Here b_0 and d_0 are the birth rate and death rate of normal individuals, respectively. As follows from the first equation in (9.2.1), the stationary number of normal individuals, $\dot{N}_0 = 0$, $N_0 = \text{const}$, is determined by the condition

$$B_0 = d_0, \qquad b_0 = d_0 e^{\mu}. \tag{9.3.2}$$

To obtain a stationary state for the number of decay individuals, $\dot{N}_n = 0$, $N_n = \text{const}$ $(n \geq 1)$, all the terms $B_n - d_n$ in Eqs. (9.2.1) should be negative, so that the appearance of decay offspring from parents of a lower level of decay could be compensated. Hence, for individuals with $n \geq 1$ the death rate δ_n should always remain higher than the birth rate β_n. One may assume, for simplicity and practically not losing generality, that

$$\delta_n - \beta_n \gg \beta_n, \qquad n \geq 1 \tag{9.3.3}$$

By doing so we consider those cases when the death rate of decay individuals is *significantly* higher than the birth rate. Under this assumption the stationary solution of Eqs. (9.2.1) assumes the following simple form

$$N_n = \frac{N_0}{(\delta_n - \beta_n)} \frac{\mu^n}{n!}, \qquad N_0 = \text{const}. \tag{9.3.4}$$

Let us now consider possible ways of ensuring the numerical superiority of normal individuals in the population, i.e. $N_0 \gg N_n$, $n \geq 1$.

As is clear from (9.3.4), when the rate of genome decay is less than unity, $\mu < 1$, there are two possible ways to meet the condition $N_0 \gg N_n$. Firstly, decay individuals may feature the same death rate as normal individuals, $\delta_n \sim 1$, whereas their birth rate is significantly lower than that of normal individuals, $\beta_n \ll 1$. Such a situation may be referred to as *regulation of birth rate of decay individuals*. In such a situation each normal individual gives birth to $d_0 e^{\mu}$ offspring per unit time. All the offspring are allowed to live in the population until their morphological properties permit (i.e. no physical elimination of decay individuals occurs), but decay individuals are prevented from reproduction.

Secondly, decay individuals may be born at the same rate as normal individuals, $\beta_n \sim 1$, whereas their death rate significantly exceeds that of normal individuals, $\delta_n \gg 1$. Such a situation may be called *regulation of death rate of decay individuals* and presumes physical elimination of decay individuals from the population at early stages of their development (i.e. high juvenile mortality).

When the rate of genome decay is equal to or exceeds unity, $\mu \geq 1$, the only possible way to meet the condition $N_0 \gg N_n$ is the regulation of death rate of decay individuals, i.e. when $\delta_n \gg 1$, $\beta_n \sim 1$ and $(\delta_n - \beta_n)^{-1} \ll 1$. As may be seen from (9.3.4), regulation of the birth rate of decay individuals in this case ($\delta_n \sim 1$, $\beta_n \ll 1$ and $(\delta_n - \beta_n)^{-1} \sim 1$) leads to the opposite situation, i.e. the number of decay individuals N_n exceeds the number of normal individuals N_0, $N_0 \ll N_n$. Taking into account that the stationary total population number N is approximately equal to $N \approx N_0 e^{\mu}$ (9.3.2), the frequency $p_0 \equiv N_0/N$ of normal individuals in the population becomes as low as $e^{-\mu}$.

The difference $\beta_n - \delta_n$ may be called relative competitiveness. In a stationary state normal individuals are characterised by the maximum competitiveness which is equal to zero, while competitiveness of decay individuals is negative. It seems more natural to consider a value of the opposite sign:

$$\gamma \equiv \delta_n - \beta_n \equiv \text{noncompetitiveness} \equiv -\text{ competitiveness} \tag{9.3.5}$$

which should then be called noncompetitiveness. As we have seen, noncompetitiveness of decay individuals for both types of regulation of their numbers remains larger than unity, $\delta_n - \beta_n \geq 1$. Hence, when we proceed from normal to decay individuals, their noncompetitiveness suffers a jump from zero ($\delta_0 - \beta_0 = 0$) to a value exceeding unity ($\delta_n - \beta_n \geq 1$). Noncompetitiveness of decay individuals cannot be much lower than unity, which corresponds to Eq. (9.3.4) remaining distant from the pole appearing at $\delta_n - \beta_n = 0$ and provides for the condition $N_0 \gg N_n$.

Conventional variables of fitness and coefficient of selection that are widely used in population genetics differ from noncompetitiveness. The absolute fitness, W, is defined as $W = b - d$, i.e. it represents the net difference between rates of birth and death, neglecting the genetic status of offspring with respect to the number of decay substitutions n. Note that in a stationary state, as follows from (9.3.2), absolute fitness W_0 of normal individuals appears to be positive, $W_0 = W_{\max} > 0$. Further, by definition, absolute fitness of normal individuals (i.e. the net increment of their number per unit of time) can be written as $W_{\max} = (dN_{0\to}/dt)/N_0$, where $N_{0\to}$ is the total number of offspring, including both normal and decay ones, produced by normal parents. The share of normal offspring is $e^{-\mu}$ and the share of decay offspring is $1 - e^{-\mu}$.

Relative fitness w is defined as $w = W/W_{\max}$ (or $W/\bar{W}$), $w_{\max} = 1$, $1 \geq w \geq 0$. Coefficient of selection s is defined as $s \equiv w_{\max} - w = 1 - w$. For stabilising selection $1 \geq s \geq 0$. It is usually assumed that the value of s is small, e.g. $s = 0.01$ (Fisher, 1930; Haldane, 1957; Ohta, 1987). The coefficient of selection against individuals with n decay substitutions, s_n, and their relative fitness, w_n, are related to the relative birth rate β_n and the noncompetitiveness, γ_n, as follows:

$$s_n \equiv 1 - w_n = 1 - \beta_n + \frac{\gamma_n}{e^{\mu} - 1} \tag{9.3.6}$$

In the situation of the death rate regulation (i.e. when $\beta_n \sim 1$) values of s_n and γ_n differ from each other by the factor $e^{\mu} - 1$, which is equal to unity at $\mu = \ln 2 \approx 0.69$.

Use of the notion of relative competitiveness $\beta_n - \delta_n = (b_n - d_n)/B_0$ is stimulated by the fact that it transparently describes the whole pattern of appearance of decay individuals in the population, which is not accounted for by the traditional definition of fitness. Relative competitiveness reaches maximum (zero) for normal individuals and is negative for decay individuals. When the rate of generation of offspring by all the decay individuals is zero ($b_n = 0$, $n \geq 1$), the numbers of decay individuals in the population may in any case remain stationary. The rate at which decay individuals die may be compensated by the rate at which they are born to normal parents. The latter produce decay individuals due to newly-arising decay mutations. In such a case, fitness of the decay individuals appears to be negative and equal to

$W_n \equiv b_n - d_n = -d_n < 0$, which cannot be explained in the traditional interpretation of fitness. For example, even if all people with hereditary diseases decided to refrain from reproduction, the corresponding genetic deviations would anyway be present in the human population as a result of reproduction of normal individuals whose germ cells underwent corresponding mutations.

To regulate birth and death rates of decay individuals, a possibility should be provided of identifying such decay individuals and of differentiating them from the normal ones. That task is solved by joining all the individuals into a population and switching on competitive interaction between them. The difference between the normal and the decay individuals is only controlled by their competitiveness. Under natural conditions the number of decay individuals in a population is suppressed in the course of their competitive interaction with normal individuals.

Superimposing conditions of stationarity, Eq. (9.3.3), which provide for the maximum competitiveness of normal individuals and for the maximum noncompetitiveness of decay individuals, actually corresponds to introduction of complex nonlinear relations between normal and decay individuals into Eqs. (9.2.1). These relations may be presented in the form of complex functional dependencies $B_n(N_0, N_k)$ and $d_n(N_0, N_k)$, which describe the differences between normal and decay individuals and which are equivalent to introducing competitive interaction between them. However, such nonlinearities may be substituted by an analogue of the boundary conditions realised in (9.3.3).

The necessity to join into a population, to switch on competitive interaction and to assign the maximum possible competitiveness to normal individuals should be genetically encoded into the normal genome. That basic programme of supporting the species stability should be safely insured against decay mutations (e.g. by multiple back-up of genetic information). Consequently, the genetic information of species should only decay in the extreme case when the number of decay substitutions approaches the lethal threshold. As a result, all individuals, independent of the number of decay substitutions their genomes contain, still enjoy the capacity to join into a population and to interact competitively. Competitive interaction between the individuals in a population always occurs with maximum intensity under conditions of the natural ecological niche of a species, and does not depend on the availability of resources (see Section 2.2).

When breeding is sexual, the genetic material in the offspring presents a random mix of parental genetic material. Decay substitutions in the genomes of the offspring individuals then happen to be distributed after Poisson (Maynard Smith, 1978; Kondrashov, 1982; 1988). Since newly-arising mutations also follow a Poisson distribution, the distribution of Eq. (9.2.5) does not change when sexual breeding is switched on. The stationary state described by (9.3.3) and all the conclusions related to it also hold when sexual breeding is switched on. Besides, Eqs. (9.2.1) may be envisaged not as equations written for numbers of individuals bearing n decay substitutions in their genomes, but as equations for numbers of genes with n decay substitutions in the whole population, independent of separate individuals bearing those genes. In such a case it becomes especially clear that the advent of sexual breeding changes nothing in either results or conclusions made in this Section.

9.4 SENSITIVITY OF COMPETITIVE INTERACTION

Determining competitiveness of individuals in the course of competitive interaction in a population is actually a process of measurement. Like any other process of measurement, competitive interaction is characterised by limited resolution (sensitivity). It means that if the number n of decay substitutions contained in the genome of a decay individual is less than the sensitivity threshold n_C of competitive interaction, $n < n_C$, such an individual cannot be told apart from a normal individual with a genome totally free of decay substitutions (see also Section 3.4 about the biotic sensitivity).

Note that a certain limit of sensitivity is inherent to any process of measurement. For example, it is impossible to weigh small loads (e.g. milligrams) using scales that are calibrated in kilograms. If each nucleotide substitution is compared to a small load weighing 1 mg, about 10^6 loads need to be put on the scales before the pointer begins to move and show non-zero weight, which corresponds to the beginning of selection process, i.e. identification of individuals with too large numbers of decay substitutions. The sensitivity of a process of measurement is determined by the properties of the applied instrument and the process itself. Similarly, sensitivity of competitive interaction is a fundamental property of life organisation and cannot be quantified *a priori*.

The competitive interaction is effectively switched on and active within the margin $n_C < n < n_L$. Here n_L is lethal number of decay substitutions. Due to the random nature of mutations, the number n_C of decay substitutions should be relatively evenly distributed along the genome, rather than concentrated in a certain part of it. Note that the limit of sensitivity of competitive interaction n_C does not depend on the total number N of individuals in the population in that sense, i.e. it has a finite non-zero value at $N \to \infty$. In larger populations individuals are given the opportunity to have more contacts with other individuals, so that the effective number of 'measurements' of their competitiveness increases. However, continuing the analogy with the scales, however many times you weigh a load on scales calibrated in kilograms, you will never be able to determine its weigh to the accuracy of milligrams, the ultimate limit to accuracy being set by the scales' calibration rather than by the number of measurements. A large number of measurements will allow you only to determine the precise number of *kilograms* in the load.

Thus, the sensitivity of competitive interaction can be represented in the form

$$n_C(N) = n_{C\infty} + (1/\sqrt{N})\,\Delta n_C, \tag{9.4.1}$$

where $n_{C\infty}$ stands for the true sensitivity of competitive interaction determined by inherent properties of individuals (it corresponds to the scales' calibration unit), whereas Δn_C describes the inaccuracy of a single process of measurement. The inaccuracy of the process of weighing a load is usually higher than the scales' calibration. For example, the person who makes the measurements may be thinking of something else when writing down the results. Also, they may inadvertently touch the scales during the process of measurement, so that the result is biased. Similarly, interactions of individuals in a population are not always characterised by the same intensity. The accuracy of determining competitiveness may depend on a number of randomly-fluctuating factors such as, for example, weather, peculiarities of the territory occupied at the moment by the population, etc. This inaccuracy, Δn_C, may be done away with by increasing the number of contacts between individuals, i.e. increasing the population number N. The multiplier $(1/\sqrt{N})$ describes the fact that with the increasing number of measurements this inaccuracy indeed decreases in accordance with the law of great numbers. Meanwhile the inherent inaccuracy, $n_{C\infty}$, of the process itself (i.e. how many decay substitutions can be discerned by the process of competitive interaction under the most favourable conditions) cannot be in principle be affected by the increasing the number of measurements N.

Under the reasonable assumption that Δn_C cannot be significantly larger than $n_{C\infty}$ (i.e. it can hardly be the case that the person weighing the load on the scales calibrated in kilograms would make a mistake in the number of tens of kilograms in the load), we come to the conclusion that the value of n_C ceases to depend on N in populations containing as few as a hundred of individuals (so that the input of the N-dependent term in (9.4.1) becomes about an order of magnitude smaller than the first constant term).

Another argument against the influence of the population number N on the limit of sensitivity of competitive interaction n_C is as follows. A single act of competitive interaction of two individuals takes a certain period of time (e.g. a fight between two bulls is necessarily followed by a long period of relaxation). Let us denote it t_{comp}. The period of the most intensive competitive activity of individuals is also limited. We denote it as T_{comp}. In many mammals this period falls during breeding seasons. Thus, during the period of the most intensive competitive activity, any individual has only time to interact with $N_{\mathrm{comp}} = T_{\mathrm{comp}}/t_{\mathrm{comp}}$ individuals. An increase of the total population number beyond N_{comp} will not have any effect on the accuracy of competitive interaction, not even in the mild form of the relation (9.4.1). One may say that N_{comp} represents the valence of individuals with respect to competitive interaction. As soon as it is saturated, i.e. when $N \geq N_{\mathrm{comp}}$, further increase in N does not produce any effect.

An increased accuracy of competitive interaction may be due to an increased length of a single act of competitive interaction t_{comp}. In other words, competitive interaction may be more efficient when few individuals compete with each other for a long time (as is the case in territorial birds) than when a large number of individuals interact competitively many times during a short time period. In such a case, even a small population may be sufficient for an acceptably accurate determination of decay individuals.

9.5 NORMAL GENOTYPES AND THE NORMAL GENOME

Genomes containing decay mutations in numbers less than n_C possess the genetic programme that ensures viability of individuals and their correct functioning within the community aimed at stabilisation of the environment. In this sense all such genomes are equivalent to each other and enjoy equal competitiveness which reaches its maximum in the conditions of the natural ecological niche of the species. In what is to follow, we call various genomes falling within the range $n < n_C$ for *normal genotypes*,[5] while those with $n > n_C$ will be referred to as *decay genotypes*. Individuals with normal and decay genotypes will be called normal and decay individuals, respectively. For simplicity we assume that n_C represents a sharp border between normal and decay genotypes. However, all the conclusions made below hold also for those cases when normal and decay genotypes are delimited by a diffusive probabilistic area. In such a case n_C may be defined, for example, as the probability that after competitive interaction with other individuals an individual with n_C substitutions will be given a status of decay individual with a probability of 0.5 (i.e. with such a probability the individual will be forced out from population, ignored by sex partners during the breeding season, or in any other way rendered low status).

Differences in localisation of decay substitutions in normal genotypes cause the observed morphological and behavioural differences between normal individuals. Obviously, only slightly deleterious substitutions, i.e. those that do not seriously affect the genetic programme of individuals, may persist in the population in relatively large numbers for a long time, remaining unnoticed by selection. Possible localisation of such substitutions is determined by the genome structure. Substitutions with a pronounced deleterious effect are immediately discovered in the course of competitive interactions of individuals, though they do not necessarily cause inviability of infertility of individuals. A single, strongly deleterious substitution may cause apparent defects in morphology or behaviour of the individual. In that sense a single, strongly deleterious substitution is equivalent to a large number of slightly deleterious mutations and may drive the genotype beyond the threshold n_C. As a result, individuals with strongly deleterious mutations are eliminated from the population in one way or another, so that the majority of strongly deleterious substitutions are not heritable. However, such substitutions are in any case present in the organism due to mutations in somatic cells of multicellular organisms. Most newly-arising mutations have a pronounced deleterious effect.

Apparently, competitiveness of individuals has no absolute value and may only be described by some relative variable. When comparing competitiveness of any two individuals, one faces only two possibilities. First, this competitiveness may remain unresolved within the resolution of competitive interaction. In such a case, neither individual is capable of forcing the other from the population. Second, the competitiveness of one may exceed that of the other. The genotype of the first individual will then inevitably force that of the second from the population, independent of the numerical difference between the competitivenesses of the two. **Thus the relative competitiveness appears to have a stepped nature. This is most vividly displayed during sexual selection. The female or the male may either accept or reject**

[5] The term 'normal' is close in its meaning to the term 'wild-type' often used in the literature. Yet the former is more clearly defined.

the mating partner. There are no intermediate possibilities. That feature comprises the main difference between the notions of competitiveness and fitness.

On this basis, if the difference in competitiveness of two genotypes exceeds the sensitivity (resolution) of competitive interaction, one may assume, to a good approximation, that the genotype of higher competitiveness has a relative fitness of unity, so that it produces the most numerous offspring in the population

$$w_{n<n_C} = 1, \qquad s_{n<n_C} = 0, \qquad \beta_{n<n_C} = 1, \qquad \gamma_{n<n_C} = 0,$$

while the genotype of lower competitiveness has a relative fitness close to zero, that is does not produce any offspring

$$w_{n<n_C} = 0, \qquad s_{n>n_C} = 1, \qquad \beta_{n>n_C} = 0, \qquad \gamma_{n>n_C} \geq 1.$$

Assume that we managed to form a clonal population of the normal genome with $n = 0$ at zero time $t = 0$. In accordance with Eq. (9.2.1) that clone will keep on decaying and accumulating decay substitutions up to $n = n_C$, since, according to the definition of n_C, for all $n < n_C$ all the differences $B_n - d_n$ coincide with each other and are equal to zero when $B_0 = d_0$. Similar to the considered case of decay down to the very lethal threshold n_L (which occurs when competitive interaction is totally absent), the population appears to be pressed to the value of n_C during the time period $t \approx n_C/d\mu$, see (9.2.6). Further increase of the number of decay substitutions is then stopped by competitive interaction, which begins to notice the decay substitutions when their number exceeds n_C. Thus, one may assume that in a stationary case all genotypes present in the population are characterised by approximately the same number of decay substitutions, namely n_C, i.e. $N_n \approx 0$ for all $n \neq n_C$ and $N_{n_C} \approx N$, where N is the total population number (see Figure 9.1).

Process of genetic decay produces genetic diversity of individuals in a population. The total number Π of different genotypes with n_C decay deviations (substitutions) from the initial clone is equal to

$$\Pi = 4^{n_C} \cdot C_G^{n_C} = (4G/n_C)^{n_C}, \tag{9.5.1}$$

where G is the total number of nucleotide pairs in the genome, $C_G^{n_C}$ is the number of combinations from G by n_C, 4 is the number of different nucleotides (genetic letters). Even when $n_C = 2$ the number of various genotypes reaches 10^{20} which by far exceeds the number of individuals in any multicellular species born during the whole lifespan of that species. Thus no two individuals in any population may happen randomly to be genetically identical to each other.

In small populations there takes place a random drift of genes in which decay mutations have occurred. A stable final state of the process of genetic drift is either loss or fixation of a given decay substitution in all the individuals of the population (Wright, 1932, 1988; Kimura, 1983; Ayala and Kiger, 1984; Altukhov, 1991). In place of the lost decay substitutions new decay substitutions appear in the process of mutagenesis, so that the total number of decay substitutions remains close to n_C. Genetic drift results in a reduction of genetic polymorphism due to fixation of some part of decay substitutions in all the individuals in the population. But it does not

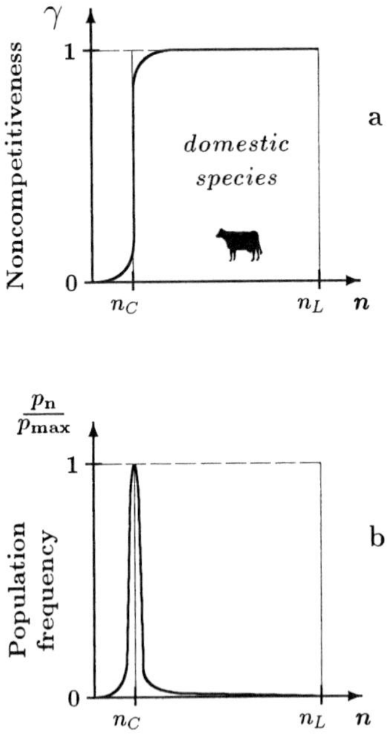

Figure 9.1. Normal polymorphism of genotypes in a population under natural conditions of the ecological niche of the species. n is the number of decay substitutions in the genotype of an individual; n_C is the number of decay substitutions characterising the sensitivity of competitive interaction and stabilising selection, which represents the limit to accumulation of decay substitutions in the genotype; n_L is the maximum number of decay substitutions compatible with viability of the individual. **a** – noncompetitiveness of individuals related to the number n of decay substitutions in their genotypes, see (9.3.5). **b** – relative population frequency of genotypes with n decay substitutions. The area $n_C < n < n_L$ unoccupied by natural species is used by humans for cultivation of domestic animals and agricultural plants.

change the statement that the average number of decay substitutions in individuals is less than n_C.

Localisation of decay substitutions in various populations originating from one and the same clone of normal individuals should be different owing to the random nature of the processes of decay and genetic drift. The value of n_C being high, genetic differences may result in the observed phenotypic (morphological, behavioural, etc.) differences between individuals from isolated populations, which are traced through many generations. Depending on the scope of these differences, they may be perceived as either national, racial or subspecific. However, nationalities, races and subspecies thus originating remain genetically equivalent: various genotypes with $n < n_C$ of individuals from whatever races are normal, and all retain the complete genetic programme of the species. Normal individuals from different races have equal competitiveness, which is higher than the competitiveness of any decay individuals from any race, the latter featuring different decay genotypes with $n > n_C$ (Figure 9.2). (Note that when two individuals from distantly-related races interbreed, the number of decay substitutions in the appearing offspring may reach $2n_C$ due to

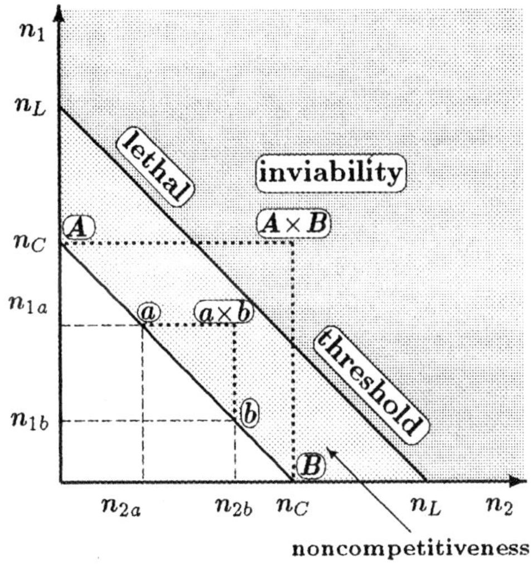

Figure 9.2. Genetic differences between different populations and subspecies of a single species. Let population A be the reference population. The species genome G is divided in a mosaic fashion into two equal parts G_1 and G_2 in such a manner that all n_C decay substitutions encountered in population A are located in the part G_1, whereas the part G_2 is free from decay substitutions. As discussed in the text, all populations and subspecies are characterised by the same sensitivity of competitive interaction, so that the average number of decay substitutions in individuals of all populations is close to n_C (see Figure 9.1). However, localisation of decay substitutions in different isolated populations is different. Let n_1 and n_2 be the number of substitutions in the G_1 and G_2 parts of the genome, respectively. While in population A all n substitutions are located in the part G_1 of the species genome, $n = n_1 = n_C$, in other populations some decay substitutions (n_1) are located in the G_1 part, while others (n_2) in the G_2 part of the species genome, $n = n_1 + n_2 = n_C$. Finally, there may be a population B where all the n_C decay substitutions are located in the part G_2, $n = n_2 = n_C$. Hence, all populations and subspecies are described by a straight line AB, each point of which corresponds to some particular localisation of decay substitutions in the genome. The genetic equivalence of all populations and subspecies is manifested by the fact that they are all equally close to the normal genome, the measure of distance being n_C. The line parallel to AB corresponds to the lethal threshold n_L. Hybridisation of individuals from two isolated populations a and b (point $a \times b$) drives the number of decay substitutions in the offspring beyond the border AB ($n > n_C$). Such offspring are viable but noncompetitive compared with normal individuals of both subspecies. Hybridisation of distant subspecies A and B (point $A \times B$) yields inviable offspring. The difference between isolated populations and subspecies is purely quantitative.

non-coinciding localisation of decay substitutions in different races. As a result, these offspring may prove to be noncompetitive and forced out from either of the two parental populations.)

A typical example of this situation is provided by Arctic gulls (*Larus* species). These birds form a set of populations which live around the Arctic Ocean (Mayr, 1963; Green *et al.*, 1989). In Western Europe the lesser black-backed gull (*L. fuscus*) is a familiar species whose range extends east into the Russian Arctic, through populations that are interbreeding but which can be arranged into several subspecies, each slightly different. The easternmost subspecies is so far east that it ranges into Western Europe as the herring gull, where it exists alongside the lesser black-backed gull. But the herring gull does not interbreed with the lesser black-backed gull in Western Europe and is called *Larus argentatus*.

The observed pattern can be explained as follows. All populations of *Larus* are predominantly composed of normal individuals with $n < n_C$, which occupy the same ecological niche. However, the localisation of the decay substitutions in genome gradually changes from one subspecies to another. When the herring gull meets the lesser black-backed gull in Europe, localisation of decay substitutions in genotypes of these two subspecies do not presumably overlap at all. Therefore the number of decay substitutions of their possible offspring would appear in the decay interval $n \sim 2n_C > n_C$, cf. cross $a \times b$ in Figure 9.2. As a result, individuals of these subspecies produce offspring that cannot stand competition with normal individuals of any of the two subspecies which, naturally, limits interbreeding between the two subspecies (Tinbergen, 1953; Ryttman *et al.*, 1979). Due to this, the subspecies that differs most from the eastern European subspecies of *Larus argentatus* is given a status of species and called *Larus fuscus*.

Within the area where the two distantly-related subspecies of gulls coexist, they still occupy the same ecological niche, i.e. individuals belonging to both subspecies interact competitively with one another. This is manifested in the observed large fluctuations in the ratio of population numbers of birds *Larus fuscus*/*Larus argentatus*. Contrary to that, different species belonging to the same type of ecological community should be characterised by a strictly specified ratio of population numbers determined by the condition of the most efficient functioning of the community as a whole (see below, Section 9.8).

In most cases, taxonomists are able to unambiguously tell apart any two closely-related species from two subspecies of the same species. If that were not the case, there would be no use in applying the notion of *species* altogether, because differences between subspecies are usually of a gradual nature. **It would be natural to define subspecies as two genetically distant populations with non-overlapping regions of the decay polymorphism n_C.** Two subspecies can be in principle reversibly transformed into each other by means of a succession of decay and reverse mutations taking place in the regions of localisation of decay substitutions of the two subspecies. (Note that speciation is an irreversible process. Two different species cannot be transformed into one another by any succession of decay and reverse mutations.) According to such a definition, *Larus argentatus* and *Larus fuscus* would be classified as two genetically distant subspecies of a single species. In fact, in spite of different taxonomic names, these populations of *Larus* are often referred to as a single *ring* species.

Morphological and behavioural geographical differences between individuals inhabiting distant regions of the species range are observed for most biological species. The uniqueness of the example of the gulls *Larus argentatus* and *Larus fuscus* lies in the fact that the two opposite ends of their range prove to be adjacent to each other. Were it possible to do the same with ranges of other species, the integrity of which is not doubted by the taxonomists, the pattern would be essentially the same.

Among substitutions appearing during the decay of a clone of a normal genome, both deleterious and neutral decay substitutions may appear (Kimura, 1989). These substitutions may be experimentally differentiated from each other in the following way. A given deleterious substitution in a given site of the genome can change the phenotype but does not change the individual's competitiveness if the total number n of substitutions remains within the margin $n < n_C$, while the same substitution lowers that individual's competitiveness if it happens to a normal individual with $n = n_C$ or to a decay one with $n > n_C$. A neutral substitution never changes the competitiveness of an individual. There is no threshold for accumulation of neutral substitutions in a genome. Those sites of the genome in which such neutral substitutions may appear (if they exist at all) do not bear any information. Thus the sequence of nucleotide pairs in a normal genome may only be defined to the accuracy of such neutral sites. On the other hand, certain substitutions may appear effectively neutral due to degeneration (i.e. existence of several meaningful variants) of certain parts of the genome. The total number of neutral sites devoid of information content is apparently much lower than the overall number of information-bearing sites of the genome.

The limit of sensitivity of competitive interaction n_C for a particular species may be estimated as the average number of nucleotide differences between genotypes of two individuals taken from two isolated natural populations of the species. The number of nucleotide differences between genotypes of two individuals from one and the same population should remain lower than n_C, since a certain portion of deleterious decay substitutions may be randomly fixed in the population. However, random fixation of one and the same decay substitution in different non-interacting populations is quite an improbable event. Therefore, if coinciding genotype sequences are found in different populations, one may safely assume that these are fragments of the normal genome of the species.

Hence, if several isolated natural populations of one and the same species are present, an experimental procedure may be suggested, capable of yielding, at least in principle, the nucleotide sequence of the normal genome to an accuracy exceeding the sensitivity of competitive interaction. Normal genome can be then determined as the genome sequence where each site is represented by a nucleotide which has the highest frequency in the whole species (i.e. when all the populations are combined). If the overall number of analysed isolated populations is large enough, that procedure may in principle identify the whole nucleotide sequence of the normal genome to the accuracy of the number of degenerative (neutral) sites present in the genome.

Localisation of neutral sites containing no information may be identified comparing homologous sequences in various individuals both within a single population and from population to population. In such sites all the four nucleotides should be discovered with equal probability, which corresponds to the state of thermodynamic equilibrium (that of maximum chaos and minimum information). Even if the initial conditions are definitely prescribed, the state of thermodynamic equilibrium in these sites will inevitably set in with time.

Neutral sites are usually related to the observed degeneration of the genetic code (when several triplet-codons of nucleotide pairs correspond to one and the same aminoacid in the protein), to the presence of non-transcribed sequences in the genome (those not coding for proteins, such as introns within genes, spacers between them, tandem repeat sequences, etc.) (Ayala and Kiger, 1984; Lewin, 1987), to the presence of functionally inactive parts in proteins.

However, ample evidence has been accumulated that suggests that these presumably neutral parts of the genome in fact play an important role in the organism's functioning, though it may remain obscure to the researchers. For example, degeneration of the genetic code is neutralised when different triplets of nucleotides correspond to different molecules of tRNA (Grosjean *et al.*, 1982; Kimura, 1983; Buckingham and Grosjean, 1986). Molecules of tRNA (transport RNA) are indispensable in the process of protein synthesis. Namely these molecules help to relate a particular aminoacid to the corresponding triplet of nucleotides. Different tRNA may be present in the cell in different abundances. Thus, if a triplet that is served by an abundant tRNA mutates to a triplet which codes for the same aminoacid but is served by a rare tRNA, the process of synthesis of the corresponding protein may slow down, leading to the shortage of the protein in the cell and to corresponding adverse changes in the competitive capacity of the organism.

Also, the number of neutral sites is by far lower than the number of noncoding sequences in the genome. For example, the telomeres, the noncoding end sequences of chromosomes in the higher species, provide for stability of chromosomes during their division. As a manifestation of such an important function, the genome sequence of telomeres remains strictly conserved from generation to generation (Moses and Chua, 1988) implying that any mutations in the telomere region have an adverse effect on the organism. There is also evidence that non-transcribed sequences probably play an important role in the process of development of a multicellular individual on early stages of embryogenesis (Maksimowski, 1988).

Differences in the effect of seemingly neutral sites on the individual's wellbeing may be detected within the resolution of competitive interaction of individuals under natural conditions and remain undetectable during laboratory experiments. Thus laboratory observations of neutral mutations and neutral sites do not necessarily mean that these correspond to equally competitive phenotypes in natural populations. The resolution of competitive interaction may be several orders of magnitude higher than that achieved in the laboratory, where only a few morphological or behavioural properties of the individual are monitored within a framework of a single experiment. The problem of the existence of genome degeneration is actually the problem of limited sensitivity of the process of competitive interaction. Neutral substitutions cannot be told apart from deleterious ones within the range of $n < n_C$.

9.6 NORMAL, DECAY AND ADAPTIVE POLYMORPHISM IN A POPULATION

Limited sensitivity of competitive interaction results in genetic diversity of normal individuals which may have different phenotypes but feature equal competitiveness under the conditions natural for the given species. Such a polymorphism may be called *normal*, despite the fact that it is determined by deleterious decay substitutions, provided the number of those substitutions n satisfies the condition $n \leq n_C$. As demonstrated above, the level of normal polymorphism in most species is very large, so that individuals with identical genotypes cannot be randomly found in practically any population (Paune and Westneat, 1988). Even when neutral substitutions are totally absent from the genome, random genetic drift taking place within the range $n \leq n_C$, where the deleterious character of mutations is not manifested, may result in random fixation of a limited number of deleterious substitutions in small populations.

'True' neutral polymorphism not related to decay mutations is also sometimes found in natural populations. A typical example of such polymorphism is the right-to-left asymmetry in the bodies of individuals. For example, the river flatfish (*Pleuronectes flesus*) is found in two forms: one with both of its eyes on the right side of its body, the other being leftsided (Andrijanov, 1954; Bisazza *et al.*, 1998). Asymmetric shape of the fish represents an adaption (Section 1.7) to sea- and riverbed life. It is reached by a correlated change in the position of the internal organs of the fish. Under natural conditions neither the rightsided nor leftsided form may have any advantage, because the environment where the fish lives is not characterised by right-to-left asymmetry.

Molecular stereoisomers in the cell are known to have identical symmetry in all living beings and are the same for the right and the left forms of the flatfish. Thus the process of formation and frequency of occurrence of the right and the left flatfish forms cannot be prescribed by two different normal genomes, independently formed in the course of evolution. Rather, they should be programmed in one and the same normal genome. The actual configuration (the right or the left form) cannot be hereditary. The only hereditary characteristic is the frequency of occurrence of the left form with respect to the right one. This frequency is apparently a neutral genetic characteristic and, as any other neutral characteristic, is subject to random genetic drift with possible fixation in an alternative position. Due to random genetic drift the frequency of occurrence of right and left forms may assume different values in different isolated populations of one and the same species (Andrijanov, 1954). In many species of the flatfish family such a drift has resulted in fixation of either the right (*Hippoglossus vulgaris*) or the left (*Rhombus maximus*) form.

Due to the continuous process of decay of normal genotypes, a certain number of decay individuals with lowered competitiveness is always present in the population, their genotypes having $n > n_C$. Genetic diversity of decay individuals results in the appearance of *decay polymorphism* in the population. In conditions of a natural ecological niche, the number of decay individuals should be kept at a low level compared with the number of normal individuals. Thus under natural conditions the decay polymorphism is much lower than the normal polymorphism and it may be safely neglected as compared with the latter.

Due to the extreme complexity of the genetic programmes of biological species, the random genetic adaptation to environmental changes appears to be improbable, be the initiating genetic change due to either a point mutation or a macromutation affecting the existing genetic program. In other words, appearance of a new, functionally-sensible genetic programme in the process of decay of an old one is highly improbable. (Such a situation is quite similar to that when some classic musical masterpiece by a genius of composition is written onto a magnetic tape. One cannot really expect, then, that the process of random erasure of the tape would result in the appearance of a new masterpiece by another genius on that tape.) The absence of random genetic adaptation agrees with the observed stability and the discrete nature of biological species and provides for the possibility of biotic regulation and maintenance of environmental stability (see Chapters 5 and 6). Absence of random genetic adaptation does not contradict the observed evolution of species, as is discussed in detail in Chapter 11.

In species with different genetic specialisation of individuals, *adaptive* genetic polymorphism is also present along with normal and decay polymorphism. The most common example of adaptive polymorphism is the genetically-encoded division of a population into females and males that is observed in a large number of species. Note that the word 'adaptive' here corresponds to the *state of adaption* of a species to the optimal environment maintained by the corresponding ecological community to which the species belongs, but not to the *process of adaptation* to changing environment (see also Section 1.7). Increase of adaptive polymorphism enhances stability of the species organisation. Contrary to that, increase in decay polymorphism represents erosion of genetic information of a species and results in shrinking of the species range, as soon as decay individuals are unable to exist sustainably within the ecological community.

All the observed adaptive polymorphism may be attributed to one and the same normal genome of the species. For example, the human genome comprises both X and Y chromosomes, though the latter is only present in male genotypes. On the other hand, adaptive polymorphism may be envisaged as a co-existence of several normal genomes (e.g. those of males and females).

Individuals with several different normal genomes may be present in the population only when they are correlated with each other, i.e. they cannot exist without each other for long periods of time and there is no competitive interaction between them. Examples of such co-existence are abundant. Several normal genomes, replicating independently of each other, are present in the nucleus of eukaryotic cells of higher organisms: one in the form of a set of nuclear chromosomes, and another as independently replicating chromosomes of mitochondria and chloroplasts (in plants) found in cytoplasm (Sager, 1972; Ayala and Kiger, 1984). Nuclear and cytoplasmatic genes cannot function independently of each other. Prokaryotic bacterial cells devoid of nuclei contain, beside their principal chromosome, autonomous ring-shaped genomes of plasmides, which replicate independently (Ayala and Kiger, 1984). Plasmids can transmit from one bacterial cell to another. Some plasmids determine resistance to antimicrobial drags and toxic metal ions in bacteria (Foster, 1983; Newbold, 1990). As already noted, many bisexual species have two normal genomes in their populations, the female and the male ones. Two different normal genomes may be present in associations of independently breeding individuals which are in a symbiotic relation to each other, as is the case for the algae and the fungus in a population of lichen. Finally, numerous normal genomes of different species are mutually correlated in the ecological community, providing for the biotic regulation of the environment.

One may speak therefore about the *collective normal genome* of all species correlated in the community in exactly the same way as one speaks about the genome consisting of several non-homologous chromosomes in the cell (which also cannot exist independently of each other), or about the combined genome consisting of a nuclear and a mitochondrial DNA, or about the combined genome of males and females in a population of a bisexual species.

The collective normal genome of a population of ecological communities remains unique for the given type of community. Rigid correlation between species in the community is often inadequately called adaptation of species to the external environment, comprising a set of physical as well as biological characteristics (e.g. other species in the community) which, as we mentioned above, should be in fact referred to as adaption, i.e. a state instead of a process. Following this way of reasoning one could speak about mutual adaption of different nuclear chromosomes in a cell.

Only a set of rigidly-correlated species combining to form an ecological community may feature the necessary wide range of reactions to changes in their environment and ensure a wide range of negative feedback responses. Such a community is characterised by a single collective normal genome comprising all the genetic information of all species belonging to the community. It is as unthinkable to compose a stable, self-sustainable, artificial community of arbitrarily picked species as to compose a viable cell on the basis of a genome combined of arbitrarily chosen genes picked from individuals of different species.

Any deviations from the normal species composition or from the normal population densities of each species in the community are decay phenomena, which are cut away in the process of competitive interaction of communities in a population of communities (Section 2.5). Relative frequency of occurrence of decay communities is controlled by the resolution of the process of competitive interaction between various communities. It may be described by equations of the type of Eqs. (9.2.1), (9.3.3) and (9.3.4).

9.7 STABILITY OF BIOLOGICAL SPECIES UNDER NATURAL CONDITIONS

Under natural conditions the resolution of the process of competitive interaction reaches its maximum, which corresponds to the minimum possible value of the sensitivity threshold n_C, which is much lower than the lethal threshold n_L (see Section 10.10), $n_C \ll n_L$. The fact that under natural conditions normal genotypes (i.e. those containing the programme of stabilisation of the environment) provide for the maximum competitiveness of their carriers is a non-trivial condition that ensures stability of the species, of the community, and of the environment (see Section 2.11). In other words, maximum competitiveness and actions of individuals on stabilisation of the environment are not always coupled. In a broad sense, competitiveness essentially means the ability of one individual to win over another individual, forcing the latter out from the population, preventing it from reproduction, depriving it of food resources or in any other way.

As noted in Section 9.3, competitiveness (9.3.5), measured for individuals within a natural community, does not coincide with fitness measured for individuals isolated from the community and hence from the population. Competitiveness drops sharply under natural conditions, while noncompetitiveness γ_n, (9.3.5), grows quickly from zero to unity with the number n of decay substitutions in the genotype entering the range $n > n_C$ (Figure 9.1a). Meanwhile relative fitness w_n (9.3.6) measured outside the population may remain practically unchanged in the range $n > n_C$ up to the lethal threshold $n \approx n_L$. Such a situation is due to the fact that decay individuals within a population are prevented from leaving offspring, not because of their physical inability to do it but because of a pressure imposed on them by normal individuals in the process of competitive interaction. Meanwhile, outside the population decay individuals with $n < n_L$ may remain perfectly viable and fertile. Such a pattern provides for the stability of the species as a whole. On the one hand, only the best individuals in the population are allowed to leave offspring. On the other hand, if the best individuals die due to certain sharp environmental fluctuations, fertile decay individuals may take part in reproduction as well, so that the persistence of the species as a whole is not threatened.

When environmental conditions deviate from the natural optimum, the resolution of competitive interaction decreases, and the value of n_C increases. In other words, competitive interaction leaves more decay substitutions unnoticed, so that competitivenesses of normal and decay individuals evens out. This can be explained as follows. Under natural conditions, the maximum competitive capacity is associated with those individuals that are able to perform strictly specified complex work on stabilisation of the environment in tight cooperation with individuals of other species of the community. Under strongly perturbed conditions no environmental regulation is possible and the regulatory abilities of normal individuals no longer impart to them competitive advantage. Similarly, some specific skills of people (e.g. knowledge of foreign languages or the ability to play the piano) that may be vitally important in some situations, remain unnoticed until only basic physical characteristics are monitored (e.g. the ability to walk, eat, reproduce, etc.). In other words, under natural conditions a huge variety of individual properties are subject to thorough control of stabilising selection, while under distorted conditions only few of them affect individual competitiveness. Hence, the huge, highly-specialised genetic information on environmental regulation is useless under strongly perturbed conditions and undergoes decay.

When environmental conditions strongly deviate from the natural ones, the genetic information of normal genotypes completely loses its meaning, and the value of n_C grows up to the lethal threshold n_L. In such a case all individuals with $n < n_L$ on average enjoy equal competitiveness (as is the case for fitness of all individuals measured outside the population). As a result, competitive interaction fails to perform its stabilising function. Decay of normal genotypes proceeds unimpeded and the decay polymorphism of the population increases. A quantitative measure of polymorphism is given by the expression $(4G/n)^n$ (9.5.1), which is equal to the number of different possible genotypes. Evidently, when $n > n_C$, decay polymorphism exceeds the normal one by many orders of magnitude and starts to dominate in the population.

Let us call the totality of decay genotypes with $n_C < n < n_L$ and corresponding decay individuals for the *decay tail* of the species.

Individuals from the decay tail of certain natural species may feature some phenotypic defects that decrease their competitiveness under natural conditions, but prove to be useful for humans. Such species served as the material for artificial selection and gave rise to all artificial plants and breeds of animals used by man. However, the overwhelming majority of species do not contain such properties in their decay tail. Artificial selection of such species proves to be useless.

Build-up of the relative number of decay individuals in a population represents erosion of the genetic stabilising programme of the species, which is otherwise responsible for correlated interaction between the species in their natural community and controls the mode of their behaviour, i.e. the necessary work of species aimed at maintaining stability of both the community and its environment.

Similarly, individuals of a given species may appear effectively deprived of the stabilising programme when placed into an alien community. Such individuals may act as 'gangsters', quickly increasing in number and destroying the correlated interaction and stability of the alien community and its environment. Among such gangsters there may be both normal and decay individuals. Reproduction of decay individuals is suppressed under natural conditions in their home community, while in the alien community they may be allowed to increase in numbers. Formally, such a situation corresponds to an increase in fitness of the corresponding decay individuals. However, it can be interpreted only mistakenly as genetic adaptation to altered environmental conditions. After the alien community is destroyed and the environmental conditions deteriorate, propagation of the 'gangsters' is also terminated, possibly ending in a complete extinction of their population. Thus, a burst of reproduction of a species in an alien environment has nothing to do with its long-term persistence in the aboriginal environment and cannot be interpreted as process of adaptation.

Insensitivity of individuals to diseases, toxins and other harmful factors not encountered under natural conditions may also appear due to decay of certain parts of the genome. This is often interpreted as the origin of meaningful information in the course of random changes of the genome. It is presumed that the same pattern may be realised in the course of evolution. In fact, however, only loss of the genetic information occurs in such cases. The corresponding parts of the genome that appear to be responsible for sensitivity to unnatural harmful factors, otherwise encode fragments playing an important role in the individual's life under natural conditions. Decay of these parts of the genome—which may be inflicted in many different ways, as any other decay—may accidentally result in higher survival of such decay individuals in the presence of unnatural toxins or infections. (Similarly, a car with a burst tyre regains some of its stability if the other three tyres are punctured as well – it 'adapts'.) In the overwhelming majority of cases the decay tail of a species contains no genotypes with such random adaptation to a prescribed change in the external conditions. However, due to the overall lack of evidence in favour of genetic adaptation, the few misinterpreted examples of the above kind are widely cited by the adherents of the genetic adaptation concept. For example, newly-acquired resistance of different organisms to artificially synthesised toxins (pesticides, antibiotics, etc.) that are absent in nature is often rendered as evolutionarily important (Sheppard, 1975; see also below).

Generally speaking, the normal genotype could be degenerated in the sense that a single normal phenotype could be encoded by a few normal genotypes. In such a case, certain (but by no means all) decay mutations could be compensated by the so-called suppressor mutations, i.e. mutations that cause the same effect as the reverse mutation (i.e. they restore the normal phenotype), being at the same time different from it. Suppressor mutations correspond to transitions from one normal genotype to another, the phenotype remaining unchanged. Degeneration of the normal genome actually means an effective increase in the number of reverse mutations in the sense that both the reverse and suppressor mutations restore the initial normal phenotype. Degeneration of the normal genome is neutralised by an increased accuracy of the process of competitive interaction. Suppressor mutations that under laboratory conditions seem to completely restore the normal phenotype, appear to be of decay nature under natural conditions of the maximum efficiency of competitive interaction. The resolution of competitive interaction being sufficiently high, all possible mutations (except for the reverse one) would lead to a decay genotype and decreased competitiveness of the individual.

9.8 STABILITY OF BIOLOGICAL SPECIES UNDER UNNATURAL CONDITIONS

When natural environmental conditions are restored, the maximum competitiveness of the normal genotypes is restored as well, so that normal individuals force all the decay individuals from the population at an exponential rate. The rate of such exponential forcing out is defined by the maximum biologically available difference between the birth and death rate of normal individuals, $B_0 - d_0$ (9.2.1), which may be called the biotic potential of the normal individuals. The time t of colonisation of the total population by descendants of a single normal individual is equal to the time during which the number of these descendants becomes equal to the total population number N, i.e. $t = \ln N/(B_0 - d_0)$. At $N \sim 10^6$ and $B_0 - d_0 \sim 1$ year, this time is equal to $t \sim 14$ years.

When external conditions have remained perturbed for a sufficiently long time, the processes of decay embrace all individuals in the population, so that not a single individual with a normal genotype is left and all individuals feature $n > n_C$. In that extreme case the population may return to its initial state after the perturbation ceases and the natural environmental conditions are restored, if competitiveness of decay individuals under the natural conditions monotonously increases with the number of decay substitutions n decreasing down to $n = n_C$. At any moment the maximum competitiveness is associated with those individuals that have the minimum number of decay substitutions, though it may still exceed n_C, $n_{\min} > n_C$. Such individuals are able to force all the other decay individuals with $n > n_{\min}$ from the population at an exponential rate. In that case we have for the noncompetitiveness (9.3.5):

$$\gamma_{n_{\min}} = 0, \qquad \gamma_{n>n_{\min}} \geq 1$$

Further process of relaxation back to the normal state of the population occurs due to reverse mutations and genetic recombination in the course of sexual breeding, which decrease the number of deleterious decay substitutions in the genotype (Section 10.1). Appearing individuals with $n_{\min} - 1$ decay substitutions in their genotype acquire the highest competitiveness and force individuals with $n_{\min}$ decay substitutions out of the population. We then have

$$\gamma_{n_{\min}-1} = 0, \qquad \gamma_{n>n_{\min}-1} \geq 1$$

and so forth until the minimum number of decay substitutions $n_{\min}$ falls down to the accuracy of the competitive interaction n_C (Figure 9.3) and the normal genetic information of the species is completely restored.

When all the fragments of the normal genome are present in the population, even though distributed over different decay individuals, genetic recombination makes it possible to reduce the minimum number of decay substitutions $n_{\min}$ by large steps from generation to generation, compared with the gradual decrease of $n_{\min}$ due to rare reverse mutations. Thus with genetic recombination operating in the population, the condition of strictly monotonous increase in competitiveness following a reduced number of decay substitutions may be relaxed.

The described pattern of relaxation of the population to the normal state does not necessarily follow a period of prolonged perturbation of the environment. The situation when the majority of the population is represented by decay individuals may occur accidentally as a result of a sharp reduction in the population number in the course of some ecological catastrophe followed by restoration of the population number owing to reproduction of the decay individuals that accidentally survived the catastrophe. Such sharp reduction in the population number is often referred to as the bottleneck effect (Nei, 1975).

Bottleneck events may result in a considerable reduction of genetic polymorphism in the population. If before the bottleneck the population consisted of normal individuals with $n \leq n_C$, after the bottleneck normal individuals will still dominate in the population, even if all the n decay substitutions are the same in all individuals, which corresponds to zero polymorphism. Thus, reduction of normal polymorphism cannot have any adverse effect on the population.

If, at the moment of the bottleneck, the population existed in perturbed environmental conditions, so that most part of polymorphism was determined by decay individuals with $n \leq n_C$, then after the bottleneck normal genotypes may appear to be absent from the population altogether. In such a case, the process of relaxation of the population to the normal state after restoration of the natural environment will take a longer time and will follow the above described pattern (Figure 9.3).

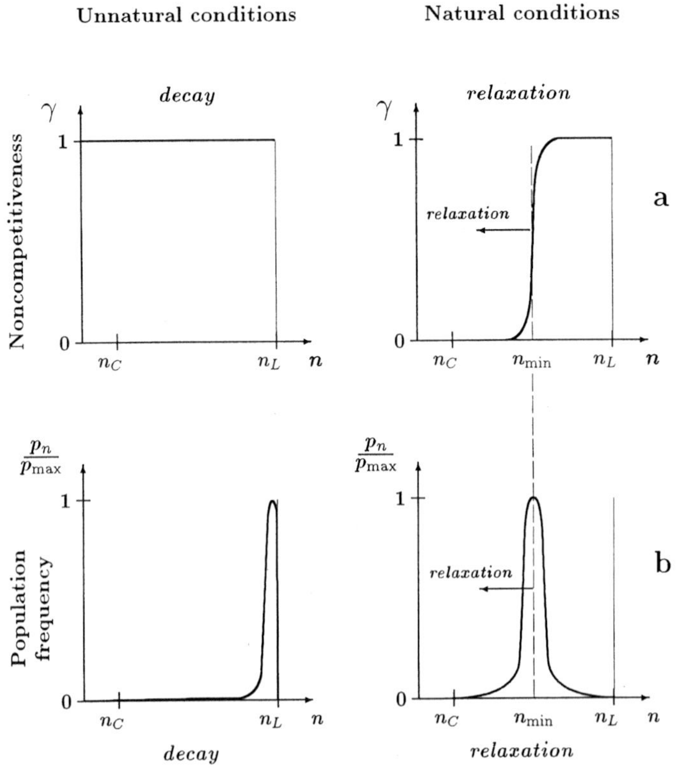

Figure 9.3. Decay polymorphism of genotypes in a population outside the natural conditions of the ecological niche of the species. n_C is the number of decay substitutions characterising the sensitivity of competitive interaction and stabilising selection, which represents the limit to accumulation of decay substitutions in the genotype; n_L is the maximum number of decay substitutions compatible with viability of the individual; n_{min} is the least number of decay substitutions encountered in genotypes at the current moment. **a** – noncompetitiveness of individuals related to the number n of decay substitutions in their genotypes, see (9.3.5). **b** – relative population frequency of genotypes with n decay substitutions. Under unnatural conditions competitive interaction and stabilising selection are switched off. Competitiveness of normal individuals decreases (and noncompetitiveness increases) down to the competitiveness of decay individuals with $n > n_C$. Decay substitutions accumulate in genotypes of individuals up to the lethal threshold n_L. When natural environment is restored, competitiveness increases with decreasing n until the limit of sensitivity of competitive interaction n_C. At $n > n_C$ individuals with the minimum number n_{min} of decay substitutions in their genotypes regain the maximum competitiveness (minimum noncompetitiveness). Relative population frequency relaxes to its normal state (see Figure 9.1).

In the absence of normal genotypes in the population, genetic relaxation to the normal genotype may follow different ways, determined by different successions of reverse mutations and acts of recombination. The process of relaxation should also depend on the pathway via which natural environmental conditions are restored for the given species, because competitiveness of individuals is a function of both their genotype and their environment (see Section 2.11). When comparing decay populations at various stages of their relaxation, it may appear that each population is best fitted to those perturbed conditions under which it exists. When transferring the decay population into different perturbed conditions it may happen to be less fitted to them than the aboriginal decay population. That phenomenon is observed in most domesticated animal and plant species which differ strongly from their natural predecessors (Begon *et al.*, 1986). Under extreme unfavourable conditions not all the individuals die simultaneously. There is always an individual (and, consequently, a genotype) that dies last. Under different unfavourable conditions, a different genotype will be the last to perish. Evidently, such a situation cannot be interpreted as a process of adaptation of the two different genotypes to different unfavourable conditions (see also Section 1.4), because neither of them is actually able to persist.

There are two very clear illustrations of the above described processes of decay and following relaxation of the genome. These are sickle cell anemia and industrial melanism, which are often cited as the textbook examples of genetic adaptation to changing environment.

Sickle cell anemia is an inherited red blood cell disorder in humans. It is caused by a single nucleotide substitution in the DNA encoding the β-globin gene. In a homozygous state (i.e. when both copies of the human genome contain the substitution) sickle cell anemia leads to death of the patient if no serious medical treatment is involved. In a heterozygous state (i.e. when only one of the two genome copies carries the decay substitution, the other retaining the normal gene) the symptoms of the disease are not so severe, though heterozygote patients may encounter problems under stress or conditions of lower oxygen abundance.

It was noticed that people who had the sickle cell mutation in the heterozygous state appeared to be more resistant to malaria than normal individuals. A certain proportion of the offspring of two heterozygous individuals always gets the sickle cell gene in a homozygous state, which under natural conditions leads to the death of such individuals and, consequently, elimination of the decay gene from the population. Thus, under natural conditions with the malaria infection absent, the population frequency of the sickle cell genes is kept at a negligibly low level. Meanwhile, under conditions of malaria epidemics, the sickle cell gene frequency increases owing to the better survival of sickle cell heterozygotes which compensates for the death of homozygotes. This is interpreted as an increase in fitness of sickle cell heterozygotes as compared to normal individuals. **However, under the unnatural conditions of malaria epidemics, fertility and, consequently, fitness of *all* the individuals (including sickle cell heterozygotes) in the population is reduced.** This is in principle equivalent to the above hypothetical situation when all individuals die under unnatural conditions irrespective of their genotypes, and there is one particular genotype that is the last to vanish. In very much the same manner as one cannot speak about adaptation of this particular genotype to the unnatural environment, one cannot speak about increased fitness of sickle-cell heterozygotes. The fact that sickle cell genes are lethal when homozygous unambiguously suggests that the natural ecological niche of humans is characterised by complete absence or extremely low rate of occasions of malaria infection. The observed high frequency of malaria cases that has led to an increase in frequency of sickle cell genes is most probably caused by the unnaturally high population density of the corresponding human settlements compared with the population density characterising the natural ecological niche of humans (see also Section 11.5).

Industrial melanism is observed in the peppered moth, *Biston betularia*, as well as in a number of other insect species. Under natural conditions the species is almost entirely composed of light grey moths. Appearance of coal-black 'melanic' forms was registered more than a hundred years ago in several industrial areas of England. In a hundred years black forms significantly increased in numbers and, in some areas, made up more than 90% of the total number of moths. It was hypothesised that, since predatory birds are more likely to eat the most conspicuous moths, melanic forms would increase as a result of industrial pollution of the environment, which leads to appearance of black polluted surfaces (e.g. tree-trunks) where melanic forms would remain unnoticed by predators (Kettlewell, 1973). It was also noted that when anti-pollution measures were taken, the number of melanic forms began to decline (Bishop and Cook, 1975).

In spite of the fact that industrial melanism has been referred to as 'evolution in action' and become the textbook example of natural selection, the observed pattern can be readily interpreted without any evolutionary implications involved. Under natural conditions the melanic form, as well as the sickle cell genes in humans, is present at a low, if any, frequency. That means that under natural conditions decay genes responsible for industrial melanism significantly deteriorate normal genetic programme of the species. In urban areas natural ecological communities are significantly disturbed by anthropogenic activities, so that the normal genetic programme of the species does not give any advantage to its carriers. Under unnatural conditions any decay individuals (including melanic ones) may randomly increase their numbers. When the natural conditions are restored, the normal individuals regain their maximum competitiveness and the normal phenotype of the population is restored. Such a process may not have any pertinence to evolution, which is known to be irreversible.

It would be also a mistake to say that while the light moths are better adapted to natural conditions, the melanic forms are better adapted to industrial conditions. Normal light moths carry a genetic programme of correct interaction with all the other species in the community aimed at long-term maintenance of both the community and its environment in a stable state. In this sense light moths are self-sustainable within their community provided that external disturbances do not go beyond a certain limit of the community's resilience. Meanwhile melanic moths are apparently deprived of the stabilising genetic programme, as soon as they appear non-competitive under natural conditions. Melanic moths are just able to exist under an unnatural industrial environment which is in itself unstable and is continuously degrading on a global scale under the growing pressure of anthropogenic activities. In other words, light moths are able to live in an environment which is characterised by a long-term stability, to which they also contribute. Melanic moths are able to exist in an environment which is inherently unstable. Thus, there is a principal difference between the two forms, the latter being but a decay state of the former, with no new meaningful genetic information gained, which is contrary to the situation taking place in the course of evolutionary process.

Environmental conditions of the natural ecological niche ascribe the maximum competitiveness to the normal genome. The relative number of decay individuals is then defined in the process of their competitive interaction with normal individuals, which takes place independently of the availability of resources (see Section 2.2 and 5.7). Population density of normal individuals is prescribed by the information contained in the normal genome and may be sustained at a level much lower than that permitted by the available resources of the ecological niche. Particularly, such a situation is realised for most large animals (see Chapter 4). That makes it possible to support stable existence not only of the species itself, but also of its environment, including all the other species in the ecological community.

Within the picture described in this section, within the natural ecological niche any biological species is unambiguously determined by its normal genome. When environmental conditions fluctuate, random oscillations occur around that normal genome without any directional changes taking place. Biological species cannot continually crawl away from the normal genome, adapting to changing environmental conditions. Only such genetically stable species and their communities are capable of stabilising the optimal state of the environment.

Accumulation of decay individuals in populations accompanying environmental perturbations should not disrupt the stabilising programme of the whole community. If a certain species completely loses its stabilising programme owing to excessive accumulation of decay individuals, such a decay species may be envisaged as an additional perturbation with respect to the other species in the community. The other species should retain their stabilising programme ensuring the community's capacity to compensate environmental perturbations (both biotic and abiotic) within some short time period. In such a case the decay species also returns to its normal state after natural conditions are restored due to the compensating reaction of the community as a whole. In the opposite case, the environment may go outside the scope of control, so that the whole community will be doomed to perish.

Thus the additional perturbation caused by incorrect functioning of the decay species is much more dangerous than complete extinction of any species in the community. Extinction of any species weakens the stabilising potential of the community, while the perturbation it has to cope with remains the same. Meanwhile incorrect functioning of a species not only impairs the stabilising potential of the community, but adds to the external perturbation. Accordingly, artificial, arbitrarily-composed communities of fields, pastures and other cultivated biological systems present much more danger to environmental stability than complete absence of biota on those territories.

It is therefore very important to ensure dominance of normal genotypes in populations of the principal species of the community, i.e. those consuming the major part of energy fluxes in the community (i.e. plants, bacteria and fungi, Section 3.7). That should be guaranteed irrespective of current perturbations of the environment. Hence, prolonged deviations from the natural environmental conditions should be prohibited for small populations, where decay individuals may quickly propagate over the whole population. Small populations may only exist under very stable environmental conditions characterised by only minor fluctuations.

This limitation allows the explanation of some observed regularities of species diversity in different ecosystems. When the biomass of the whole community is fixed, an increase in species diversity of the community naturally results in a reduction of population numbers of species. Thus communities of high species diversity may only exist under quite stable external conditions (cf. Section 3.6).

The highest species diversity is found in tropical forests on land, and in the coral reefs at sea. Communities of tropical forests and coral reefs indeed exist under extremely stable environmental conditions. In the absence of any sharp perturbations of abiotic nature which otherwise are uncommon for these ecosystems, these communities maintain their environment in a stable state, quickly compensating all spontaneous fluctuations of the environment in accordance with the biotic Le Chatelier principle (Section 5.5). Communities of rainforests feature the highest productivity and greatly provide for stabilisation of the environment for the whole of the biosphere on the global scale. At the same time, these communities appear 'unprepared' for sharp anthropogenic perturbations. Such large-scale perturbations are not encountered in their natural environment. Hence, the normal genome of such communities cannot contain a programme of compensation for such large-scale disturbances. These communities rapidly degrade when exposed to anthropogenic influence. As a result, the global environment also appears to be destabilised.

Communities of the moderate and polar areas exist under strongly fluctuating environmental conditions. Contemporary terrestrial biota is apparently incapable of suppressing these fluctuations, stabilising the environmental conditions in these zones to such an extent that communities with high species diversity, low numbers and low stability of each separate population would become possible. Communities in those zones feature low species diversity, high population numbers and, respectively, high stability of the genomes of all the species. These communities may suffer stronger perturbations, anthropogenic included, without losing their stability. However, their role in stabilising the overall environment for the biosphere is significantly smaller than that of the tropical and subtropical communities, since their production contributes less to the overall production of the biosphere (Whittaker and Likens, 1975).

9.9 BIOLOGICAL SPECIES: DEFINITION

To give a clear and non-contradictory definition of species has been one of the challenges in biological science. Biologists describe species on the basis of differences in morphology and behaviour. The number of characteristics involved may be so high and the differences themselves so exotic that taxonomy may sometimes be compared to an art. However, it is taxonomy that is traditionally considered as the highest instance in doubtful cases.

The attempts to give a formal unifying definition of a species as an array of populations within which sexual reproduction takes place and which are reproductively isolated from populations of any other species, encounter many contradictions. Many apparently different species may produce viable and fertile offspring. Impossibility of interspecific breeding between congeneric species is an exception rather than a rule (Raven and Johnson, 1988). This fact is often used by humans in conservation programmes. For example, the present-day population of Przewalski horses (*Equus przewalskii*) has been restored via breeding of the remaining few Przewalski horses with domestic horse *Equus caballus* (Bowling and Ryder, 1987). The same pattern was followed in restoration of some of the present-day populations of European bisons (*Bison bonasus*) that were bred with American bisons (*Bison bison*).

On the other hand, in some cases crosses between individuals from different populations apparently belonging to the same species may produce inviable offspring or remain infertile altogether, as is the case with the so-called ring species, which represent an array of slightly different populations where each two adjacent populations freely interbreed, while 'the ends' of the species range appear to be so genetically distant from each other that they do not interbreed (see Section 9.5).

Finally, many species in the biosphere are asexual. Though such species are not very numerous compared with sexual ones, some of them (e.g. bacteria) play a principal role in the overwhelming majority of ecosystems being responsible for decomposition of more than 90% of the primary production (see Section 3.7). The fact that these organisms are completely ignored in the above formal definition of species points to sufficient flaws in the latter.

At present it is not possible to give a quantitative genetic definition of species either, in spite of a great number of well-documented differences in DNA of different species. In some close species the number of identifiable genetic differences does not exceed the intraspecific genetic diversity. For example, the average number of randomly differing proteins between two individuals belonging to the same *Drosophila* species is of the order of 12% (Nevo *et al.*, 1984), while the number of fixed protein differences between some *Drosophila* species may be less than one percent (Coyne and Orr, 1989). Thus, the interspecific genetic difference may not be taken for the basis of definition of species.

Most difficulties are encountered when it is necessary to delimit species and the so-called subspecies, which approximately correspond to human races. On the basis of the biotic regulation concept, we propose a constructive definition of biological species that makes it possible to differentiate between subspecies of a single species and different species.

We have noted repeatedly that any biological species is determined by the information contained in the normal genome. The normal genome only remains functionally sensible under conditions of the natural ecological niche of the species and within the ecological community to which the species belongs. Level of intrapopulational genetic variability is determined by the sensitivity of the process of competitive interaction. Individuals with genotypes containing a number of deviations from the normal genetic programme less than the critical threshold value feature equal competitiveness. All populations of the species are characterised by the same genetic programme encoded in the normal genome. Genetic differences between populations are due to different localisation of decay deviations from the normal genetic program caused by random nature of the processes of genome decay and genetic drift. **Thus, all functionally sensible fragments of the normal genome ('genetic signal') are the same in all populations of the species. Decay deviations from the normal genome ('genetic noise') are different in isolated populations.**

On this basis one may formulate a criterion which allows us to discriminate between two different species and two subspecies of a single species coexisting on a given territory. This criterion does not depend on the mode of reproduction of the species and may be used for asexual organisms as well.

Consider two populations coexisting on the same territory. If the population density of individuals from both populations may be characterised by a relatively constant time-independent value, the two populations belong to different species and occupy different ecological niches in the same community. If the population densities strongly fluctuate with time and cannot be characterised by a certain definite value, individuals from the two populations belong to one and the same biological species and interact competitively with each other.[6]

Each species performs a strictly specified amount of unique work on stabilisation of the environment, so that the population density of its individuals is of immediate importance to the community. In fact population density of each species is an important environmental characteristic for all the species of the community, as well as such abiotic characteristics as temperature, humidity, air pressure, etc. Due to this fact population densities of different species should retain their optimal values corresponding to the most efficient functioning of the community as a whole. Under natural conditions they may not undergo random changes; in particular, one species cannot expand at the expense of the other. Such a situation could be compared to substitution of certain parts of a working mechanism by some other parts of the same mechanism, which are also important when in appropriate numbers. For example, a car works well when it has four wheels and one motor, while one wheel and four motors make quite a different situation.

Meanwhile two different subspecies carry essentially the same meaningful genetic programme the difference lying only in the localisation of decay substitutions. Thus, changes in the ratio of population numbers of different subspecies make no difference for the community. As a result, population densities of co-existing subspecies may fluctuate quite arbitrarily provided that the cumulative number of individuals of that species remains approximately the same. Random fluctuations in population numbers of different subspecies is caused by competitive interaction of their individuals.

This is particularly well illustrated by the situation with human races. World history may be envisaged as a number of large-scale fluctuations of population numbers of people belonging to different races, which are caused by competitive

[6] If gradual morphological and behavioural changes are observed along the species range, all populations (both sympatric and allopatric) inhabiting that range should be classified as a single species. This follows from the observed discreteness demonstrated by the majority of extinct as well as extant species.

interaction between different races and nations. The very fact of existence of such fluctuations and of competitive interaction between people belonging to different races and nations points unambiguously to the genetic equivalence of all races and nations and proves that they all belong to a single species.

10

Genetic Bases of Biotic Regulation and Life Stability: Analysis of Empirical Evidence

All the known features of the genome structure, the characteristics of life cycle, the morphology and behavioural traits of individuals work to enhance the genetic stability of species and to speed up the process of relaxation back to the normal genome after external perturbations of natural environmental conditions. The available quantitative data are consistent with the statement that under natural conditions intraspecific genetic variation is allowed to accumulate up to a certain limit prescribed by the sensitivity of competitive interaction and stabilising selection.

10.1 GENETIC RECOMBINATION

As noted in the previous chapter, competitive interaction of conspecific individuals prevents genetic information of the species from decay. Efficiency of competitive interaction is at its maximum under conditions of the natural ecological niche of the species. In a perturbed environment competitive interaction becomes weaker and deviations from the normal genetic programme may accumulate in genotypes of individuals. Restoration of the normal environment switches on competitive interaction which ensures relaxation of the genotypes to the normal genome (see Section 9.7 and Figure 9.3).

Relaxation to the normal genome may occur as a result of reverse mutations that lower the number of decay substitutions in the genotype, which results in a stepwise growth in competitiveness of the corresponding individuals and following exponentially quick capture of the whole population by individuals with the lowest number of decay substitutions. However, reverse mutations occur very rarely. For example, to observe a single reverse mutation in a particular genome site in a population of mammals, $(\nu k)^{-1} \sim 10^9$ individuals should be accumulated, i.e. 10^4 generations should pass in a population of 10^5 individuals. Here ν is the rate of mutations per site per cell division, k is the number of divisions in the germ line (see Section 9.1).

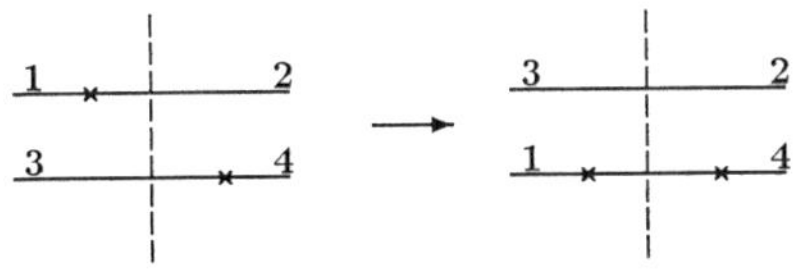

Figure 10.1. Genetic recombination—exchange of genetic material between chromosomes. Crosses indicate deleterious substitutions in the two chromosomes. After recombination one of the new chromosomes contains two decay substitutions, while the other does not contain them at all. If the cut (dashed line) happens to the one side of both decay substitutions, recombination does not change the number of decay substitutions in the chromosomes. Genetic recombination significantly accelerates the process of relaxation of a population to the normal genome.

Genetic recombination (crossover) may drastically accelerate the rate of genetic relaxation of the population to the normal state. Basically, recombination represents exchange of portions of genetic material between homologous chromosomes when the latter are engaged in the production of gametes.

Assume that there are two homologous chromosomes, each containing one decay substitution. The probability of random coincidence in localisation of these substitutions is negligibly small. Let us cut both chromosomes at some arbitrary but identical point and connect the right-hand part of one to the left-hand of another and vice versa (see Figure 10.1). The obtained two new chromosomes get the head of one parent chromosome combined with the tail of the other. If the cut (the chiasma) passes between the decay substitutions, one of the two new chromosomes will contain no decay substitutions, that is, it will be normal, while the second one will contain two decay substitutions. If both decay substitutions lie to one and the same side of the chiasma, the recombination changes nothing.

In the course of gamete production a number of chiasmata occur in the process of each act of recombination (Cano and Santos, 1990; Burt *et al.*, 1991). Contact (syngamy) of two chromosomes belonging to different genomes is always needed for genetic recombination to occur, which is attained in the course of sexual breeding of individuals. Note also that in the course of the individual's life cycle there should be a period of diploid phase in which the haploid genomes (or haploid parts of the genomes as is the case with bacterial recombination) of two different individuals combine into one. Different species allocate the most part of their life cycle to either diploid or haploid phase (see Section 10.3, Maynard Smith, 1978; Kondrashov, 1988; Michod and Lewin, 1988).

In a general case genetic recombination between homologous chromosomes results in the appearance of genomes with both higher and lower numbers of decay substitutions, as compared with the parent genomes. Recombination is a random process which brings about a Poisson distribution of the number of decay substitutions in the recombinant genomes of the offspring (Maynard Smith, 1978).

The Poisson multiplicity, Eq. (9.1.6), i.e. the average number of mutations in the recombinant genomes should, apparently, coincide with the average number of decay substitutions in the genomes of the parent population, as soon as genetic recombination does not change the total number of decay substitutions in the population. If decay substitutions already followed a Poisson distribution in the parent population, recombination would not change anything if no reverse mutations took place in the parent genomes. However, if the parent population is characterised by a narrow peak in the $n > n_C$ range (Figure 9.3), recombination leads to appearance of normal individuals with $n < n_C$ as early as in the first generation of the offspring. **Hence, genetic recombination accelerates the process of genome relaxation by several orders of magnitude, as compared with reverse mutations.**

If all the fragments of the normal genome are present in the population, which otherwise consists of decay individuals each carrying some fragment of the normal genome, sexual breeding and genetic recombination may result in recovery of normal individuals as early as in the first generation of offspring. Recombination makes it possible to bring together all the fragments of the normal genome into a single genome of an individual in the offspring.

Thus genetic decay of a sexual population (or even a whole species) placed under unnatural conditions accompanied by loss of all normal individuals with $n \le n_C$ and accumulation of decay individuals with their numbers of decay substitutions close to the lethal threshold $n \sim n_L$ presents no danger for survival of the species. Thanks to sexual breeding and genetic recombination, restoration of the natural environment and normal process of competitive interaction is almost immediately followed by recovery of the high population frequency of normal individuals. Note that sexual breeding does not necessarily imply sexual dimorphism, i.e. splitting of the population into males and females. The only essential thing is that exchange of genetic material between different individuals must be ensured. Thus hermaphrodites (i.e. organisms that are able to perform functions of both male and female) are also able to make use of genetic recombination.

To support a stationary state of a continuously decaying genome, condition (9.3.2) should be satisfied, which equals the rate B_0 of appearance of normal individuals in the population with their death rate d_0, $B_0 = d_0$. Note that the rate of appearance of normal individuals is not equal to the total number b_0 of progeny produced by normal individuals of the former generation, because a certain part of this progeny is composed of individuals that carry additional decay substitutions absent in the genomes of their parents, see (9.3.2). With μ being the average number of mutations per genome per generation, the number of normal individuals appearing in the population during a single generation is given by $B_0 = b_0 e^{-\mu}$ (Section 9.3). When μ is small ($\mu \ll 1$) the stationary condition $B_0 = d_0$ means that each normal individual should produce about one offspring during its lifetime, $b_0 = d_0 e^{\mu} \approx d_0$. However, to have at least one normal individual among the offspring of each normal parent at high values of μ, $\mu \ge 10$, the total number of offspring produced by a single (pair of) normal parent(s) should exceed $e^{\mu} \ge 2 \cdot 10^4$.

Many species are indeed characterised by a very high fecundity (number of offspring produced by an individual during its lifetime). The number of seeds in many higher plants satisfies that condition (Wilson, 1975). The number of eggs spawned by a single fish during its lifetime often exceeds 10^6 (Wilson, 1975). Both higher plants and fishes are on average characterised by large genome sizes and, consequently, by large values of μ (9.1.3). (Note that high fecundity is not necessarily related to high values of μ. It may result from ecological reasons as well, when the main ecological work specific to the given species is carried out by the numerous juvenile individuals, while the few adult ones are only responsible for reproduction of those juveniles. Such is presumably the situation with fishes.) However, in both mammals and birds their progenies cannot be numerous because of their endothermic (warm-blooded) nature. There are limits imposed on body sizes of endothermic animals. Very small animals cannot in principle sustain constant body temperature (Gorshkov, 1995). While the ability to have numerous offspring is necessarily characterised by small individual body size (e.g. fish roe), this means that fecundity of endothermic animals is limited.

A stationary state of genetic organisation in organisms with low fecundity and large value of $\mu, b_0/d_0 \ll e^{\mu}$, may be ensured if a large number of offspring characterises the haploid phase of otherwise diploid organisms. For example, in humans ($\mu \approx 10$) the number of haploid spermatozoids in sperm ejaculate exceeds 10^8 (Vogel and Rathenberg, 1975), while only one of them fertilises the ovule. Decay mutations affecting germ cells may be thus selected out at the haploid stage.

10.2 SEXUAL DIMORPHISM AND REGULATION OF BIRTH RATE OF DECAY INDIVIDUALS

The considered advantages of sexual breeding and genetic recombination relate to both single sex (hermaphrodite) and bi-sexual (sexually dimorphic) species.

In bi-sexual species only part of the population (the females) is capable of actually producing the offspring, while in single-sex species all individuals in the population are capable of producing offspring (such as hermaphroditic and parthenogenetic species). This leads to a reduced biotic potential of the population (i.e. of the maximum possible value of fitness, $W_0 \equiv b_0 - d_0$, which defines the maximum possible rate of exponential growth of the population), which is considered to be a significant drawback of the bi-sexual strategy of existence (Wilson, 1975; Maynard Smith, 1978; Bell, 1982; Hamilton *et al.*, 1990). Large non-productive losses of both pollen and sperm are envisaged as another drawback of sexual breeding. However, no population ever increases its numbers at a rate close to the biotic potential under natural conditions. In some species of invertebrates the sex ratio of emerging adults can vary from nearly 100% males to nearly 100% females (Clutton-Brock, 1982). Therefore, halving the biotic potential in a bisexual population with an equal number of males and females against that of a hermaphroditic unisexual or an asexual population can never manifest itself under the natural conditions and will never lead to an advantage of unisexual species over bisexual ones. Energy losses associated with futile expenditure of either pollen or sperm are negligible in comparison with natural variations of individual productivity. Therefore, genetic recombination, based on unisexual (hermaphrodite) and bisexual (sexually dimorphic) modes of breeding are practically equivalent to each other in all the characteristics so far considered.

However, the bisexual strategy of genetic recombination is a much more complex system, which requires additional genetic information to form two different types of individuals in the population—the male and the female. In this respect the bisexual mode of existence may indeed be considered to be more costly compared with the unisexual one. Therefore the observed wide spread of sexual dimorphism in the biosphere requires an explanation.

As demonstrated in Section 9.3, it is only possible to prevent genetic degradation of the population and support it in a stationary state when the death rate of decay individuals is forcefully increased or their birth rate is decreased compared with that of normal individuals.

When the strategy of regulation of the birth rate of decay individuals is realised, the average lifetime of both normal and decay offspring (born to normal parents due to the inevitable processes of genetic decay) may be kept the same. The alternative strategy (that of regulation of the death rate of decay individuals) is physical elimination of all the decay individuals at early stages of their development, so that their average lifetime is low compared to that of normal individuals. Such a strategy requires additional energy expenditures from normal individuals. Within the first strategy, juvenile mortality is low. Normal individuals do not spend their energy on forced elimination of decay individuals. Decay individuals are allowed to exist in the population until their life capacity permits. Such a strategy permits the identification of normal and decay individuals with high precision. The competitive capacity of an individual and the quality of its genotype are assessed during the individual's lifetime. It cannot be predicted or estimated *a priori* due to a huge number of characteristics involved. The longer individuals live, the more precise determination of their competitiveness is possible. Thus it proves to be advantageous for a species to keep the juvenile mortality at a low level allowing all born individuals to manifest themselves during their normal lifetime. Meanwhile the low birth rate of decay individuals necessary for the stationarity of population is maintained by normal individuals that prevent decay individuals from reproduction.

However, that is not an easy task to perform. By all appearances, there is only one way to regulate the relative birth rate of decay individuals (briefly discussed in Section 2.3). That is polygamy based on sexual dimorphism, during which two different but correlated normal genomes—of the male and the female—appear in the population. Sexual dimorphism is often understood as apparent morphological differences between the male and the female. However, the main difference between them is functional. The male is incapable of reproduction by himself. The female is only capable of reproduction after sexual contact with the male.

In such a situation it is possible to translate all the processes of competitive interaction into sexual relations between individuals. The stabilising function will then be performed by sexual selection. The normal male may prevent one female (a decay one) from reproduction, while stimulating another female (a normal one) into it. Similarly, the normal female may accept for mating the normal male and reject the decay male. Competitive interaction between males is thus aimed at winning the normal female instead of being determined by availability of environmental resources. In one way or another, competitive interaction between males results in suppressing sexual activity of the decay males. As a result, most offspring in the population appear to be produced by normal females and males, while reproduction of all decay individuals is suppressed (see Figure 2.2). Competitive interaction between females may be either completely switched off or minimised (Partridge and Harvey, 1986), which in many cases allows the female to completely devote her time to parental care. All these advantages have worked to make sexual dimorphism and sexual selection widely spread throughout the animal world independent of species ecology (Emlen and Oring, 1977; Bradbury and Andersson, 1987).

Sexual dimorphism in plants (diclinous plants) also provides for regulation of birth rate of decay individuals: specific aerodynamic properties of pollen, specific structure of the female sexual organs to catch the normal pollen and other similar means may be here employed. However, that does not exclude competitive interaction between all the individuals within the population of plants, so that the principal mechanism for genetic stabilisation of the species in the plant world remains that of regulating the death rate of the decay individuals. Thus sexual dimorphism among plants is a feature much less developed than in the animal world. The relative number of diclinous plant species constitutes not more than several per cent of the total number of plant species (Maynard Smith, 1978; Geodakyan, 1987).

All the advantages of sexual dimorphism are only manifested in polygamy, be that polygyny (one male having many female sex partners) or polyandry (one female having several male sex partners). In fact, many cases of polyandry may actually represent hidden polygamy: decay males either totally or partially barred from sexual contacts are used to help the females to bring up the offspring (Gorshkov, 1995; Andelman, 1987), while normal breeding females keep coupling with normal males.

Meanwhile monogamy loses all the advantages of sexual dimorphism. In monogamous bisexual species, as well in hermaphrodite and parthenogenetic ones, regulation of the death rate of decay individuals is only possible when the number of decay individuals in the population is reduced via increased juvenile mortality. When the monogamous pattern is governing the population structure (i.e. when every two individuals unite for life to produce offspring), it becomes practically impossible to prevent two decay individuals from reproduction. Meanwhile under conditions of polygamy normal males control reproduction of all the females in the population and vice versa. In its stabilising strategy monogamy does not differ from hermaphroditism at all. Monogamous couples of sex partners may be envisaged as an analogue to a single hermaphrodite individual capable of self-fertilisation. All the population consists then of internally correlated associations—families—composed of monogamous couples and their offspring. Competitive interaction and stabilising selection then operate at the level of such associations. At low population densities hermaphroditism and parthenogenesis may appear even more advantageous than monogamy, since there is no need to waste energy seeking for a sexual partner.

As indicated in Section 9.3, see Eq. (9.3.4) and below, regulation of the birth rate of decay individuals and sexual dimorphism may only be completely effective at low values of genome decay rate μ. In such a situation the population overwhelmingly consists of normal individuals, so that the presence of a small number of non-breeding decay offspring of normal parents does not perturb the normal genetic information of the population. In an opposite case, when $\mu \gg 1$, i.e. each normal individual needs to produce a huge number of offspring in order to ensure that at least one of them is normal, only regulation of the death rate of the decay individuals may be effective. That is seemingly testified to by the data on plants for which $\mu \gg 1$ (Williams, 1975). The majority of plants exhibit no sexual dimorphism (Maynard Smith, 1978) and hence do not ensure regulation of birth rate of decay individuals.

Meanwhile in animals, including those with large values of μ (mammals, birds), sexual dimorphism is widely spread and, hence, the strategy of regulation of the birth rate of decay individuals and maintenance of relatively low juvenile mortality is followed. This is made possible owing to the fact that in these species regulation of the death rate of decay individuals (which is inevitable at large values of μ) is transferred from the diploid phase to the haplophase stage of the life cycle. In haplophase it is possible to attain huge population numbers of haploid individuals (germ cells) and select out the majority of decay individuals. As already noted, in humans one spermatozoon out of 10^8 present in the ejaculate fertilises the egg cell. Strong selection of decay germ cells in the haplophase makes it possible to keep the juvenile mortality at a negligibly low level in the diploid phase, where the organisms spend most part of their life cycle.

10.3 HAPLOIDY AND DIPLOIDY

Most species in the biosphere including humans are diploid, i.e. they carry two copies of their genetic material in most of their cells. Certain organisms (e.g. bacteria) are haploid, i.e. they carry a single copy of their genome. Finally, some species (some plants, amphibians, fishes) display polyploidy, which means that they have more than two copies of their genetic material in their cells.

In a multicellular organism, genetic information pertaining to individual development and complex concerted action of its cells and organs must be stored. Consequently, multicellular organisms have large genomes with the minimum size of 10^8 bp (base pairs or nucleotide pairs); most multicellular animals have genomes of 10^9 to 10^{10} bp (White, 1973). For comparison, bacteria are characterised by genome sizes of the order of 10^7 bp (Lewin, 1987).

Large genomic and body size entails the problem of somatic mutations. Somatic mutations are those that are not inherited from parents but occur during the lifetime of an organism in its somatic cells.[1] Deleterious *germline* mutations are eliminated from the population by selection of individuals. However, deleterious *somatic* mutations cannot be thus eliminated because cells of different organs in a multicellular organism are highly differentiated and associated and cannot substitute each other like individuals in a population. Thus, a multicellular organism carries a load of somatic mutations. We demonstrate below that this load is proportional to the genome and body size. One can assume that, if the critical genomic and body sizes are exceeded, the existence of a multicellular organism in the haploid phase is impossible and the organism becomes diploid. Diploidy (i.e. presence of two homologous copies of the genome in each somatic cell) ensures protection from

[1] A somatic cell is any cell of the body except for germ cells (sperm cells and egg cells) and their precursors.

the deleterious mutation effects (Efroimson, 1932; Orr, 1995; Gorshkov and Makarieva, 1999): the copy that is not mutationally affected can compensate defective function of the mutant copy.

The mean number of somatic mutations μ_S per cell of an organism with the genome size G and body mass M can be estimated as follows. The mutation rate per one nucleotide pair (bp) per cell division (d) is of the order of $\nu \sim 10^{-10}(\text{bp})^{-1}\,\text{d}^{-1}$ (see Section 9.1). We can estimate the number of cell divisions in a somatic line k_S using the dichotomous approximation, i.e. assuming that the whole body of the multicellular organism is built on the basis of dichotomous cell divisions only. In this case, an organism that consists of n cells will have $k_S = \log_2 n$ somatic cell divisions. The value of n can be estimated as $n = m/m_{\text{cell}}$, where m is the body mass, and m_{cell} is the mean mass of cells of the organism. Taking mean mass of an eukaryotic cell to be 10^{-9} g and converting logarithms from binary to decimal ones, we obtain

$$k_S = \log_2 10 \cdot \lg(m/m_{\text{cell}}) = 3.3(\lg m + 9) \tag{10.3.1}$$

where m is the body mass in grams. Thus,

$$\mu = \nu G\, k_S = 3.3\nu G(\lg m + 9). \tag{10.3.2}$$

For small organisms having body mass of about 0.1 g and the genome size not exceeding that of haplodiploid insects $G_1 \approx 2 \cdot 10^8$ bp (Jordan and Brosemer, 1974; Rasch *et al.*, 1977) we obtain from (10.3.1)

$$\mu_1 = 3.3 \cdot 10^{-10} \cdot 2 \cdot 10^8 \cdot (\lg 0.1 + 9) \approx 0.5 < 1 \tag{10.3.3}$$

This means that the mean number of somatic mutations arising in each cell during development of the organism is less than unity, i.e. somatic mutations are absent or rare. This value of μ_1 can be regarded as the critical value, which determines the transition to diploidy, because males of haplodiploid insects apparently are the largest haploid animals having the largest genome.

Note that some haploid algae are large in size. This is explained by a weak association between their cells because algae lack the vascular system. The appearance of cells with an unacceptable number of somatic mutations may result in the death of these cells, but the normal functioning of other cells is not impaired. In animals, which have strongly associated cell systems, this course of events is impossible because the death of cells of any specialised organ may lead to death of the individual. Consequently, in algae, the correlation radius (the area within which the death of given cells results in the death of all cells associated with them) is small, not exceeding the size of strongly correlated small haploid organisms (Section 3.3). Thus, relationship (10.3.3) apparently holds for all haploid algae as well.

Large animals, e.g. mammals of about 10 kg (10^4 g) in mass, which have the 1 C (one copy) genome size $G_2 \approx 4 \cdot 10^9$ bp (Orlov and Bulatova, 1983), would have contained about 20 expressed somatic mutations in each cell, if the genomes of these cells were haploid. This is 40 times higher than the corresponding value for haploid insects μ_1 (10.3.3):

$$\mu_2 = 3.3 \cdot 10^{-10} \cdot 4 \cdot 10^9 \cdot (\lg 10^4 + 9) \approx 20 \gg 1 \tag{10.3.4}$$

It is known that about 1% of the genome is not polymorphic in any individual of any population (Lewin, 1987). This means that any mutations in this part of the genome lead to lethality or sterility of their carriers. Thus, if the mean number of mutations per cell is higher than unity (10.3.4), up to 1% of cells of an individual would contain mutations in impermissible regions, which would result in mortality of large haploid organisms having large genomes. Thus, diploidy seems to be a necessary prerequisite for the existence of large organisms with large genomes. The fact that $\mu_1 \ll \mu_2$ quantitatively demonstrates the significance of diploidy in the protection of an organism from deleterious somatic mutations.

This role of diploidy is confirmed by the results of experiments on irradiation of haploid males and diploid females of haplodiploid insects (Rasch *et al.*, 1977). Irradiation in these experiments serves as an artificial source of mutations. In haploid males, cells of many organs are known to acquire diploidy or even polyploidy in ontogeny. In adult males, most cells have ploidy of 2, 4 and more. However, at early stages of development, most somatic cells in these males remain haploid. In females, somatic cells have ploidy of 2 or higher throughout their life. When males and females are exposed to radiation at early developmental stages, males whose cells are still haploid die far more often than females. At later stages of development, this difference in mortality rates decreases and ultimately disappears when diplo- and polyploidisation of most male cells is completed. Finally, after irradiation of adult males and females, their mortality is equal. Thus, diploidy of somatic cells can have a substantial effect on individual mortality.

Estimation of the division number k_S in the somatic line on the basis of dichotomous approximation (10.3.1) is not exact because of the fact that after termination of individual development, stem cells in many tissues continue to divide and old cells are substituted by new ones although the organ and the entire body do not grow. However, only part of the genome, which is several times smaller than the total genome size, functions in all specialised organs. Totipotency, i.e. functioning of the entire genome required for development of the organism, is preserved only during the first cleavage divisions of the zygote. Consequently, the account for stem cell divisions in an adult individual is unlikely to change the order of magnitude of the mean number of mutation substitutions in somatic cells [(10.3.3.) and (10.3.4)] estimated using the dichotomous approximation.

Another argument in favour of using dichotomous approximation (10.3.1) is low variation of k_S in different organisms as compared with the variance in genome size G. Indeed, the genome of multicellular animals varies in different taxa by several orders of magnitude (from 10^8 bp in sponges to 10^{11} bp in some fish species and tailed amphibians) (White, 1973). By contrast, the minimal number of cell divisions in somatic line k_S, which is proportional to the logarithm body mass (10.3.1), changes in those organisms by less than an order of magnitude (from $k_S \approx 10$ in the smallest to $k_S \approx 50$ in the largest multicellular animals). Thus, the primary relationship is that between μ and G (10.3.2), whereas possible bias in the k_S estimation is not significant.

In diploid organisms, somatic mutations are fully expressed only in the case when they simultaneously occur in the same site of the two copies of the diploid genome. The probability of this double event is proportional to the product of the

Table 10.1. Dependence of the coding region of the genome on the total genome size in different organisms. After (Szathmary and Maynard Smith, 1995). The values were approximated by the power function $G_g = aG^b$, which gave $b = 0.58 \pm 0.06$ ($r = 0.98, p < 0.001$ with 5 d.f.).

Organism	Total genome size ($G \equiv 1$ C)(10^9 bp)	Relative size of the coding region (%) ($G_g/G \cdot 100$)	Absolute size of the coding region G_g (10^9 bp)
Bacterium (*E. coli*)	0.004	100	0.004
Yeast (*Saccharomyces*)	0.009	70	0.0063
Nematode (*Caenorhabditis*)	0.09	25	0.0225
Fruitfly (*Drosophila*)	0.18	33	0.0594
Newt (*Triturus*)	19.0	3	0.57
Man (*Homo sapiens*)	3.5	18	0.63
Lungfish (*Protopterus*)	140.0	0.8	1.12

probabilities of the single events (ν^2), i.e. this probability is very small. Thus, the appearance of diploidy opened the possibility for constructing organisms as large as is wished.

Equation (10.3.2) may be written for the coding region of the genome G_g:

$$\mu_g = \nu G_g k_S = 3.3\nu G_g(\lg m + 9), \tag{10.3.5}$$

where μ_g is the mean number of somatic mutations in the coding region of the genome. According to the data reviewed by Szathmary and Maynard Smith (1995), the size of the coding region of the genome G_g increases with the total genome size G following the relationship $G_g \sim G^{0.6}$ (Table 10.1). Because of this, the difference between μ_g values estimated similarly to (10.3.3) and (10.3.4) for large and small organisms having respectively large and small genomes, persists in being approximately $1.8 \cdot 20^{0.6} \approx 10$. Thus, these values differ approximately 10 times: by a factor of 1.8 due to the difference in k_S values between small and large organisms and by a factor of 6 due to the change in the coding genome regions, which corresponds to the 20-fold change in the total genome size:

$$\frac{\mu_{g2}}{\mu_{g1}} = \frac{G_2^{0.6} k_{S2}}{G_1^{0.6} k_{S1}} = \frac{(4 \cdot 10^9)^{0.6}}{(2 \cdot 10^8)^{0.6}} \cdot \frac{(\lg 10^4 + 9)}{(\lg 0.1 + 9)} \approx 20^{0.6} \cdot 1.8 \approx 10.$$

Thus everywhere below we operate with characteristics pertaining to the coding region of the genome, since this region is studied in most detail in a great number of organisms.

10.4 EFFECTIVE HAPLOIDY: AUTOSOMAL HETEROZYGOSITY AND SEX HEMIZYGOSITY

In all diploid organisms, a small portion of the genome can be haploid not violating condition (10.3.3). This possibility is used in nature for chromosomal determination

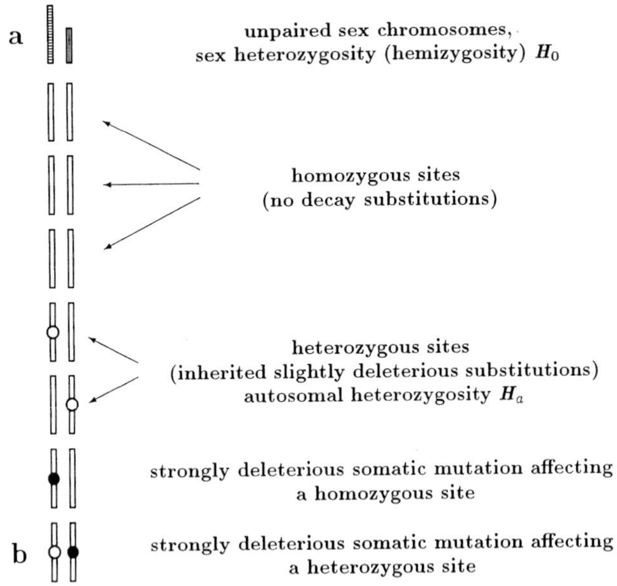

Figure 10.2. Schematic description of heterozygous and homozygous sites of a diploid genome of a heterogametic individual. In homogametic individuals sex chromosomes are paired.

of sex. Heterogametic males in mammals (and females in butterflies and birds) carry unpaired sex X and Y (in birds and butterflies, W and Z) chromosomes. Being unpaired, sex chromosomes are referred to as hemizygous and represent a haploid part of the diploid genome. In mammals, these chromosomes constitute 5% of the total genome (Orlov and Bulatova, 1983).

In addition to sex chromosomes, the diploid genome contains a small heterozygous part, in which alleles of the diploid set of paired chromosomes (autosomes) do not coincide because of accumulated decay substitutions (Figure 10.2). This part of the genome can be described by the value of autosomal heterozygosity H_a, which for the protein coding region of the genome is determined as the ratio of heterozygous (non-coinciding) genes to the total number of studied genes and is called protein heterozygosity. Different variants of one and the same gene locus are called alleles.

If a certain heterozygous locus contains a normal and a decay allele (i.e. an allele carrying decay substitutions), the decay allele may not be manifested at all in the phenotype, or its manifestation may be significantly masked by functioning of the normal allele. In such a case one speaks about the dominant (normal) and the recessive (decay) alleles.

Dominance of the normal allele of a gene owes itself to the fact that the normal allele encodes enzymes that are able to join the concert of correlated biochemical reactions of the organism, which are enzyme-controlled. By contrast, enzymes transcribed from the recessive decay allele do not fit into the general set of the normal biochemical reactions of the organism because its coding sequence is randomly 'spoilt' by mutation. Decay mutations affecting the normal allele of a heterozygous gene reveal the masked dysfunction of the decay allele.

The heterozygous part of the diploid genome actually represents a set of normal genes in a single copy. It may be thus envisaged as the effectively haploid part of the genome, because mutations affecting the normal alleles of heterozygous genes remain unmasked as well as in truly haploid genomes. Note that the same situation is realised for the unpaired sex chromosomes.

We define total effective heterozygosity H, which characterises the effectively haploid part of the genome, as the number of non-coinciding sites of the diploid genome related to the total genomic site number. In this case, sites of unpaired sex chromosomes are supplemented by null sites. The total effective heterozygosity H is thus equal to the sum of random autosomal heterozygosity H_a determined by randomly distributed decay mutational substitutions and fixed sex heterozygosity (traditionally called as hemizygosity) H_0 (see Figure 10.2):

$$H = H_a + H_0. \tag{10.4.1}$$

Comparing effects of somatic mutations in the hemizygous sex part and the heterozygous autosomal part of the genome, it should be taken into account that the observed protein heterozygosity H_a mostly corresponds to the presence of slightly deleterious inherited mutation substitutions (Ohta, 1987) that were able to pass through the 'selection sieve' which eliminates from the population all the extremely deleterious variants. Meanwhile, as noted above, newly-arising somatic mutations escape selection. Consequently, deleterious somatic mutations in heterozygous autosomal loci may have a weaker phenotypic expression than corresponding mutations in the effectively haploid hemizygous parts of the genome. This is due to the fact that in heterozygous autosomal loci somatic mutations affecting the normal allele (Fig. 10.2b) unmask the slightly deleterious substitution, while the only existing copy of the effectively haploid sex chromosomes (Figure 10.2a) displays severe dysfunction when affected by somatic mutation.[2]

Sex heterozygosity (hemizygosity) H_0, associated with unpaired sex chromosomes, exists only in the heterogametic sex, i.e. in males of mammals and females of butterflies and birds (Haldane, 1922; Charlesworth *et al.*, 1987; Figure 10.2a). In homogametic sex, the sex chromosomes are paired, $H_0 = 0$, and the total effective heterozygosity H practically coincides with conventional autosomal heterozygosity H_a. In mammals, autosomal heterozygosity H_a is on average 4% (Nevo *et al.*, 1984), whereas the relative length of the X and Y chromosomes (that determine the value of sex heterozygosity H_0) is on average 5% of the haploid genome (Orlov and Bulatova, 1983).

[2] This fact can be formally accounted for by introducing a coefficient α $(0 < \alpha < 1)$ before H_a in all expressions. In other words, the fact that manifestation of somatic mutations is milder in the heterozygous autosomal part than in the effectively haploid part may be accounted for by effective reduction of the value of H_a as compared to H_0, which essentially corresponds to the reduction in the number of mutations themselves. However, as this substitution does not alter any further results and conclusions, we omit coefficient α in all further intermediate calculations, i.e. for simplicity we consider $\alpha = 1$.

Thus, the total effective heterozygosity H is at least twice as low in the homogametic as in the heterogametic sex:

$$H = (H_a + H_0) \leq 9\%$$

for the heterogametic sex and

$$H \approx H_a \leq 4\%$$

for the homogametic sex. Here, we neglected the contribution of X-chromosome heterozygosity in the homogametic sex, which provides a correction of not more than 5%. Thus, the homogametic sex is at least twice better protected from the effect of deleterious somatic mutations. This can account for the fact that mammalian males have a higher mortality rate than females (Loudon, 1985). In birds, females (the heterogametic WZ sex) should have a higher mortality rate than males, which is actually observed (Payevski, 1985; Curio, 1989; Blondel *et al.*, 1992).

10.5 THRESHOLD HETEROZYGOSITY VALUES AND HALDANE'S RULE

In large animals having large genomes, the effective haploid part of the diploid genome H cannot increase indefinitely due to the rapid accumulation of deleterious mutational substitutions in ontogeny. The larger the animal, the more cell divisions required for its development, and the more new mutational substitutions contained in each cell of its body.

This limitation determines two threshold values of heterozygosity: H_C, which characterises the sensitivity of competitive interaction of individuals and H_L (the lethal threshold). H_C and H_L have the same qualitative meaning as the threshold numbers of decay substitutions n_C and n_L (Section 9.4). All individuals with total effective heterozygosity H that does not exceed H_C are equally competitive in a natural population. Individuals whose total effective heterozygosity H is higher than H_C are characterised by low competitive capacity and are forced out from the population by normal individuals with $H \leq H_C$. However, such decay individuals retain their viability up to $H_L > H_C$. Outside their natural ecological niche and in the absence of competitive interaction, the number of decay individuals can increase. All individuals with heterozygosity $H > H_L$ are inviable or sterile, i.e. their genotypes are in any case eliminated from the population.

The existence of the lethal threshold heterozygosity H_L makes it possible to explain the famous Haldane's rule, according to which the heterogametic sex in interspecific hybrids is often absent, inviable, or sterile (Haldane, 1922; Laurie, 1997). Haldane's rule holds for a wide range of organisms (mammals, birds, butterflies, *Drosophila*) irrespective of which sex (male or female) is heterogametic, i.e. carries unpaired sex chromosomes.

The diploid genome of an interspecific hybrid is composed of two different haploid genomes of the parental species. In this case, autosomal heterozygosity is determined not only by random mutational substitutions accumulated through generations in the genome of each parental species, but also by fixed differences between the normal parental genomes that appear in strictly defined locations irrespective of the presence or absence of random intraspecific mutational substitutions. These differences can be termed as hybrid heterozygosity H_h in contrast with random heterozygosity H_a of each parental species.

The total effective heterozygosity of a homogametic hybrid H_{hom}, which is equal to the sum of hybrid and autosomal heterozygosities, appears to be larger than the total effective heterozygosity of homogametic individuals of each parental species and can be close to the lethal threshold H_L:

$$H_{\text{hom}} = H_a + H_h \leq H_L.$$

The addition of the contribution of sex chromosomes H_0 to the total autosomal heterozygosity may drive the total effective heterozygosity of the heterogametic hybrid H_{het} beyond the lethal threshold and cause the hybrid lethality or sterility:

$$H_{\text{het}} = H_a + H_h + H_0 > H_L, \tag{10.5.1}$$

which corresponds to Haldane's rule (Gorshkov and Makarieva, 1999).[3]

In simple words, Haldane's rule is observed due to the fact that heterogametic hybrids are more exposed to the deleterious effects of somatic mutations having a larger haploid part of their genome due to the presence of the unpaired sex chromosomes.

Note that either the fact that in some cases patterns of hybrid lethality or sterility are highly predictable (e.g. when hybrids always die at one and the same stage of development) or the fact that one is often able to map Haldane's rule inviability genes (i.e. genes where artificial changes cause inviability of heterogametic hybrids) do not contradict the proposed explanation of Haldane's rule based on the effect of spontaneous and, hence, unpredictable somatic mutations.

All genes perform some function in the organism. Mutations in these genes have unequal effects on fitness. If we take humans, for example, there are many genes where mutations are very deleterious and lead to drastic reductions in fitness or even to death. Meanwhile there are loci where mutations are more or less tolerable. But one cannot say that humans die only from the first type of mutations, because all people die sooner or later while not all of us carry this crucial type of mutations. The same is true for hybrids. One may find a gene in one of the two hybridising species which plays a very important role for hybrid viability. If one introduces deletions or insertions into this gene, its function will be most likely damaged and the hybrid will die. But it does not mean that hybrids cannot die due to other reasons. The same is true for sterility loci.

We do not argue that all those loci that may cause hybrid sterility, are affected by somatic mutations. We do not actually think that decrease in hybrid fitness is due to mutations in some particular loci, because somatic mutations are spontaneous and

[3] Taking into account that H_a enters into expressions (10.5.1) instead of αH_a (see the previous footnote), note that the result holds true for any possible values of α including $\alpha = 0$, since Haldane's rule is determined by the α-independent term H_0 which describes the difference between total effective heterozygosites of the hetero- and homogametic sexes.

thus unpredictable. One may then ask why the pattern of hybrid inviability is often very predictable in that sense that for a given cross the developmental stage at which hybrids die is almost always the same. The explanation goes as follows.

Due to incompatibilities between genomes of the two different hybridising species, hybrids always possess a number of 'weak' incompatible sites—loci where normal functioning of genes is disrupted. Random accumulation of somatic mutations decreases the organism's fitness and results in a hybrid breakdown. But what exactly happens with the hybrid organism may to a considerable degree depend on the quantity and quality of these weak sites. The following analogy may be helpful here. Compare hybrid fitness to a cord, weak sites to incisions in this cord (more or less deep) and somatic mutations to a load fixed to the cord. This load is heavier for males than for females, so that the male cord is more often to break. But *where* it breaks does not depend on the load, it depends on the properties of the cord, that is, hybrid genome properties. If, for example, loci responsible for larvae development are the weakest, hybrids will die at the larvae stage.

A hybrid differs from both parental species in a variety of properties. Particular properties of hybrids that are absent in the parental species allow one to speak about 'hybrid force' with respect to this property and use it in artificial selection. Thus one may often hear about advantages of hybrids as compared to their parents. However, such logic completely ignores the adverse changes in a vast majority of other important characteristics that are due to incompatibilities of the genomes of the parental species.

In very much the same manner, a heterozygous state of a gene when the dominance of one of the alleles is not complete may differ from the two corresponding homozygous states. In artificially changing environmental conditions one may encounter a situation where the heterozygous individuals will display higher viability and fitness than homozygous ones. This phenomenon is known as *heterosis*. Under natural environmental conditions the normal genome ensures long-term sustainable existence of the species via correlated interaction of the species individuals with other species of the ecological community. Under randomly distorted conditions the stabilising programme of the normal genome 'goes to waste' and the long-term stability of the species is undermined. Competitive capacities of normal and decay individuals even out and the latter begin to accumulate in the population, which is continuously degrading. In such a situation some decay individuals (including heterozygous ones) may for some period of time become the most fit. However, further maintenance of the distorted environment finally leads to death of all individuals including the fittest ones. In this sense one cannot speak about their adaptation to a changed environment.

This situation is realised with sickle cell anemia in the presence of malaria infection (see also Section 9.8). Very roughly, this disease can be characterised as follows. Sickle cell gene homozygotes die from the sickle cell anemia. Normal homozygotes die from malaria. Meanwhile sickle cell gene heterozygotes (i.e. those individuals who have one normal and one sickle cell gene) survive malaria better. The more often the population experience malaria attacks, the more pronounced the advantage of the heterozygotes. In the extreme case of a continuous malaria epidemic the advantage of heterozygotes becomes overwhelming and they are indeed the last to die.

Apparently, such a situation cannot be interpreted as genetic adaptation of the sickle cell gene heterozygotes. In other words, it is useless to speak about better or worse adaptation, when the absolute survival (or any other absolute characteristic of the population wellbeing) is continuously decreasing nearing zero. When things similar to the example of the sickle cell anemia are observed, it suggests that the population is placed under distorted environmental conditions and the stability of its existence is undermined. **In the case of the sickle cell anemia, the most likely factor responsible for the unnatural situation is the unnaturally high population density of people in the malaria-affected regions.**

10.6 ESTIMATES OF LETHAL AND HYBRID HETEROZYGOSITIES

On the basis of the absence (or, more precisely, relatively low abundance) of hybrids in natural environments (Raven and Johnson, 1988) and Haldane's rule for artificially produced hybrids, it is possible to estimate values of threshold heterozygosities H_C and H_L. Under natural conditions the population is mostly composed of normal individuals with $H \leq H_C$, which is equal to the relation

$$H_0 + H_a \leq H_C < H_L. \tag{10.6.1}$$

Under natural environmental conditions, interspecific hybrids of both sexes are noncompetitive compared with normal individuals from the corresponding species. It means that the total effective heterozygosity of homogametic hybrids $(H_h + H_a)$ already exceeds the limit of sensitivity of competitive interaction H_C, i.e.

$$H_h + H_a > H_C \tag{10.6.2}$$

Note that the fact that hybrid zones in nature are strictly limited provides an independent argument in favour of the protective function of diploidy with respect to accumulation of phenotypically manifested somatic mutations in the process of development and functioning of multicellular organisms.

From inequalities (10.6.1) and (10.6.2), we obtain

$$H_h \geq H_0 \tag{10.6.3}$$

This means that hybrid heterozygosity (i.e. the relative number of fixed (regular) differences in genes of related hybridising species) exceeds the relative length of sex chromosomes H_0. Hybrid heterozygosity H_h can be estimated from the data on Nei's genetic distance D available for a great number of species (Nei, 1975). For two species 1 and 2 Nei's genetic distance is determined as

$$D \equiv -\ln\left[(1 - H_{12})/\sqrt{(1 - H_1)(1 - H_2)}\right],$$

where $H_{12} \equiv H_h$ is the hybrid heterozygosity, and H_1 and H_2 are the random differences in individuals genomes within each population, i.e. H_a. With $H_1 \approx H_2 = H_a$ we have[4]

$$H_h = 1 - e^{-D}(1 - H_a) \tag{10.6.4}$$

Coyne and Orr (1989) list values of Nei's genetic distance D for 101 pairs of hybridising *Drosophila* species with an average of $\bar{D} = 0.78$. Nevo and colleagues (1984) estimated the average autosomal heterozygosity to be about $\bar{H}_a = 0.12$ for 34 species of *Drosophila*. The average hybrid heterozygosity $\bar{H}_h$ for *Drosophila* estimated using these values according to (10.6.4) appears to be 0.60. The relative length H_0 of sex chromosomes in Drosophila varies around 25% with a maximum of 40% (Turelli and Begun, 1997). Thus, we have for *Drosophila* $0.60 > 0.25$ in complete accordance with (10.6.3). In mammals, the average value of D for 144 pairwise congeneric species comparisons was equal to 0.30 (Avise and Aquadro, 1982), whereas the average autosomal heterozygosity H_a is about 0.04 (Nevo *et al.*, 1984), giving $\bar{H}_h \approx 0.29$. Relative length of sex chromosomes in mammals is about 0.05. Thus, inequality (10.6.3) holds true for mammals as well [$0.29 > 0.05$].

Haldane's rule makes it possible to estimate the lethal heterozygosity H_L. The existence of homogametic hybrids and effective absence (inviability or sterility) of heterogametic hybrids corresponds to inequalities

$$H_h + H_a \leq H_L, \quad \text{(homogametic hybrids are viable)} \tag{10.6.5}$$

$$H_L \leq H_h + H_a + H_0 \quad \text{(heterogametic hybrids are inviable)} \tag{10.6.6}$$

which on average gives for mammals ($\bar{H}_h \approx 0.29, \bar{H}_a \approx 0.04, \bar{H}_0 \approx 0.05$),[5] see Figure 10.3:

$$0.3 \leq H_L \leq 0.4. \tag{10.6.7}$$

Note that values of total effective heterozygosity H close to the obtained estimate of H_L (10.6.7) are practically never observed in natural biological species of mammals (see Figure 10.9 in Section 10.10). This clearly demonstrates that under natural conditions the heterozygosity is limited by competitive interaction of viable and fertile individuals, i.e. total effective heterozygosity of natural species is determined by the limit of sensitivity of competitive interaction H_C. Meanwhile under distorted environmental conditions where competitive interaction of individuals may be completely switched off, individuals with a number of decay substitutions in their

[4] Note that namely the hybrid heterozygosity $H_{12} \equiv H_h$ characterises non-random genetic distance between the two species. The intraspecific heterozygosity of parent species H_1 and H_2 are due to random accumulation of decay substitutions the number of which is limited by the sensitivity of competative interaction. Combining these two principally different variables (Fuerst and Ferrell, 1983; Kellog and Appels, 1995) into one, D, the only peculiarity of which is the fact that it turns to infinity at $H_h = 1$, seems unreasonable.

[5] The lower limit is obtained when H_a in (10.6.5) is substituted by αH_a with $\alpha = 0$, while the upper limit is obtained when H_a in (10.6.6) is substituted by αH_a with $\alpha = 1$ (see footnote 3 on page 291). Note that the unknown true value of α does not significantly affect the obtained result, because the main contribution to the upper and lower limits in (10.6.5) and (10.6.6) comes from $H_h, H_h \gg H_a$. The unknown value of α results only in larger error of H_L.

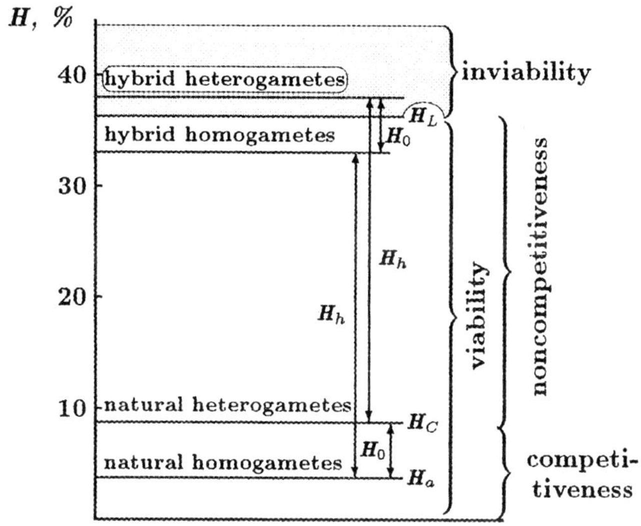

Figure 10.3. Haldane's rule. Total effective heterozygosity of the homogametic sex (females in mammals) is determined by autosomal heterozygosity H_a. Heterozygosity of the heterogametic sex (males in mammals) H_C exceeds H_a by the relative value of the length of sex chromosomes, H_0. The sum $H_a + H_0 = H_C$ is equal to the maximum value of heterozygosity tolerated by stabilising selection. Individuals with $H > H_C$ are noncompetitive. Hybrids possess additional heterozygosity H_h determined by regular differences between genomes of the hybridising species. If the value of H_h is large, the total effective heterozygosity of homogametic hybrids may approach the lethal threshold H_L. Heterogametic hybrids with their additional heterozygosity H_0 may go beyond H_L and become inviable or sterile. The predominant sterility and inviability of heterogametic hybrids as compared to homogametic ones constitutes the essence of Haldane's rule.

genomes usually unacceptable in nature may accumulate. Heterozygosity of such individuals may increase up to the lethal threshold H_L (10.6.7). To our knowledge, the highest values of heterozygosity ever encountered in mammals characterise Przewalski horses, for which only zoo populations are preserved, and domestic horses. Total effective heterozygosity of horses is close to 40%, which, at least in Przewalski horses, is accompanied by increased juvenile mortality and decreased life-span (Bowling and Ryder, 1987). Other domestic animals (cows, sheep) also demonstrate elevated heterozygosity values (see Figure 10.9 below).

10.7 BRIEF ACCOUNT OF DIFFERENT VIEWS ON THE NATURE OF INTRASPECIFIC VARIABILITY

Under natural conditions the limit of sensitivity of competitive interaction determines the observed value of the total effective heterozygosity H_C of individuals. The value of H_C is equal to the sum of fixed value of sex heterozygosity (hemizygosity) H_0 and random (mutational) heterozygosity H_a (10.4.1). The sex heterozygosity H_0 is equal to the relative length of chromosomes and appears to be an evolutionary conserved parameter. Most mammals have their H_0 values close to 5% (Orlov and Bulatova, 1983). Thus, the only characteristic that may be controlled in the process of competitive interaction of individuals is the random autosomal heterozygosity H_a, which we will below call simply heterozygosity[6] and omit the low index '*a*'. It is natural to denote the value of H that is under natural environmental conditions tolerated by competitive interaction due to its limited sensitivity as *ecological threshold heterozygosity*. Heterozygosity is one of the possible quantitative characteristics of intraspecific genetic variability.

The question of the nature and maintenance of intraspecific genetic variability is widely discussed in the literature (Nei, 1984; Nevo *et al.*, 1984; Ayala and Fitch, 1997). There are two polar views on the problem. According to the first one (the older of the two) intraspecific variability represents the adaptive potential of species. In other words, the presence of genetic variability has a direct relevance to the population wellbeing. This view is often referred to as selectionist's. The more genetic variability a species possesses, the more easily it can adapt to an unpredictably changing environment. That is, the more genetic variability, the more chances to survive in a changing environment, because when there are many genetic variants, the probability that one of them will fit a future environment is higher than when there is only one genetic variant (i.e. when all individuals are genetically uniform). Thus, many conservation biologists are seriously concerned about the fact that some endangered species exhibit very low levels of genetic variability. Accordingly, for conservation breeding programmes populations with high genetic variability are generally preferred (Cohn, 1986; O'Brien, 1994).

We have repeatedly noted that the concept of genetic adaptation to changing environment contradicts a variety of the observed data, among which the most vivid is the incompatibility of genetic adaptation based on intraspecific genetic variability and the observed discreteness of extinct and extant species. More concrete critique pertains to the absence of a general view on the problem. Single cases are usually cited as if they were representative of the whole picture. Contrary to common statements, there are many examples of perfectly prosperous species with very low intraspecific variability (Merola, 1994) and many cases where endangered species are characterised by a very high level of genetic variability. For example, among more than 340 mammalian species studied by the present authors (see Figure 10.4), over 40 species were characterised by protein heterozygosity values not exceeding one per cent, which is four times less than the average of the considered data set. Among these, there were many populations of widely spread rodents (e.g. genera *Dipodomus, Peromyscus*, *Spermophilus*, *Rattus* and others) which may in no way be characterised as endangered. Among non-rodent species, there were indeed some endangered mammals, e.g. the famous cheetah *Acinonyx jubatus*, but also many 'safe' species like, for example, the northern elephant seal *Mirounga angustirostris* which at the time of measurements numbered more than 30,000 individuals (Bonnel and Selander, 1974), which led the researchers to conclude that 'genic variability is not

[6] In mammals autosomes account for about 95% of the genome size. Thus, autosomal heterozygosity coincides to the accuracy of 5% with the random heterozygosity of the whole genome.

essential for the continued existence of animal species'. Meanwhile the critically endangered greater one-horned rhinoceros *Rhinoceros unicornis* exhibits a very high level of protein heterozygosity, namely 10%, which is more than twice the mean value for mammals (Dinerstein and McCracken, 1990). We have already cited the example with Przewalski horse which exhibits the highest value of protein heterozygosity ever recorded in mammals (about 40%) and is characterised by high juvenile mortality and decreased lifespan (Bowling and Ryder, 1987).

Thus, in accordance with the biotic regulation concept, the available empirical data show that a low level of intraspecific genetic variability in no way threatens the species existence. Rather, the observed accelerating process of species extinctions (with both high and low genetic variability) is apparently due to the increased anthropogenic pressing imposed on natural habitats and the on-going, large-scale degradation of the biosphere as a whole (Merola, 1994). Meanwhile, too-high values of intraspecific may indeed be harmful for species, whereas lowered level of protein heterozygosity prove to be normal for large mammals in their natural ecological niches (see Section 10.10 and Figure 10.9).

The second view about the intraspecific genetic variability consists of the statement that most part of the inherited variation is neutral, i.e. it does not have any effect on the individual phenotype. This neutralist's view is now shared by the majority of population geneticists, especially those who work on developing mathematical models for understanding genetic variation. Due to the extreme complexity of calculations that emerges when a non-zero selection coefficient, see (9.3.6), is assigned to mutational substitutions, the only approach that lends itself to be relatively easily modelled is that of neutral mutations theory first introduced by Kimura (Kimura, 1968; 1983). Kimura stated that most of the observed variation does not have any considerable impact on the organism's fitness.

Such an assumption leads to a drastic simplification of all calculations related to population genetical models, as one may cancel the selection coefficient making it equal to zero. As a result, the neutral approach is extensively used in population biology. The main conclusion yielded from it is that the observed variation appears as a result of balance between newly-arising mutations and random genetic drift that tends to decrease the variation. Genetic drift means random changes of allele frequencies between successive generations. In a relatively small population, one may sometimes observe that some allele become fixed solely due to stochastic fluctuations in frequency. Very simply, the fluctuation in frequency may be so high that the frequency becomes equal to unity and the mutational substitution is fixed.

An essential condition for application of the neutral theory is that the mutation rate should be considerably lower than the rate of fixation, so that a mutation has time to be fixed until another mutation occurs, exactly the same. This is not true for huge populations (e.g. bacterial ones), which makes the nature of genetic variation there not very easy to explain in terms of neutral theory. However, it is helped by stating that bacteria actually do not represent huge populations, but exist in the form of isolated, not very numerous, populations. In other words, their effective population size is small. The effective population size N_e is very difficult to be measured directly, as it involves the unknown coefficient of mixing among sympatric populations. Thus, in most cases when the neutral expectations do not fit the observed results, it is very convenient to modify namely N_e and state that the observed poor fit is due to the poor knowledge of N_e.

A direct test of the neutral theory and its relevance to explaining the nature of the existing variation can be made comparing changes in genetic heterozygosity H of populations in relation to the effective population size N_e. The value of H should grow roughly proportionally to $\nu_g k N_e/(1 + \nu_g k N_e)$ approaching unity at large N_e, where $\nu_g k$ is a very small value of mutation rate per gene per generation (see Section 9.1). As soon as N_e remains practically unknown, the only way to perform such analysis is to assume a proportional relationship between the N_e and the real population size N and test the dependence of H on N.

To our knowledge, such a test was performed only once with a sufficiently large number of species (Nei and Graur, 1984).They indeed found that, despite sharp fluctuations, genetic variation H tends to grow with increasing population size. However, the observed dependence was very far from the predicted $\nu_g k N_e/(1 + \nu_g k N_e)$ dependence, which was related to the poorly known N_e. Since then no such testshave been performed, so that the statements about the explained nature of genetic variation on the basis of the neutral theory remain unproved.

An indirect way of testing the neutral theory and its relevance to the problem of the intraspecific genetic variability is to compare evolutionary genetic distances between different species. The neutral theory predicts that genetical distance between species grows linearly with the time of their evolutionary separation, the proportionality coefficient being again the mutation rate per gene per generation which is expected to vary not very significantly among species, being determined by the universal biochemical nature of life. This statement constitutes the essence of the well-known molecular clock (Zuckerkandl and Pauling, 1965; Zuckerkandl, 1987).

It is indeed observed that species that separated a longer time ago are more genetically different than closer relatives (in terms of species). The coefficient proportionality varies greatly from gene to gene and from one group of species to another, leaving one to doubt: whether there is a neutral molecular clock or the observed trend is simply due to the fact that evolution is a time-dependent process and, naturally, the more time has elapsed, the more difference is accumulated, the proportionality coefficient being in each particular case determined by some specific reasons (see, e.g. Ayala, 1999 for this point of view).

But even if molecular clock does exist in the neutral understanding of it, there is ample evidence showing that the observed *interspecific* variation (that refers to evolutionary changes) has nothing to do with *intraspecific* variability (see, e.g. Fuerst and Ferrell, 1983; Kellog and Appels, 1995). Thus, no evidence gathered on the evolutionary field may support the neutral theory statements when the nature of the observed intraspecific genetic variation is concerned, rendering the present-day situation with regard to explaining the nature and maintenance of intraspecific genetic variability as confused as ever.

Within the biotic regulation concept, the phenomenon of intraspecific genetic variability acquires a clear meaning. The observed intraspecific genetic variability represents 'genetic noise', i.e. loss of the meaningful species–specific genetic information. Thus, in its very nature the intraspecific variability is by all means *harmful* to the species (contrast to the assumptions of both the neutral and selectionist theories). However, a certain amount of variation is allowed to persist in natural populations. This is not due to a balance between genetic drift and mutation, but to the fact that the process of competitive interaction that performs stabilising functions in the population is not (and cannot be) infinitely precise. In what is to follow, we show that the observed patterns of intraspecific genetic variability in different taxa are in accordance with such view.

10.8 POISSON DISTRIBUTION OF THE NUMBER OF POLYMORPHIC LOCI

To date, ample evidence on heterozygosity of the protein coding region of the genome has been accumulated for a great number of species (Nevo *et al.*, 1984). Measurements of the protein heterozygosity H are based on differences in electrophoretic mobility of proteins encoded by different alleles of the same gene. This is one of the oldest methods of measuring genetic variability and has been widely used in genetics since the end of the 1960s. At present, owing to rapid development of measuring techniques, many studies assess the DNA variability directly (e.g. by DNA sequencing) rather than by proteins encoded by it. However, development of DNA technique during the last several years has not yet led to creation of a comparable DNA variability data set for different species, especially taking into account the growing bias in molecular studies towards medicine and, consequently, the single species *Homo sapiens*. Thus, the protein variability data still remain unique due to their extensiveness with respect to the number of species studied. Here, mammals are one of the best studied taxa.

Due to the continuous process of mutagenesis the intraspecific genetic variability (and, hence, protein heterozygosity as one of its quantitative characteristics) should increase up to a certain threshold level, which is determined by the sensitivity (accuracy) of competitive interaction. It is natural to expect that the threshold values of genetic variability are similar inside groups of evolutionary close organisms. Meanwhile the observations show that intraspecific genetic variability is remarkably different even among evolutionary close species, e.g. within the class of mammals (Nevo *et al.*, 1984). In what is to follow, we show that the observed different values of protein heterozygosity in different species and the wide spread of the observed values of protein heterozygosity around the mean are explained by the small number of loci studied. With increasing number of studied loci the scatter of heterozygosity values decreases in accordance with a Poisson distribution while the mean does not change. **This suggests that all natural species of mammals are characterised by nearly the same level of protein heterozygosity, which can be thus considered the ecological threshold heterozygosity common for the whole class of mammals.**

The coding region of the mammalian genome is known to contain about 100,000 genes. Such numbers of genes cannot be studied in any modern experiment. In most studies, sets of $L \sim 20 \div 30$ gene loci are investigated. For each individual in the population the number of the observed heterozygous loci (i.e. loci where two different alleles of the gene are present) is related to the total number of studied loci yielding the value of the heterozygosity. These individual heterozygosity values are further averaged over all individuals studied in the population. The obtained average population heterozygosity H_{Lo} characterises genetic variability of the chosen group of L loci and may differ significantly from the value of heterozygosity H of the total protein-coding region.

Gene loci that are heterozygous in some individuals in the population (i.e. those loci that have at least two different alleles) are called *polymorphic*. Polymorphic loci appear in the population due to mutations. The new alleles may further spread over the population in a number of generations, provided that the new mutational substitutions are not very deleterious and may remain unnoticed by competitive interaction. Note that a polymorphic locus characterises population as a whole, while heterozygous locus is a state of the polymorphic locus in a particular individual. In other words, polymorphic loci are heterozygous in some individuals and homozygous in the others, while monomorphic loci are homozygous in all individuals of the population.

Localisation of polymorphic loci in the diploid genome of the population should be random, as the appearance of polymorphic loci is determined by the random process of mutagenesis. If this is the case, the number l of polymorphic loci observed in different sets of L loci studied in a population should follow the Poisson distribution. The probability $p(l)$ to observe l polymorphic loci in a given set of L loci is equal to

$$p(l) = e^{-\bar{l}} \frac{\bar{l}^l}{l!}, \qquad (10.8.1)$$

where $\bar{l}$ is the average number of polymorphic loci observed in sets of L loci.[7] The number of polymorphic loci observed in the population is related to the population heterozygosity H_{Lo} (see below). If the ecological threshold heterozygosity is the same for all mammals, then all mammals should be characterised by the same average proportion of polymorphic loci. In such a case the number l of polymorphic loci observed in sets of L studied loci *in different species* will follow the same Poisson distribution as the number of polymorphic loci in sets of L loci chosen in different parts of the coding region *of the same species*. We collected published data on heterozygosity of 343 mammalian species (Figure 10.4). Let us now show that the observed distribution of heterozygosity values (Figure 10.5) indeed agrees with the Poisson distribution of the number of polymorphic loci l observed in different species.

For each polymorphic locus, the relative frequency h of its occurrence in a heterozygous state is determined as

[7] To be precise, formula (11.8.1) holds true for cases when $\bar{l} \ll L$, i.e. the proportion of the observed polymorphic loci is not large. This is usually the case for mammals where the average polymorphism (i.e. the relative number of polymorphic loci) does not exceed 0.20 (Nevo *et al.*, 1984).

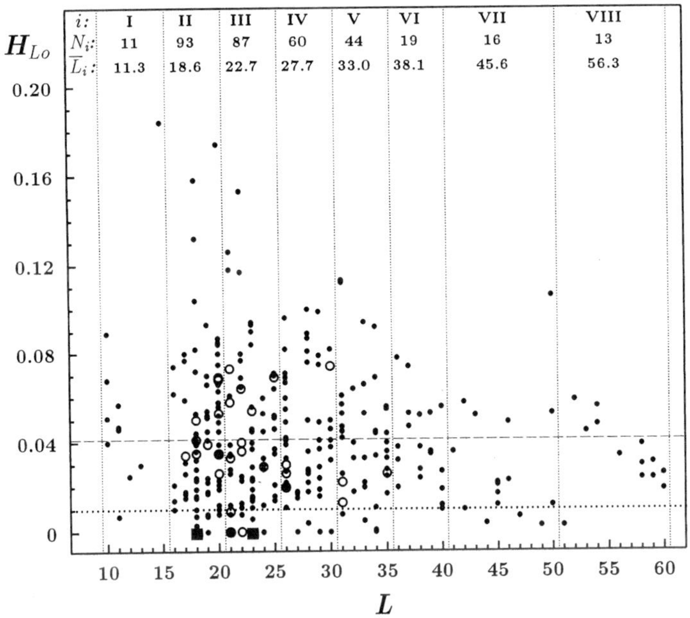

Figure 10.4. Heterozygosity H_{Lo} of natural species of mammals with respect to the number L of loci studied. Points represent values for one population each. Large empty circles, large filled circles and large filled squares represent randomly coincided values of heterozygosity for two, three and four populations, respectively. Vertical dotted lines divide the total range of L values into eight intervals. The proposed division is arbitrary. It was chosen to minimise differences between lengths of intervals and at the same time between numbers of heterozygosity values observed in each interval. It can be shown, however, that the results of the study do not depend on different ways of choosing intervals. For each i-th interval the number of heterozygosity values N_L and the average number of loci $\bar{L}$ is shown. Horizontal dotted line corresponds to $H_{Lo} = 0.010$ and delimitates over 40 species with low heterozygosity. Horizontal dashed line corresponds to the average heterozygosity of natural mammals $\overline{H_{Lo}} = 0.041$.

$$h = \sum_{i \neq j} 2q_i q_j = 1 - \sum_i q_i^2$$

where q_i is the population frequency of the i-th allele of the considered locus, $\sum_i q_i = 1$. If all the l polymorphic loci were characterised by the same value of heterozygosity h, each individual would on average have hl heterozygous loci among the L loci studied. The mean value of heterozygosity of individuals in the population could be then determined as

$$H_{Lo} = hl/L. \tag{10.8.2}$$

From this expression we obtain the following relationship for the average values $\bar{l}$, $\overline{H_{Lo}}$, $\bar{L}$ and $\bar{h}$, where the averaging is made over populations of different species of mammals:

$$\bar{l} = \overline{H_{Lo}}\bar{L}/\bar{h}. \tag{10.8.3}$$

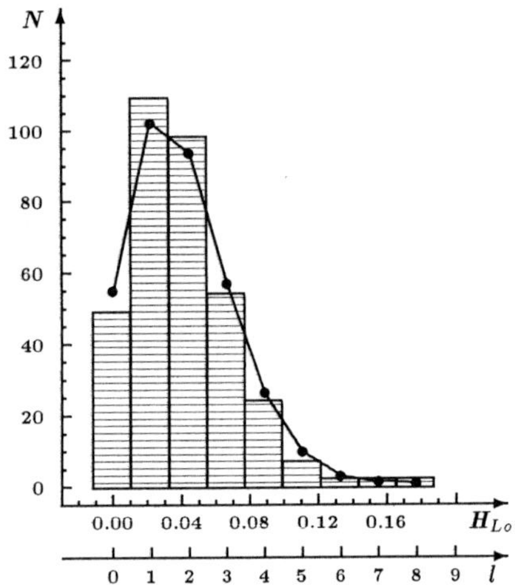

Figure 10.5. Frequency distribution of heterozygosity values in natural species of mammals. Cross-hatched diagram is drawn on the basis of observed values represented in Figure 10.4. Area enveloped by the histogram is equal to the total number of species studied (343). The polygonal line with dots shows the best approximation (χ^2 significance level $S = 0.75$ with 4 degrees of freedom) of the observed values by a Poisson distribution of the effective of polymorphic loci l (the lower horizontal axis), see text for the procedure of relating heterozygosity H_{Lo} to the effective number l of polymorphic loci.

The Poisson distribution is characterised by the equality between the variance and the mean. Thus, if the number of polymorphic loci l follows a Poisson distribution, the following equality should hold:

$$\bar{l} = \overline{(l - \bar{l})^2} \equiv \sigma_{\bar{l}}^2, \tag{10.8.4}$$

where $\sigma_{\bar{l}}^2$ is the variance of the number of polymorphic loci l. Using (10.8.2) we may express the variance $\sigma_{\bar{l}}^2$ using the variance of heterozygosity H_{Lo} and neglecting the non-zero variances of L and h, i.e. considering these values to be the same for all species and all loci in all species, respectively:[8]

$$\bar{l} = \sigma_{\bar{l}}^2 = \frac{\sigma_{H_{Lo}}^2 \bar{L}^2}{\bar{h}^2}. \tag{10.8.5}$$

Equating the right-hand parts of relationships (10.8.3) and (10.8.5) we thus obtain:

$$\bar{h} = \sigma_{H_{Lo}}^2 \frac{\overline{H_{Lo}}}{\bar{L}}. \tag{10.8.6}$$

Values of $\sigma_{H_L}^2 = 0.00090$, $\overline{H_{Lo}} = 0.041$ and $\bar{L} = 26.7$ are known from the collected data set (Figure 10.4). Using the relationship (10.8.6) which was obtained under the

[8] Variance of the number of the studied loci L may be kept at a low level if approximately the same number of loci is studied in all species. Variance of heterozygosity of polymorphic loci h may be obtained empirically studying the distribution of h values in different loci of different species. According to this distribution (data not shown), the non-zero variance of h does not violate expression (11.8.4) and all the further results by more than 30%.

assumption of a Poisson distribution of the number of polymorphic loci l we get $\bar{h} = 0.59$ and estimate the effective average number[9] of polymorphic loci $\bar{l}$ using (10.8.5) to be $\bar{l} = 1.85$. Using this value we may transform the observed continuous distribution of heterozygosity values into a discrete distribution and further compare it with a Poisson distribution. This is done as follows. As soon as the average effective number of polymorphic loci $\bar{l} = 1.85$ corresponds to the average heterozygosity $\overline{H_{Lo}} = 0.041$, it means that one effective polymorphic locus corresponds to heterozygosity $\overline{H_{Lo}}/\bar{l} = 0.022$. The whole range of heterozygosity values was thus divided into intervals corresponding to integer numbers of effective polymorphic loci. Heterozygosity values less than $0.5\overline{H_{Lo}}/\bar{l}$ (i.e. less than a half of effective polymorphic loci) were assigned the integer value 0 (i.e. such values roughly correspond to complete absence of polymorphic loci). Heterozygosity values falling within the interval from $0.5\overline{H_{Lo}}/\bar{l}$ to $1.5\overline{H_{Lo}}/\bar{l}$ were assigned the integer value 1 (i.e. such values roughly correspond to one polymorphic locus), and so on, Figure 10.5.

The obtained distribution of integer values was then compared to Poisson distribution using the χ^2-criterion (Figure 10.5). It was shown that, indeed, it nicely agrees with the Poisson distribution with the average multiplicity $\bar{l} = 1.85$ (significance level $S = 0.75$ with four degrees of freedom). Poisson distribution with low average multiplicity of the order of unity is characterised by a high frequency of zero values. Thus, the obtained low value of the average multiplicity $\bar{l} = 1.85$ explains the high frequency of occurrence of zero heterozygosity values in natural biological species, which repeatedly causes concern to conservation biologists who consider the intraspecific genetic variability as the adaptive potential of species (O'Brien, 1994).

Thus we have shown that the observed distribution of heterozygosity in different species is well described by a Poisson distribution of an effective number of polymorphic loci averaging $\bar{l} = 1.85$. This effective value is three times lower than the observed average number of polymorphic loci $\bar{l}_S = 5.3$ and corresponds to a three times higher single-locus heterozygosity $\bar{h} = 0.59$ than the actually observed value of $\bar{h}_S = 0.21$, which is related to $\bar{l}_S = 5.3$ by the relation (10.8.3).

Poisson distribution describes a multiplicity of independent events. The observed situation indicates that loci in the studied sets are not independent but form internally-correlated groups. Within groups all loci are characterised by specific heterozygosity values. The appearance of particular groups in the studied sets of loci is independent and conforms to the Poisson distribution. Thus, the effective number of polymorphic loci $\bar{l}$ actually stands for the number of independent groups of polymorphic loci, the average number of loci in a group being close to three. Within a polymorphic group, each locus is characterised by the averaged single-locus heterozygosity $\bar{h}_S = 0.21$, while the group as a whole naturally makes a three times higher contribution equal to $\bar{h} = 0.59$.

The existence of correlation between the studied gene loci is confirmed by independent analysis of the empirical evidence. Most studied proteins are coded

[9] This value appears to be three times lower than the observed average number of polymorphic loci l, see below.

by two, three or more genetic loci. If a given protein is studied, all its loci are studied as a rule, such pattern being determined by the applied techniques. These groups of loci are characterised by significantly different values of heterozygosity (Selander and Johnson, 1973; Powell, 1975; O'Brien *et al.*, 1980). For example, in the studied species the average heterozygosity of esterases and phosphoglucomutases exceeds the overall average heterozygosity of all loci more than twofold, while lactate dehydrogenases, malate dehydrogenases and hexokinases have average heterozygosity which is 4–5 times less than the overall average value.

Another reason for correlation between the studied loci is the fact that different researchers may use different techniques when analysing protein heterozygosity. If one researcher uses a more sensible technique than the others, he may discover higher levels of heterozygosity in all studied loci. This effectively resembles the situation when some researchers study only polymorphic loci, while others study only monomorphic ones. In such a case the average number of correlated loci in a group can theoretically increase up to the total number of studied loci L.

A good analogy seems to be a situation when a person has a broken telephone which rings in response to a call but makes no connection. If a colleague rings this person, and the connection is lost, the colleague dials the number again and again until he realises that the line is bad and gives up. Thus, rings from each colleague form correlated groups and follow each other in a non-random manner. In other words, if one hears a ring, one is nearly sure to hear at least one ring more during a short period of time. Thus, separate rings do not follow the Poisson distribution. Meanwhile, groups of rings from different colleagues are evidently independent and do conform to the Poisson distribution.

A group of loci characterised by high heterozygosity may be thus compared to a number of correlated telephone signals. They are considered against the background of loci with low heterozygosity which constitute the majority of the loci studied and correspond to absence of calls (negligibly low heterozygosity). Thus, the observed fact that the effective number of polymorphic loci which corresponds to the Poisson distribution (Figure 10.5) is three times lower than the actual number of the observed polymorphic loci is explained by the peculiarities of the applied techniques and properties of the studied loci.

Distribution of heterozygosity values shown in Figures 10.4 and 10.5 is obtained for natural species of mammals only. Domestic and zoo populations were ignored in this study. The obtained Poisson approximation of the observed heterozygosity distribution makes it possible to determine the probability of natural occurrence of high values of heterozygosity observed in mammalian species living under unnatural conditions (horses, cows and humans themselves).

As we have seen, the Poisson distribution (10.8.1) describes the observed heterozygosity distribution (Figure 10.5) if we understand l as the number of groups of polymorphic loci with approximately three loci in each group. Let us denote for m the number of such groups observed in each species. If the actual number of polymorphic loci observed is l, the number of triplets is apparently $m = l/3$. Using (10.8.2) we may thus express heterozygosity H_{Lo} as $H_{Lo} = mh/(3L)$, where h is the observed average heterozygosity of polymorphic loci ($h \sim 0.21$), L is the total number of loci studied. According to (10.8.3), the average number of observed polymorphic triplets $\bar{m}$ is naturally proportional to the total number L of loci studied and is equal to $\bar{m} = \overline{H_{Lo}}L/(3\bar{h}) = 0.065L$. The probability $p(H)$ of observing heterozygosity H in a species where L loci is studied is equal to the probability of observing $m = HL/(3h)$ triplets of polymorphic loci. As the number of triplets of polymorphic loci m follows the Poisson distribution with $\bar{m} = 0.065L$, we may write for $p(H)$:

$$p(H) = p(m) = e^{-\bar{m}}\frac{\bar{m}^m}{m!}, \qquad m = \frac{HL}{3h}, \qquad \bar{m} = 0.065L \tag{10.8.7}$$

For humans we have $H = 0.125$, $L = 107$ (Nevo *et al.*, 1984), which gives us $m \approx 21$ and $p(H) = p(m) < 10^{-5}$. *Homo sapiens* is the only species studied with $L > 70$. Thus, the observance of the value $H = 0.125$ within the natural distribution (10.8.7) is highly improbable. Heterozygosity of humans apparently differs from the natural patterns observed in other mammals living under natural conditions of their ecological niches. This means that the genetic information of *Homo sapiens* has considerably degraded during the time that people have spent living in the artificial conditions of civilisation outside the natural ecological niche. The obtained result indicates that, under the existing conditions, competitive interaction of individuals within the human population is no longer able to support genetic stability of the species.

Similar results are obtained for the other species living under artificial conditions outside their natural ecological niche. For horses (*Equus caballus*) $H \approx 0.36$, $L = 18$ (Bowling and Ryder, 1987), and $p(H) < 10^{-6}$. Given that the total number of the studied species is less than 500, the probability to observe $H \approx 0.36$ within the studied data set is less than 10^{-3}, which indicates that genetic variability of horses is also in excess as compared with what is tolerated in nature. For various breeds of cows (*Bos taurus*) $H \approx 0.30$, $L = 11$ (Bannikova and Zubareva, 1995), which gives $p(H) < 10^{-3}$, which is also on the verge of real probabilities. Although the above estimates are tentative, as they are based on the assumption that all polymorphic loci exhibit similar values of heterozygosity $h \approx 0.21$ (while in highly heterozygous species greater values of h may be encountered), they do show that species living under artificial conditions exhibit significantly elevated levels of heterozygosity. **This is in full agreement with the prediction of the biotic regulation concept, according to which the accuracy of competitive interaction becomes low in artificial environments, so that individuals with high numbers of decay substitutions in their genomes may accumulate and the intraspecific genetic variability increase (see Figure 9.1a).**

Note that the fact that some domestic species do not show high heterozygosity values (e.g. for some domestic goats $H = 0.030$ (Randi *et al.*, 1990)) does not contradict the above way of reasoning. The elevated level of heterozygosity in domestic species is not necessarily due to accumulation of deleterious substitutions during the process of mutagenesis which, as we have seen in Section 9.2, operates very slowly. Rather, it is due to the breeding practice to which the domestic animals are exposed. For example, increased variability may be due to crosses between genetically distant breeds, whereas in populations where no such crosses are performed and, besides, inbreeding is present, intrapopulational variability may sustain its natural low level. (High variability may also characterise a combined population composed of all individuals from different breeds of the same species.) The most important thing here is that if individuals with high heterozygosity do accumulate, it may only occur under artificial environments, where competitive interaction is weak or absent. Meanwhile under natural conditions competitive interaction of individuals does not tolerate high heterozygosity.

10.9 NATURAL LEVEL OF HETEROZYGOSITY IN MAMMALS

Values of heterozygosity observed in different natural species of mammals are characterised by a rather wide spread around the mean (see Figure 10.4). If the sensitivity of competitive interaction is indeed similar in all mammals (which would mean that all mammals are actually characterised by approximately the same heterozygosity), the observed scattering should be solely due to the small number of loci studied. In other words, in some species many polymorphic loci were found occasionally and, consequently, high value of heterozygosity, while in other species a small number of polymorphic loci and low heterozygosity were found. In such a case, variance of heterozygosity should decrease inversely proportionally to L in accordance with relationship (10.8.6), which may be written in the form

$$\sigma^2_{H_{Lo}} = \bar{h}\frac{\overline{H_{Lo}}}{\bar{L}}. \tag{10.9.1}$$

If, on the other hand, the observed wide scattering (Figure 10.4) reflects true differences in heterozygosity values of different species of mammals, variance of heterozygosity cannot depend on the number of studied loci L.

More generally, dependence between $\sigma^2_{H_{Lo}}$ and L may be written in the allometric form as follows:

$$\lg \sigma^2_{H_{Lo}} = a \lg(\overline{H_{Lo}}/\bar{L}) + b. \tag{10.9.2}$$

For the Poisson distribution of polymorphic loci we should obtain in accordance with (10.9.1):

$$a = 1, \qquad b = \lg \bar{h} = -0.68. \tag{10.9.3}$$

To test this prediction, the whole range of L values (Figure 10.4) was divided into eight intervals, each of them containing not less than 10 values of heterozygosity H_{Lo}. For each interval the average number of studied loci $\bar{L}$, the average heterozygosity $\overline{H_{Lo}}$, and variance of heterozygosity $\sigma^2_{H_{Lo}}$ were calculated. In accordance with (10.9.1), logarithms of $\sigma^2_{H_L}$ and of the ratio $\overline{H_{Lo}}/\bar{L}$ were then tested with a linear regression, which gave the following results (Figure 10.6):

$$a = 1.1 \pm 0.2, \quad b = 0.1 \pm 0.7 \quad (r = 0.88, P = 0.004 \text{ with 6 degrees of freedom})$$

The obtained estimate of b exceeds significantly the expected value (10.9.3), though the latter still falls within the 95% confidence interval of the obtained estimate. Such

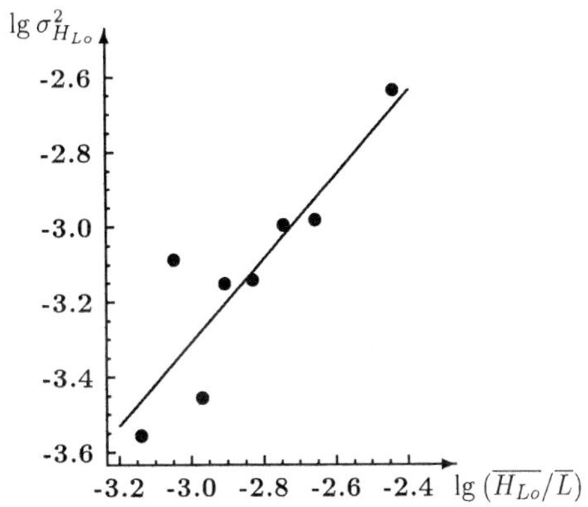

Figure 10.6. Dependence of heterozygosity variance $\sigma^2_{H_{Lo}} \equiv \overline{(H_{Lo} - \bar{H}_{Lo})^2}$ of heterozygosity H_{Lo} of different species of mammals on the number of studied loci L in the logarithmic form (10.9.2). Parameters of the linear regression: $r = 0.88$, $P = 0.004$ with six degrees of freedom.

discrepancy is due to the peculiarities of the studied loci, which were discussed in the previous section. Meanwhile, the obtained value of a agrees very well with the corresponding relationship (10.9.3). The high correlation coefficient and low probability level both testify in favour of the statement that variance of heterozygosity decreases inversely proportionally to L.

Thus, we have seen that the observed wide scattering of heterozygosity values obtained when measuring a small number of loci does not contradict the prediction of the biotic regulation concept that all mammals are likely to be characterised by a similar level of heterozygosity that is determined by the accuracy of the process of competitive interaction in natural mammalian populations. With increasing number of studied loci the observed heterozygosity distribution (Figure 10.5) should shrink to a narrow peak around the average value of $H = 0.04$, which thus represents a quantitative estimate of the average sensitivity of competitive interaction in natural species of mammals.

As noted in the previous section, heterozygosity values of domestic animals do not belong to the natural heterozygosity distribution and are thus unlikely to change significantly with an increased number of studied loci, as was demonstrated for humans.

10.10 HETEROZYGOSITY DEPENDENCE OF BODY MASS AND GENOME SIZE

As we have seen, protein heterozygosity may serve as a quantitative characteristic of the sensitivity of competitive interaction in a population. The fact that the sensitivity of competitive interaction is likely to be conserved in evolutionarily close species determines the fact that all natural species of mammals are characterised by approximately the same average heterozygosity.

However, competitive interaction is a process of comparison of *phenotypic* characteristics of individuals, meanwhile heterozygosity characterises an individual's *genotype*, i.e. the number of accumulated decay substitutions. In diploid organisms only a certain part of the decay substitutions present in the genome is manifested. The relative number of manifested decay substitutions is dependent on heterozygosity, but not equal to it. Thus, heterozygosity represents an indirect characteristic of the limit of sensitivity of competitive interaction.

The sensitivity of competitive interaction can be measured directly as the permissible total number of phenotypically manifested decay substitutions per individual genome, which is influenced (beside heterozygosity) by such parameters as the mutation rate per nucleotide per cell division, genome size, number of divisions in the germ line (and, hence, individual body size). Let us now discuss how the number of phenotypically manifested decay substitutions may be expressed using these parameters (as above, we only consider the protein coding region of the genome).

The number of heterozygous (and, hence, effectively haploid) gene loci of the genome is equal to

$$G_g H, \tag{10.10.1}$$

where G_g is the total number of gene loci in the protein coding region of the genome, H is the heterozygosity. We defined heterozygosity as the relative number of non-coinciding loci in a diploid genome. Similarly, heterozygosity may be defined for a haploid population as the average relative number of non-coinciding loci between any two randomly compared haploid genotypes. As soon as in haploid phase all decay substitutions are manifested, $G_g H$ represents the total number of decay substitutions (i.e. defective genes) manifested in each cell of a haploid unicellular organism.

Random mixing of genetic material in a sexually breeding population due to genetic recombination results in the fact that in any individual certain parts of gene loci contain decay substitutions in both copies of the diploid genome. Such loci are deprived of normal alleles that could ensure a masking effect, so that the decay substitutions in these loci are manifested. The number of such loci in a unicellular diploid organism is naturally given by

$$G_g H^2, \tag{10.10.2}$$

because the probability of occurrence of two decay alleles in a single site is equal to the squared probability of occurrence of one decay allele, which is equal to H.

In multicellular organisms, beside slightly deleterious inherited substitutions, strongly deleterious somatic mutations are also present. Somatic mutations are manifested if they affect the normal copy of a heterozygous locus (Figure 10.2b). The number of phenotypically manifested somatic mutations per cell is equal to

$$G_g H k_S \nu, \tag{10.10.3}$$

where $k_S = 3.3(\lg m + 9)$ is the average number of divisions in the somatic line (m is the body mass in grams), see (10.3.1), $\nu = 10^{-7}\ (\text{gene})^{-1}\ \text{d}^{-1}$ is the rate of mutations per gene per cell division.

Accumulation of phenotypically manifested decay substitutions is limited by competitive interaction of individuals. Individuals with an intolerable number of such substitutions are forced out from the population by normal individuals.[10] The limit of sensitivity of competitive interaction determines the permissible number of decay substitutions that may remain unnoticed in the individual. In other words, it represents the permissible level of erosion of species-specific information that yet does not disintegrate correct behaviour of the species in the community. It is natural to assume that this fundamental characteristic should be similar in evolutionarily close species. In such a case, evolutionary increase in the genome size G_g in haploid organisms should be accompanied by an inversely proportional decrease in species heterozygosity H, so that the total number of manifested decay substitutions G_gH (10.10.1) remained the same.

Similarly, in diploid organisms heterozygosity should decrease proportionally to $G_g^{-1/2}$ with increasing genome size G_g of the organisms, so that the total number of manifested substitutions G_gH^2 remained the same (10.10.2). However, in multicellular organisms increase in the genome size is also accompanied by an increase in the number of somatic mutations (10.10.3), which grows proportionally to G_g.

The average number of cell divisions in the somatic line does not significantly exceed 40 for most multicellular organisms. The ratio of the number of inherited to somatic decay substitutions is equal to $G_gHk_S\nu/G_gH^2 = \nu k_S/H \sim 10^{-4}$ at $H \approx 0.04$ (as in mammals) and $k_S \sim 40$. Under the assumption that diploidy evolved to protect multicellular organisms from deleterious effects of somatic mutations (Section 10.3), we have to conclude that the negative phenotypic effect of not numerous but strongly deleterious somatic substitutions in most multicellular organisms exceeds the cumulative phenotypic effect of more numerous but only slightly deleterious inherited substitutions. Otherwise diploidy would make no sense. From this we obtain that phenotypic effect of a single somatic substitution is comparable to the phenotypic effect of 10^4 inherited, slightly deleterious substitutions. In other words, a somatic substitution is 10^4 times more deleterious than an inherited one. Thus, for natural species the value (10.10.2) can be considered as a lower estimate of the permissible number n_C of decay substitutions tolerated by stabilising selection under natural conditions, whereas for domestic species this value represents a lower estimate of the maximum number n_L of decay substitutions compatible with viability (the lethal threshold), see Sections 9.5–7. Hence, it is possible to estimate the ratio between n_C and n_L from the known values of heterozygosity of natural species (e.g. mammals, $H = 0.04$) and the maximum recorded heterozygosity values still compatible with viability, $H_L \approx 0.40$ (for horses, see above). We thus obtain

[10] The statement that it is the total number of manifested decay substitutions per genome rather than their genome density that is stabilised by competative interaction, is suggested by the rigidly correlated structure of the genome. This property can be compared to a complex computer program, where a single mistake may lead to disintegration of the whole programme irrespective of its length and the absolute number of blocks in it.

from (10.10.2) that $n_L/n_C \sim (H^2/H_L^2) \sim 100$, i.e. the lethal threshold number of decay substitutions is about a hundred of times larger than what is tolerated by stabilising selection under natural environmental conditions.

Qualitatively, the necessity to limit the number of phenotypically manifested decay substitutions with increasing genome size and body mass of multicellular organisms can be described by the following pattern. Small multicellular organisms ($m \leq 0.1$ g) with small genomes ($G \leq 10^8$ bp) can be haploid, which is exemplified by the existence of haploid males in haplodiploid insects[11] (see also Section 10.3). In haploid organisms, all somatic mutations are phenotypically manifested. Their number increases proportionally $G_g \nu k_S$ irrespective of the value of heterozygosity H_1 which for haploid organisms decreases as $H_1 \sim G_g^{-1}$. When the critical values $G \sim 10^8$ bp and $m \sim 0.1$ g are reached, the negative phenotypic effect of somatic substitutions exceeds that of inherited slightly deleterious substitutions. As a result, existence of larger multicellular organisms with larger genome sizes becomes impossible.

Appearance of diploidy results in a sharp stepwise reduction of the number of manifested somatic substitutions due to the masking effect of the normal copy of the diploid genome unaffected by mutations. The number of manifested inherited substitutions becomes equal to $G_g H_2^2$, where H_2 is heterozygosity of diploid organisms. Thus, relatively small diploid organisms with relatively small genomes can afford an elevated level of heterozygosity as compared to haploids of similar body and genome sizes, so that

$$G_g H_2^2 \sim G_g H_1, \qquad (10.10.4)$$

i.e. the number of manifested decay substitutions remains approximately the same). In other words, a diploid organism is allowed to accumulate more decay inherited substitutions as compared to a haploid organism, because the deleterious effect of most decay substitutions in the diploid organism is masked. The observed distributions of heterozygosity values of haplodiploid (H_1) and diploid (H_2) insects with similar genome sizes are indeed remarkably different, with $H_1 \ll H_2$ as predicted by (10.10.4), although the relationship (10.10.4) does not hold exactly (Table 10.2). This may be due to significant evolutionary distance between the orders of haplodiploid (*Hymenoptera, Thysanoptera*) and diploid (*Diptera, Coleoptera, Heteroptera, Lepidoptera*) insects, which may lead to different values of the sensitivity of competitive interaction in these taxa and violation of the exact equality (10.10.4).

Unpaired sex chromosomes in sexually breeding species can be considered as another example of haplodiploidy, although exotic. In haplodiploid insects, haploid males have only one copy of the genome, while diploid females carry two copies. Similarly, in many other species (e.g. mammals) males carry only one copy of X chromosome, while females carry two copies of it. Y chromosome is present in males only. Thus, all decay inherited substitutions in sex chromosomes are manifested in

[11] Diplo- and polyploidisation of somatic cells of certain organs in haploid males of haplodiploid insects at late stages of ontogeny (Rasch *et al.*, 1977) suggests that genome and body size of these organisms may already exceed the critical threshold value beyond which haploid multicellular organisms are inviable.

Table 10.2. Heterozygosity and genome size in different insect orders with respect to ploidy level (published data collected by the present authors). G – haploid genome size (1C) (10^9 bp) (± standard deviation). N – number of studied genera (species). H – average heterozygosity (± standard deviation). Note the difference between heterozygosity values of diploid and haploid species with similar genome sizes.

Order	*G*	*N*	*H*	*N*
Diploid				
Coleoptera (beetles)	0.6 ± 0.6	54 (92)	0.13 ± 0.07	12 (15)
Diptera (flies, mosquitoes)	0.8 ± 0.8	26 (60)	0.11 ± 0.03	10 (84)
Heteroptera (bugs)	1.1 ± 0.4	5 (14)	0.13 ± 0.09	5 (19)
Lepidoptera (butterflies)	0.8 ± 0.3	12 (13)	0.14 ± 0.07	22 (62)
Orthoptera (locusts, grasshoppers)	10 ± 4	34 (57)	0.06 ± 0.05	10 (33)
Haplodiploid:				
Hymenoptera (bees, ants, wasps)	0.22 ± 0.07	4 (6)	0.05 ± 0.04	37 (64)
Thysanoptera (thrips)	?	–	0.06 ± 0.02	4 (4)

males. According to the above logic, autosomes that are always diploid should be characterised by higher heterozygosity than the sex chromosomes that are effectively haploid in males. This is indeed observed. Whereas approximately 1 in 560 bp is variant in autosomal DNA, the variability of X-chromosomal DNA is only about 1 bp in 2100 bp, while viability of the Y chromosome (which is never diploid) is less than 1 in 48,000 bp (Hofker *et al.*, 1986; Jakubiczka *et al.*, 1989; Malaspina *et al.*, 1990; Dorit *et al.*, 1995).

Further increase of genome size in multicellular organisms is accompanied by proportional growth of the number of manifested somatic substitutions $G_g H_2 k_S \nu$ (10.10.3). However, in small organisms with relatively low values of k_S the cumulative phenotypic effect of somatic substitutions may remain lower than that of inherited substitutions, $G_g H_2^2$. Constancy of the phenotypic effect of inherited substitutions is then ensured by the dependence

$$H_2 \sim G_g^{-1/2} \tag{10.10.5}$$

Qualitatively, this dependence is observed in diploid insects (see Figure 10.7 and Table 10.2). Among the five insect orders studied, four insect orders (*Diptera, Coleoptera, Heteroptera, Lepidoptera*) have their genome sizes clustered around 10^9 bp and heterozygosity values exceeding 10%. The fifth order (*Orthoptera*) is characterised by significantly larger genomes (about 10^{10} bp) and significantly lower values of heterozygosity. The observed drop of heterozygosity is approximated as $H_2 \sim G^{-1/3}$. This may be explained by the fact that the protein coding region of the genome G_g grows proportionally to $G^{2/3}$ (where G is the absolute genome size, see Table 10.1 in Section 10.3), which makes $H_2 \sim G^{-1/3} \sim (G_g^{3/2})^{-1/3} \sim G_g^{-1/2}$, as predicted by (10.10.5).

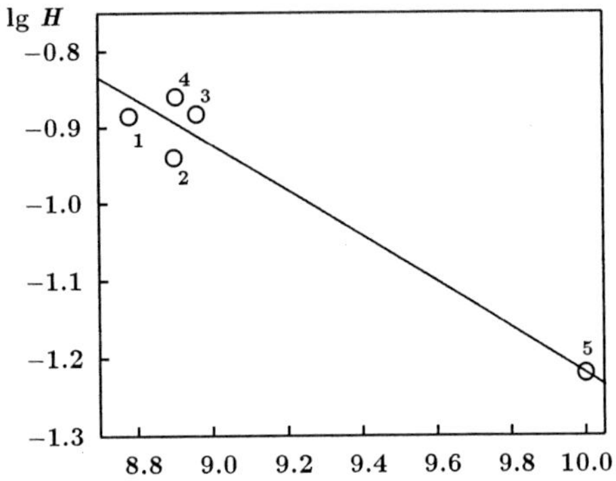

Figure 10.7. Heterozygosity dependence on genome size G in different orders of insects. 1—*Coleoptera*, 2—*Diptera*, 3—*Heteroptera*, 4—*Lepidoptera*, 5—*Orthoptera*, see also Table 10.2. Parameters of the regression $\lg H = a + b\lg(G/G_0)$: $r = 0.96$, $P = 0.01$ with 3 degrees of freedom, $b = -0.3 \pm 0.05$ ($a = 0.1 \pm 0.01$).

Finally, further increase of genome size and body size in multicellular organisms may lead to the situation when the major contribution into phenotypically manifested substitutions is made by somatic substitutions, $G_g H_2 k_S \nu$. To prevent this contribution from further growth heterozygosity H_2 should decrease proportionally to $(G_g k_S)^{-1}$. In such a case the contribution of somatic substitutions $G_g H_2 k_S \nu$ remains constant, while the contribution of inherited substitutions $G_g H_2^2$ decreases inversely proportionally to G_g $[G_g H_2^2 \sim G_g (G_g k_S)^{-2} \sim G_g^{-1} k_S^{-2}]$ and becomes negligibly small as compared to the contribution of somatic substitutions.

Heterozygosity dependence $H_2 \sim (G_g k_S)^{-1}$ is indeed observed in urodelous amphibians and mammals. Species of urodelous amphibians are characterised by drastically different genome sizes and relatively similar body sizes, corresponding to a relatively constant value of k_S number of divisions in the somatic line (see Section 10.3). The observed change of heterozygosity is satisfactorily described by dependence $H_2 \sim G^{-1}$ (see Figure 10.8).

In contrast, mammals are characterised by only slightly varying genome sizes not correlated with body mass. Thus the genome value G_g may be considered constant for the whole class of mammals with a sufficiently high accuracy. Body mass of mammals varies from 1 g (some shrews) up to 100 t (some whales), which corresponds to the change in the average number of divisions in the somatic line k_S from $k_S \approx 30$ in the smallest up to $k_S \approx 56$ in the largest mammals, see (10.3.1). This allows to study the dependence of heterozygosity H_2 in mammals on the inverse number of somatic divisions k_S^{-1} (Figure 10.9).

Values of heterozygosity in mammals were plotted against the inverse number of somatic divisions k_S^{-1}, calculated using published data on body mass m according to (10.3.1). The whole range of k_S^{-1} values was divided into 14 equal intervals. Within each interval the average value of heterozygosity was calculated to reduce, in accordance with the results of the previous section, the effect of the small number

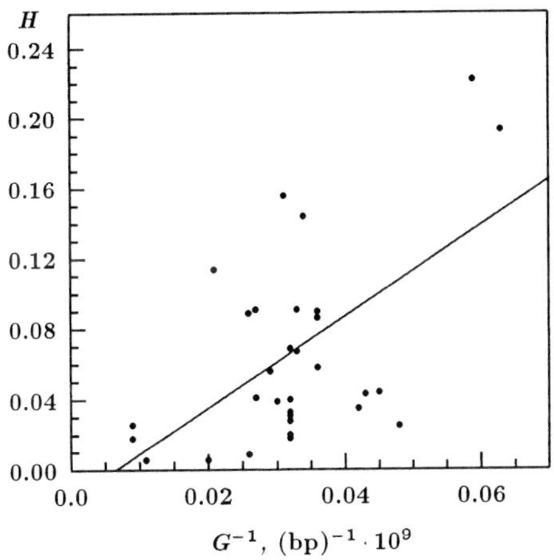

Figure 10.8. Heterozygosity dependence on genome size G in urodelous amphibians. Heterozygosity values taken from Nevo *et al.*, (1984), genome size is the average of four sources: (Pierce and Mitton, 1980; Ginatulin, 1984; Larson, 1984; Session and Larson, 1987). For some species with unknown genome size it was estimated as the average for the genus. Parameters of regression $\lg H = a + b \lg(G/G_0)$: $r = -0.52$, $P = 0.003$ with 29 degrees of freedom, $b = -1.1 \pm 0.3$ ($a = 0.3 \pm 0.5$).

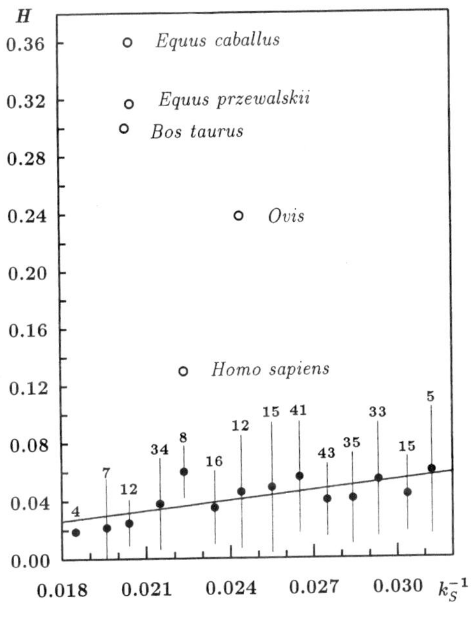

Figure 10.9. Heterozygosity dependence on the number k_S of divisions in the somatic line in natural species of mammals. Published data collected by the authors (after Gorshkov and Makarieva, 1997). Filled circles represent average heterozygosity values for each interval of k_S^{-1}, figures indicate the number of species in each interval, thin vertical lines represent standard deviation. Open circles stand for heterozygosity values of mammals living in unnatural conditions: horses *Equus caballus* and *Equus przewalskii* (Bowling and Ryder, 1987); cow *Bos taurus* (Bannikova and Zubareva, 1995); domestic sheep *Ovis* (Wang *et al.*, 1990); man *Homo sapiens* (Nevo *et al.*, 1984). Parameters of regression $H = a + bk_S^{-1}$: $r = 0.79$, $P = 0.008$ with 12 degrees of freedom, $b = 2.2 \pm 0.5$ ($a = -0.01 \pm 0.01$).

of studied loci. The obtained values were then used in the linear regression analysis, which gave the following results:

$$H_2 = a + bk_S^{-1}, \quad b = 2.2 \pm 0.5; \quad a = -0.01 \pm 0.01$$
$$(r = 0.79; p < 0.001 \text{ with 12 d.f.}) \qquad (10.10.6)$$

As follows from (10.10.6), the desired inverse proportionality between H_2 and k_S is indeed discovered, which means that the value of H_2k_S is conserved in all mammals, being determined by the evolutionarily conserved value of sensitivity of competitive interaction in natural populations. Note that the obtained result does not contradict the statement made in the previous section that all mammals are characterised by approximately the same heterozygosity. One may say that this statement describes the general pattern for the whole class of mammals, while results (10.10.6) reveal a more refined structure. The observed dependence (10.10.6) describes a less than twofold change in heterozygosity values between the largest and the smallest organisms. Both very large and very small mammals that are largely responsible for dependence (10.10.6) constitute a small part of the total number of mammals (see size distribution of the number of mammalian species in Figure 3.4 in Section 3.8) and cannot significantly influence the overall distribution of heterozygosity values discussed in Section 10.9.

Summing up, the performed analysis of the available empirical evidence unambiguously testifies in favour of the fact that the number of phenotypically manifested mutational substitutions (both inherited and somatic) is equally limited within large taxa of organisms with different genome and body sizes. This points to the existence of an ecological threshold of permissible level of erosion of genetic information of the natural species of the biosphere. The observed intraspecific genetic variability is a manifestation of the limited sensitivity of the process of competitive interaction of individuals, that allows slightly deleterious inherited substitutions to accumulate up to the ecological threshold.

The question of whether these substitutions may serve as the genetic basis of the evolutionary process is addressed in the following chapter.

11

Evolution

It is shown that all the available data on biological evolution are consistent with the biotic regulation concept.

11.1 EVOLUTION AND ENVIRONMENTAL CHANGE

As we have seen in the preceding chapters, all the peculiarities of organisation of living objects are aimed at conserving the existing level of order and preventing spontaneous decay of life organisation. The described mechanism of maintaining stability of biological organisation, i.e. stabilising selection based on competitive interaction of individuals in a population, may work to sustain life at a prescribed level of organisation associated with the maximum possible competitiveness without any evolutionary changes. However, if that maximum level is either not reached or does not exist, the same mechanism may bring about evolutionary changes. Random changes in genotypes, associated with loss of competitiveness, are cut off in the process of competitive interaction of individuals in the population. Random changes in genotypes, associated with higher competitiveness, provided that such take place, spread through the whole population. Progeny of more competitive individuals force individuals with unchanged competitiveness out from the population.

In other words, stabilising selection ensures propagation of the most competitive communities of species, i.e. those that are best able to stabilise their environment and are the quickest to compensate for any adverse changes in it. If the existing communities are characterised by the maximum possible competitiveness, no evolutionary changes are possible. However, the Earth's environment is exposed to the influence of various abiotic factors of cosmic and geophysical nature (e.g. changes in solar activity, rock weathering, meteorite fall, etc.), which cannot be controlled by the biota. These factors slowly change the average external conditions under which the biotic regulation of the environment takes place, making possible evolutionary changes in the biosphere.

For example, suppose that at a given level of solar activity a given ecological community is characterised by the maximum possible competitiveness, i.e. it stabilises the environment in the most efficient way. It means that no other community's organisation would ensure a quicker compensation for spontaneous environmental perturbations. When the solar activity (or any other abiotic parameter) changes, a different strategy of environmental stabilisation becomes most efficient. As a result, there opens a possibility for evolutionary changes. Evolutionary origin of new species contributes to competitiveness of the community and enhances its regulatory potential. Thus, evolution results in improved environmental regulation with respect to changing abiotic conditions uncontrollable by the biota. Note that at any time during the evolutionary process the environment remains under biotic control, be the latter ensured by either the old or a new evolutionary type of community.

The traditional interpretation of evolution is seemingly based on the same succession of events. First, the environment changes. Second, there appears a genotype which under new environmental conditions imparts increased competitiveness to its carriers and allows them to produce most progeny, forcing out other genotypes from the population. Third, this genotype gives rise to a new species (and, consequently, to a new community with altered species composition).

The principal additional statement made within the biotic regulation concept is that increased competitiveness (and, hence, increased number of progeny) is only possible when the new species that appeared in the course of evolution enhances the regulatory potential of the community. In other words, within the biotic regulation concept, competitiveness of individuals and the regulatory potential of the community to which they belong are tightly coupled (see Section 2.11). By contrast, within the traditional paradigm, an increased competitiveness (or fitness) is considered to be a sufficient condition for evolutionary changes *per se*.

Genetic information of closely related species differs by about 1% (Lewin, 1987). The average time of species existence is of the order of 10^6 years (Avise *et al.*, 1998). Consequently, during the time period of the order of 10^8 years the genetic information of the biosphere is completely renewed. This 'upgrades' the biota's ability to regulate the environment under the changing external impact of geophysical and cosmic factors. On this basis, evolution may be envisaged as another mechanism of maintenance of the regulatory potential of the natural biota, along with other stabilising characteristics of the biota that were considered in the preceding chapters.

11.2 ORIGIN OF NEW SPECIES

As shown above (Section 9.8), when environmental conditions deviate significantly from the natural conditions of the ecological niche of the species and stabilising selection ceases to operate efficiently, decay genetic polymorphism in the population may increase up to the lethal threshold. The totality of possible decay genotypes corresponding to defective but viable individuals we called the *decay tail* of the species (Section 9.7).

We have seen (Section 10.10) that the decay polymorphism and the decay tail of the species by far exceeds the normal polymorphism. Under the assumption that the appearance of a new species' genotype is approximately the same over the whole space of viable genotypes, that means that a new species' genotype will most likely appear within the decay tail of the old species. In other words, the first individual of a new species will be with a high probability a decay individual from the point of view of the old species. Hence, such an individual will be noncompetitive under the natural conditions of the ecological niche of the old species.

Meanwhile, it is evident that an individual carrying the genotype of a new species may only survive in the population of individuals of the old species if this individual, as well as its offspring, possesses higher competitiveness compared with normal individuals of the old species. **Hence, we come to the conclusion that survival and propagation of individuals of the new species may only be ensured if the environmental conditions of the natural ecological niche of the old species are seriously distorted and stabilising selection weakened, so that both normal and decay individuals of the old species become equally competitive. In such a case, individuals of the new species (decay individuals from the point of view of the old species) will have a chance to force out all individuals of the old species, provided that the new species' individuals possess just a slight advantage in competitiveness.** This conclusion agrees with the available paleodata, according to which bursts of speciation often follow considerable environmental restructuring, i.e. transformation of the existing natural ecological niches, e.g. changing sea level, appearance of physical barriers like the Panama Isthmus, etc. (Knowlton *et al.*, 1993; Jackson, 1994).

Increase in competitiveness is associated with a higher degree of order in life processes (see Section 2.11). This factor controls spontaneous evolution of life towards its higher organisation. Possible decay changes in genotypes resulting in the decrease of order in life processes outnumber by many orders of magnitude possible changes of genotypes resulting in a higher degree of order. This statement constitutes the essence of the notion of orderliness. It is why the rate at which life orderliness decays when competitive interaction and stabilising selection are switched off should be many orders of magnitude higher than the rate of evolutionary changes.

The observed difference between the normal genomes of two closely related eukaryote species in the number of substitutions of nucleotide pairs is about 1% of the size of the genome (Lewin, 1987; Ayala and Kiger, 1984). The average difference between normal genotypes of diploid individuals of one and the same species due to limited sensitivity of stabilising selection (normal polymorphism, see Section 9.5) is also of the order of 1% of the genome size (Hofker *et al.*, 1986; White and Lalouel, 1988). Hence, the observed difference between two individuals in the number of the non-coinciding nucleotide pairs in their genomes does not contain information as to whether or not these two individuals belong to the same species. Assuming that the genome of the two kin species contains 10^9 bp, we find that the difference in the number of nucleotide pairs between both the individuals from different species, and individuals within one and the same species, amounts to $m = 10^7$ bp. The number of different genotypes, which may be constructed on the basis of that difference, is

$$4^m \approx (10^6)^{10^6} \tag{11.2.1}$$

These genotypes correspond to normal, decay but viable and simply inviable phenotypes of individuals of both species. 'Inviable' genotypes comprise the overwhelming majority of all the possible genotypes (11.2.1), in spite of the fact that the total number of genetic substitutions does not go beyond 10^7, which corresponds to the sensitivity of competitive interaction within the population and characterises normal viable individuals in the population. The observed number of genetic substitutions tolerated by stabilising selection under natural environmental conditions is predominantly composed of slightly deleterious substitutions that have passed through the 'selection sieve' (Section 10.4). Thus, these substitutions are not random in contrast to the majority of variants of genotypes described by (11.2.1). Among the latter there are many strongly deleterious changes that are not compatible with viability of individuals.

Evidently, evolutionary transition from one species to another may only occur if the evolutionary succession of genotypic changes does not lead to loss of individual viability at any stage of the evolutionary process. One of the major questions of the evolutionary theory is whether there exists a viability 'bridge' connecting decay tails of two closely related species. In other words, the question is whether there exists a succession of separate point mutations and macromutations that could ensure transition from the initial species to a new one without loss of individual viability at any stage. If such a viability bridge does exist, gradual evolution of species is possible based on endogenic genetic modifications of the species.

If there is no such bridge between species, evolution may only occur by 'jumps' over the inviability precipice separating the two species. Such jumps mean that a certain genotype of the old species incorporates a genetic macrofragment that ensures a saltatory transition to a genotype of a viable and fertile individual belonging to the decay tail of a new species. Such a macrofragment is absent from the genome of the old species and cannot be obtained by the result of a succession of single mutational changes of the genome of the old species without loss of viability of its individuals. Such evolution by genetic 'jumps' would be manifested as punctuated equilibrium on the morphological level (Gould and Eldridge, 1993). It means that individuals of a given species retain constant morphology over the most period of the species' lifetime over geological time scale and undergo rapid morphological changes during the speciation burst corresponding to acquisition of the new macrofragment and subsequent restructuring of the genome (see below).

The new macrofragment can only be formed as a result of a multiple mutation event, when several germ-line mutations occur simultaneously in the same genome region of a newly-born individual. The probability of such a multiple event is proportional to the product of probabilities of single mutational events. Owing to this fact there exists a high probabilistic barrier that hampers evolutionary transitions between species and, by doing so, increases the stability of organisation of the existing species. The existence of such a probabilistic barrier may explain the observed discreteness of the extant species.

Genetic information of all the extant species was once formed in the course of the evolution of genomes of species-precursors. The unique genetic information of newly-appearing species that was absent in genomes of the old species should be also written as a row of genetic letters—the four different nucleotides. The low probability of multiple mutational events imposes limits on the maximum length of a new unique macrofragment that can be obtained within a given species. Naturally, the probability to observe several mutational events within a given region of the genome is proportional to the mutation rate per nucleotide per generation and the number of individuals in the population. Accordingly, maximum macrofragments can be formed within huge populations of rapidly mutating viruses.

Let us now estimate the length of such a unique fragment formed in a population of viruses. In other words, let us discuss the following situation. Suppose that there exists a virus species which we call species A. This species can form another species, species B, if n mutations occur simultaneously in particular sites of its genome, i.e. if n particular sites of the genome of the old species are somehow changed. These n mutational substitutions represent the viability bridge between the two species. If this multiple mutational event occurs, there appears an individual belonging to species B. The question is, how long such a bridge can be, i.e. what is the value of n? If it is too long, the probability of the multiple mutational event becomes too low, and during the whole lifespan of the species A there appear no individuals with the n mutational substitutions needed to ensure the evolutionary transition to species B.

The highest mutation rate characterises RNA-viruses that cannot make use of the proofreading system of DNA repair of their hosts. RNA-viruses mutate at a rate of 10^{-4} nucleotide substitutions per site per replication (Holland *et al.*, 1982). The probability of observing n mutational substitutions in a given site of a single viral genotype is equal to $(10^{-4})^n$. Thus, to observe a genetic 'word' consisting of n new genetic letters in a given site of the viral genome we need to consider about 10^{4n} individuals of viruses. We assume the total number of viruses in the biosphere to be of the order of 10^{29}, i.e. an order of magnitude higher than that of bacteria (see Section 3.8). The average time of reproduction of viruses is of the order of 10^{-5} years. Thus, during the time period of the order of million years (the average lifespan of eukaryotic species) there appear about 10^{40} individuals of viruses. It means that the maximum length of a new genetic word prepared by viruses during this period of time is equal to $n = 10$ genetic letters ($10^{4n} = 10^{40}$). In other words, longer fragments could in principle be prepared by the use of macromutations combining unique genetic 'words' obtained by point mutations into longer 'phrases'. However, the probability of macromutations that could compose a long fragment from a non-random set of genetic words should be very low. One may therefore assert that preparation of a fragment with a prescribed succession of nucleotide pairs longer than 100 bp is absolutely improbable. Hence, all new genes (and other genome parts) appearing in the course of punctuated evolution are likely to be constructed on the basis of incorporation of short unique fragments from 10–100 bp in length into the genome of the old species.

To investigate the question about the gradual or saltatory character of genetic evolution, let us now compare evolutionary rates in eukaryotes and prokaryotes.

11.3 EVOLUTION OF PROKARYOTES AND EUKARYOTES

According to paleodata, the average lifespan of eukaryote species is about 10^6 years (Simpson, 1944; MacFadden and Hubbert, 1988; Avise *et al.*, 1998). With the observed relative constancy of the number of species in the biosphere (Raup and Sepkoski, 1982; Wilson, 1989) extinction of every species should be compensated by the appearance of a new one which, among the eukaryotes, would differ from the old one by a change of about 1% of its genome.[1] If the genome is 10^9 bp in size, that corresponds to a change of 10^7 bp. It is natural to call these substitutions *positive*, as far as they lead to the formation of a new species and, hence, increase the level of genetic orderliness. By contrast, the overwhelming majority of possible substitutions are of decay nature and increase genetic disorder.

Estimating the number of individuals in a eukaryote species as 10^9, and assuming the time of generation to be equal to one year (e.g. a small mammal or bird), we find that there appear about 10^{15} individuals during the whole period of the species' existence. That is, about 10^8 individuals fall for every of the 10^7 positive substitutions leading to formation of a new species. Following the analogy with the rate of genome decay μ we may thus introduce the generative rate μ^+, i.e. the rate at which positive (+) substitutions are generated:

$$\mu^+ \sim 10^{-8} \ (+) \text{ substitutions per individual}$$

The rate of genome decay μ (9.1.3), i.e. the number of decay substitutions per individual (or per genome per generation, which is the same) is of the order of unity for eukaryotes (we denote μ as μ^- below):

$$\mu^- \sim 1 \ (-) \text{ substitutions per individual}$$

The ratio between the decay rate μ^- and the generative rate μ^+ for eukaryotes is of the order of

$$\theta_E = \mu^-/\mu^+ \sim 10^8, \tag{11.3.1}$$

which means that during the species lifespan there appear 10^8 decay substitutions per each positive substitution out of those 10^7 leading to formation of a new species.

Now we may estimate a similar value for prokaryotes, θ_P. The average lifespan of a prokaryote species is of the same order of magnitude as that for eukaryotes, that is,

[1] Note that the nucleotide differences between any two species may be of two types. Differing sites can be dispersed over the whole genome or concentrated in some place. Knowledge about the compact type of interspecific difference is scarce because, being concentrated in some concrete places of the genome, sequences of that type are difficult to discover.

$\sim 10^6$ years (Ochman and Wilson, 1987). The size of the bacterial genome is of the order of $\sim 10^6$ bp. Kin bacterial species differ by approximately 10% of nucleotide substitutions in their genome, that is by $\sim 10^5$ bp (Ochman and Wilson, 1987). A single bacterial species contains about 10^{24} individuals (Section 3.8). During the species lifespan there appear 10^{33} bacterial individuals. Hence, there are about 10^{28} bacterial individuals per one positive nucleotide substitution, that is, the generative rate for bacteria is equal to

$$\mu^+ \sim 10^{-28} \text{ (+) substitutions per individual.}$$

The rate of genome decay for bacteria is of the order of

$$\mu^- \sim 10^{-3} \text{ (−) substituions per individual,}$$

(see Section 9.1). Hence, the ratio between decay and positive mutations for bacteria θ_P is of the order of

$$\theta_P = \mu^-/\mu^+ \sim 10^{25} \tag{11.3.2}$$

which is 17 orders of magnitude higher than the respective value for eukaryotes (11.3.1).

The difference is striking. If we envisage evolution as a continuous test of newly-appearing genotypes for their ability to give rise for a new species, we have to conclude that prokaryotes have to test almost a billion of billions of times more genotypes than eukaryotes do to form a new species.

Meanwhile, due to the universal biochemical nature of life organisation, it is natural to expect that the ratio between the numbers of decay and positive substitutions should also be a universal characteristic of life as a whole. In any case, these ratios cannot differ by those 17 orders of magnitude from eukaryotes to prokaryotes. Therefore we consider two possibilities: first that θ_E is universal, see Eq. (11.3.1), and, second, that θ_P is universal, see Eq. (11.3.2).

If we take the value of θ_E to be universal, we find that for bacteria $(\mu^+)^{-1} \sim 10^{12}$ individuals per single positive substitution.[2] Recalling that any two bacterial species differ by approximately 10^5 (+) substitutions per species, we find that 10^{17} individuals need to accumulate within a single bacterial species before an evolutionary transition to another species becomes possible. Hence, during the lifetime of a bacterial individual which is of the order of $\sim 10^{-3}$ years, each bacterial species containing 10^{24} individuals would produce $10^{24}/10^{17} = 10^7$ new prokaryote species. Given that the total number of bacterial species in the biosphere is of the order of 10^4 (Chislenko, 1981), we obtain for the evolutionary rate of prokaryotes a value of about 10^{14} new species per year, or 10^7 new species per second.

Global population numbers may amount to 10^{18} individuals in unicellular eukaryotes and up to 10^{15} individuals in small invertebrates (Section 3.8). Given the average lifetime of about 10^{-2} years and the total number of lower eukaryote species of about 10^5, unicellular eukaryotes would evolve at a rate of 10^8 species per year. Small invertebrates constituting up to 10^7 species of the biosphere (Thomas,

[2] This is obtained from (11.3.1) using $\mu^- \sim 10^{-3}$ decay substitutions per individual.

1990) would evolve at a rate of about 10^7 species per year, were the value of θ_E universal. All such species would be genetically unstable and would not be capable of preserving the observed discreteness for any noticeable period of time, that is, they would not exist as biological species.

Being in drastic contrast with the available empirical data, the possibility of the universality of the value of θ_E should be rejected. We are left with the second possibility.

The universal value may thus only be that of θ_P, Eq. (11.3.2). In that case, assuming that the evolution of eukaryotes occurs exclusively by the internal (endogenic) mutational transformations of nucleotide pairs and of their macrofragments in the genome, we find that $(\mu^+)^{-1} \sim 10^{25}$ individuals per single positive substitution need to be produced on average by every eukaryote species in order to ensure an evolutionary transition to a new species.[3] However, such a great number of individuals is not produced in any eukaryote species within the lifespan of that species.

This leads one to the unambiguous conclusion that endogenic evolution of eukaryotes is impossible. Evolution (and, apparently, the origin of eukaryotes (Margulis, 1975)) may only be explained on the assumption that genomes of eukaryotes incorporate exogenic genetic material in the form of functionally sensible genetic fragments from virus and bacterial genomes (Margulis, 1975; Martin and Fridovich, 1981; Kordum, 1982; Hesin, 1984; Syvanen, 1984, 1987, 1989; Moses and Chua, 1988; Gorshkov, 1995; Amabile-Cuevas and Chicurel, 1992; Martin *et al.*, 1993; Marienfield *et al.*, 1997). Without assuming the existence of a flow of genetic information from viruses and prokaryotes to eukaryotes, there is no possibility of explaining the observed evolution of eukaryotes.

In other words, population numbers of eukaryotes are by far insufficient to ensure a gradual evolutionary transition from one species to another on the basis of a succession of endogenic mutations, even if such succession (i.e. a viability bridge between the two species, see the previous section) does exist. Thus, speciation of eukaryotes is not gradual, but occurs by a saltatory transition from one species to another. Such a transition is initiated by the acquisition of a new genetic macrofragment, which increases competitiveness of its carriers as compared with individuals of the old species. As a result, progeny of the individual that incorporated the new macrofragment exponentially forces out individuals of the old species from the population.

Note that the new, functionally-sensible macrofragment that ensures an evolutionary transition should incorporate as a single unit in an individual genotype, since separate parts of that fragment may be deprived of functional meaning and unable to impart additional competitiveness to their carriers.

Although sexual breeding may occasionally combine genetic fragments from different genotypes into a single genotype, it cannot accelerate evolution. Consider two neutral or slightly deleterious genetic 'semifragments' that produce a positive, functionally-sensible macrofragment when combined with each other. A single

[3] This is obtained from (11.3.2) using $\mu^- \sim 1$ decay substitutions per individual.

semifragment appears in the genome in the process of mutagenesis with a given probability p. The probability of the appearance of two semifragments within a single genotype is equal to p^2. It may seem that the two semifragments may unite with each other in the course of sexual breeding of their carriers with a higher probability than the probability p^2 of the double mutational event.

However, sexual partners do not know that they carry semifragments that could be combined into a single positive fragment because, as stated above, separate parts of the positive fragment do not have any effect on the competitiveness of their carriers. Thus, the two individuals carrying parts of the positive fragment may only unite by chance, the probability of that event being of the order of $1/N$, (i.e. the individual carrying the first semifragment chooses its mating partner by pure chance out of the N individuals in the population). If the population number is equal to $N = p^{-1}$, so that the probability of the appearance of a single semifragment in that population approximates unity, the probability of double event, i.e. of appearance of the second semifragment in the same genotype where the first semifragment appeared, is equal to $p = 1/N$, which is exactly the same as in the presence of sexual breeding. Thus, sexual breeding of individuals has no accelerating effect on formation of positive macrofragments that initiate speciation.

As discussed above, the average length of positive macrofragments does not exceed one hundred nucleotide pairs. Meanwhile the observed nucleotide differences between related species amount to about 1% of the total genome size, which by many orders of magnitude exceeds the length of the positive macrofragment. That means that acquisition of the positive macrofragment is equivalent to a genetic jump from the old species to the decay tail of a new species (Figure 11.1).

The first individual of the new species, the founder, is a decay individual of the new species, because its genotype differs from the normal genome of the new species in about 1% of nucleotide sites. Thus, acquisition of the positive fragment is further followed by a process of endogenic genetic changes in genotypes of the new species aimed at formation of the normal genome of the new species. This process may be called *evolutionary tuning* of the normal genome of the new species. It is practically identical to the process of genetic relaxation to the normal genome, which occurs in a population that has spent long time under distorted environmental conditions, lost some fragments of the normal genome but was then returned to its natural ecological niche (see Section 9.8 and Figure 9.3).

The only difference lies in the fact that, during the process of relaxation, the normal genome is *restored*, whereas in the course of evolutionary tuning the normal genome of the new species is *formed* for the first time, its formation being predetermined by the acquired positive macrofragment. If we represent species as potential pits (Figure 11.1), incorporation of the positive macrofragment corresponds to a jump from the initial pit corresponding to the old species onto the slope of an empty pit corresponding to a new species. The process of evolutionary tuning of the genome (i.e. acquisition of dispersed genetic differences) can be then compared with rapidly sliding down the slope of the new pit to the very bottom, where the stable normal genome of the new species resides.

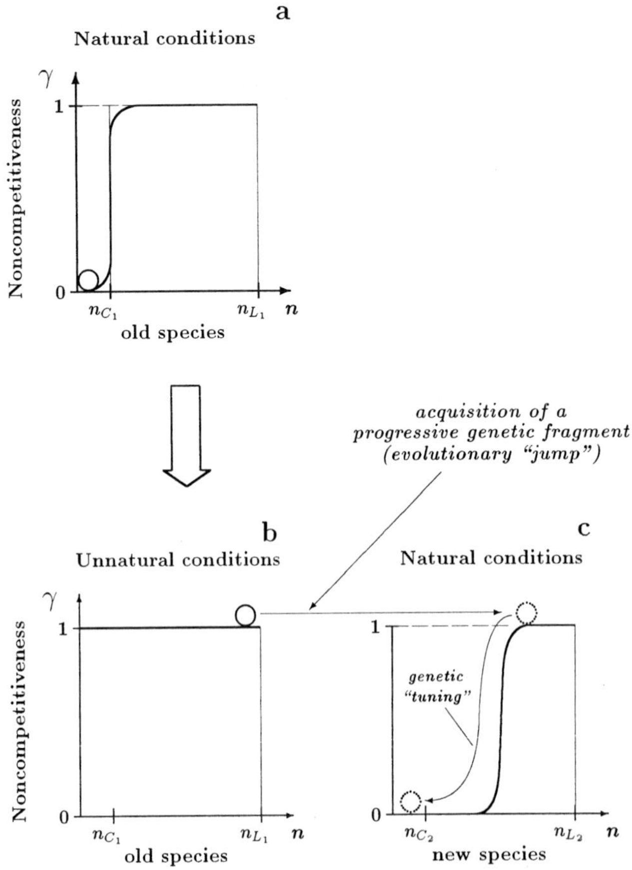

Figure 11.1. Appearance of a new species in the course of evolution. **a** – under natural conditions genotypes of the initial species randomly fluctuate around the stable normal genome of the species, $n < n_C$. **b** – under unnatural conditions normal individuals lose competitiveness, so that their competitiveness becomes equal to that of decay individuals populating the decay tail of the species, $n_C < n < n_L$. **c** – acquisition of a progressive genetic fragment by one of the individuals of the old species drives this individual to the decay tail of a new species. If the environmental conditions prove to be optimal for the new species, the normal genotype is formed in the process of genetic tuning which is analogous to the process of genetic relaxation to the normal genome (see Section 9.8 and Figure 9.3). Genetic tuning is due to increased competitiveness associated with genotypes that are most close to the normal genome. In a sexual population genetic tuning occurs very rapidly, see text.

Let us now consider the time scale of evolutionary tuning. Suppose that one individual in a population incorporates a positive macrofragment into its genome, acquires high competitiveness and becomes the founder of a new species. Its highly competitive offspring exponentially quickly force out individuals of the old species from the population. Hence, the population is soon composed of decay individuals of the new species, with genotypes including the positive macrofragment, but differing by at least 10^7 nucleotide substitutions from the normal genome of the new species.

Let us estimate the average time of tuning for a population of 10^9 individuals with genome size of the order of $G \sim 10^9$ bp and the number of divisions in the germline $k \sim 10$. Such parameters are common for many plants, invertebrates and small vertebrates. As far as the probability of a single mutation per nucleotide site per cell division is of the order of $v = 10^{-10}$, each individual in this population will carry $\mu = Gkv \sim 1$ new mutational substitution, absent from the genotypes of its parents. Thus, in a population of 10^9 individuals any mutational substitution possible in the genome (including the 10^7 positive substitutions) will be present with a high, close to unity, probability in at least one genotype. Being closer to the normal genome of the new species, individuals carrying positive substitutions become more competitive than other individuals in the population. Such highly competitive individuals tend to force out less competitive individuals from the population and breed with each other. Genetic recombination that accompanies sexual breeding results in the appearance of two positive substitutions in genomes of some of their offspring. Individuals with two positive substitutions become the most competitive, force out other individuals and breed with each other. Some of their progeny will contain four positive substitutions in their genomes and so on. As a result, normal individuals of the new species (i.e. those containing all the 10^7 positive substitutions) will dominate in the population in n generations, where $n = \log_2 10^7 \approx 23$.

Twenty-three generations of any species represent an infinitely small time period on the evolutionary time scale. Thus, we have seen that evolutionary tuning occurs nearly instantaneously, so for the majority of the time of the species' existence, its individuals retain morphological and genetical stasis (Haldane, 1954), which perfectly agrees with the punctuated equilibrium pattern of speciation observed in the paleodata (Gould and Eldridge, 1993).

The obtained result could be predicted from the biotic regulation concept. The main condition of the species' persistence and prosperity in the biosphere is its ability to stabilise the environment within the ecological community to which it belongs. Meanwhile during evolutionary tuning the species is composed of decay individuals with an impaired regulatory genetic programme. Communities with slowly evolving species with a prolonged tuning would be less efficient in environmental stabilisation and, hence, less competitive. As a result, evolution of life would be directed towards minimisation of the time of evolutionary tuning, so that any species could devote most part of its lifespan to regulation of the environment.

In smaller populations of larger organisms evolutionary tuning takes longer time. Most time is spent on the appearance of the necessary positive substitutions. In a population of 10^8 individuals, appearance of the needed substitution in a given site of the genome takes on average 10 generations, in a population of 10^7 individuals—100 generations, etc. When all the necessary positive substitutions have arisen, further tuning takes about 20 generations, as discussed above. Thus, for all species evolutionary tuning occurs nearly instantaneously compared with the average time of species existence of the order of 10^6 years, which comprises at least 10^5 generations of larger animals. Note also that all the above estimates were done under the assumption that each individual in the population produces on average one offspring, which leads to overestimation of the time of evolutionary tuning. In reality, individuals in most species produce a large number of offspring, most of which is very short-lived (e.g. spawn in fish species). This increases the probability of the appearance of the positive substitutions in the population and shortens the time period necessary for evolutionary tuning. In endothermic animals, that are characterised by relatively low fecundity compared with other species of the biosphere, appearance of positive substitutions can be accelerated by production of a huge number of sperm cells, that may compete with each other for the possibility of egg fertilisation. Provided that spermatozoa carrying positive substitutions are more competitive, the rate of accumulation of positive substitutions in the population of mammals can be substantially enhanced.

The high competitiveness of individuals carrying single positive substitutions makes it possible for them to recognise each other and allow genetic recombination to combine all the positive substitutions in a single genotype of their offspring. Due to this fact, sexual breeding significantly accelerates evolutionary tuning. In the absence of genetic recombination, evolutionary tuning in an asexual population consisting of 10^9 individuals takes 10^7 generations (one generation per appearance of one positive substitution). Thus, asexual species deprived of genetic recombination may only appear in the course of evolution as sexual species that used to reproduce sexually but lost that ability later. Independent evolution of purely asexual species is strongly suppressed.

11.4 UNIFORMITY OF EVOLUTIONARY TEMPO IN DIFFERENT BIOLOGICAL TAXA

According to the proposed evolutionary pattern, functionally-sensible genetic macrofragments that initiate speciation in eukaryotes are prepared by rapidly-evolving viruses and bacteria. **This explains the observed uniformity of evolutionary tempo in different taxa of eukaryotes, i.e. the fact that the average lifespan of species is of the order of several million years irrespective of the population numbers, metabolic rates or body sizes of species' individuals.**

Generally, the contradiction discussed in the previous section stems from this uniformity and can be formulated in the following simple form. Why does a prokaryote species have to test almost a billion of billions of times more genotypes than a eukaryote species does, until a genotype coding for a new species is found? This striking contradiction is relaxed, when the horizontal transfer of genetic material from viruses and bacteria to eukaryotes is taken into account as the major driving force of the evolution.

This can be explained as follows. High mutability makes it possible for viruses to rapidly create positive viral fragments. Integrating into genome sequences of their hosts, viruses give them a possibility to test new positive viral fragments for compatibility with the host organism. If a new fragment appears to be positive for the host species as well, the individual with this fragment will have advantage and the fragment can be retained in the population (Marienfield *et al.*, 1997). Thus, evolution of viruses is endogenic in the sense that viruses create their positive fragments themselves.

Higher organisms mutate and reproduce much more slowly. We have seen in Section 11.3 that if evolution of eukaryotes and prokaryotes were endogenic as well, there would be a 17 orders of magnitude's difference between the average lifespans of eukaryote and prokaryote species, which is not the case. It means that higher organisms find their positive genetic fragments much more quickly among genetic fragments delivered to their genomes by viruses than if they created these fragments themselves. In other words, concentration of positive eukaryotic and prokaryotic fragments proves to be higher in an array of positive viral fragments delivered by viruses into eukaryotic genomes than in an equally large array of all possible decay fragments generated by higher organisms themselves in the process of mutagenesis. This is not surprising. Positive viral fragments have been already tested for life compatibility in the living cell and proved to be functionally sensible, as soon as they were able to ensure formation of a new virus species. Meanwhile endogenic genetic fragments formed during mutagenesis in genomes of higher organisms represent decay of genetic information and loss of function of previously meaningful genetic fragments.

Hence, the limiting factor of evolution in both prokaryotes and eukaryotes is the creation of functionally sensible genetic fragments by viruses. As soon as the limiting factor is the same, the evolutionary tempo is the same as well.

The proposed evolutionary pattern explains the observed relatively constant rate of genetic substitutions in homologous genes of different eukaryote as well as prokaryote species, i.e. the so-called 'molecular clock' (Zuckerkandl and Pauling, 1965; Kimura, 1987; Ohta, 1987; Zuckerkandl, 1987; Ochman and Wilson, 1987; Ayala, 1999). In many cases, the time various contemporary species have taken to diverge from a common ancestor is known from paleodata. The observed differences in the nucleotide sequences of homologous genes of the two species, divided by the time of their independent evolution, gives the doubled rate of molecular evolution of each species, i.e. the absolute rate of appearance of positive substitutions in the species' genomes. Comparing several pairs of such species with different times of divergence, one may find the average rate of molecular evolution during various time periods. These rates appear to be roughly the same for different evolutionary lineages, and that effect is essentially what the 'molecular clock' is about.

Within the proposed evolutionary pattern, such an effect is simply due to the fact that the average relative genetic distance between closely-related species appears to be approximately the same in different taxa; meanwhile the average lifespan is approximately the same for all species being determined by the rate of appearance of new positive genetic fragments in viral genomes. Dividing approximately equal values of evolutionary genetic distance (i.e. the number of positive genetic substitutions) by approximately equal values of the species' lifespan, we naturally arrive at approximately equal values of the absolute rate of appearance of positive substitutions per year. Note, however, that taking into account the saltatory character of the evolutionary process, the absolute rate of appearance of positive substitutions appears to be a non-informative characteristic when averaged over the whole time of species existence. The overwhelming majority of positive substitutions appear in the course of evolutionary tuning (Figure 11.1) shortly after acquisition of a positive macrofragment that initiates speciation, while during the remaining time of the species' existence no directional changes take place in the genome. Thus it makes no sense to speak about the average number of substitutions per year during the whole lifespan of a species.

11.5 CONCLUSIONS

Taking into account the fact that for the majority of the time of their existence all species remain in a state of evolutionary stasis, the problem of tempo and mode of evolution is of but secondary importance to the biotic regulation concept. The biotic regulation concept pertains mostly to the everyday existence of species and their everyday work on stabilisation of the environment.

However, we considered it necessary to briefly discuss in this chapter the major empirical data on evolution. This was done in order to show that all of them find an explanation within the biotic regulation concept. In other words, the existence of evolution neither discredits the biotic regulation concept nor provides evidence for genetic adaptation of species to the changing environment. Moreover, the proposed evolutionary pattern that arises from the central genetic issue of the biotic regulation concept, namely that of genetic stability of species which is necessary for stabilisation of an optimal environment, provides explanation for some evolutionary questions (like that of uniform evolutionary rate in prokaryotes and eukaryotes) that still remain even unaddressed within the traditional biological paradigm.

12

Conclusions: Can the Biosphere Be Treated as a Resource?

Humans are unique in the living world. Unlike all other species, humans are able to accumulate cultural heritage, which is transmitted to following generations. Accumulation of cultural heritage is not only a unique ability of *Homo sapiens* individuals, but one of their natural needs encoded into the species genome. The development of cultural heritage cannot be stopped. Were the cultural information cease to accumulate, that would mean termination of the existence of the *Homo sapiens* species.

An inherent component of the development of cultural heritage is scientific and technological progress. Its modern rate can be characterised by the average time of replacement of major technologies, which is now of the order of several years. The corresponding natural process of replacement of biotic 'technologies' (i.e. the ways of the biota's organisation and functioning) is characterised by the average evolutionary time of complete change of the genetic programmes of the extant species, which is of the order several hundred million years (Chapter 11). Thus, the contemporary rate of scientific and technological progress exceeds the rate of evolutionary progress by 7–8 orders of magnitude. Any possibility of sustainable development of the biosphere in pace with the development of humanity is therefore unthinkable. The misleading idea of such a possibility is based on erroneous identification of relatively rapid processes of decay and relaxation of the genetic information of natural biota (including artificial selection of domestic plants and animals) with natural evolutionary processes (Chapters 9, 10 and 11).

Thus, humanity is facing two contradicting phenomena—the inevitability of development of civilisation and the impossibility of an equally rapid development of the biosphere. In other words, while civilisation develops, the biosphere degrades. To solve the contradiction, one needs to confine civilisation's development within such limits that would make it possible to ensure the safe existence of the biosphere, i.e. the biota and its environment. In order to do it, one needs to quantify the permissible threshold of anthropogenic perturbation of the biosphere beyond which the biotic regulation of the environment ceases to operate on a global scale and global environmental changes commence. This threshold, in the first approximation, was found in this book in several independent ways (Sections 3.7; 4.6; 6.4; 6.5). It was shown that anthropogenic environmental impact is completely determined by the world human population number (Figure 6.5). Thus, the only limit that should be imposed on the development of civilisation is that the world population number of humans should not exceed the threshold value beyond which the biosphere starts to degrade on a global scale.

The existence of the modern world population is owed completely to the continuous progress of science and technology, which enables humans to accelerate the process of cultivation of the biosphere envisaged by humanity as but one of the environmental resources. With growing population density the *per capita* energy consumption inevitably increases (see Figure 12.1). It is therefore impossible to achieve a considerable reduction of anthropogenic energy consumption on a global scale at fixed population density. This means that the way towards sustainability does not lie through reduction of consumption of developed countries. Division of the modern world into developed countries with high *per capita* consumption and anthropogenic environmental impact and developing countries with lower *per capita*

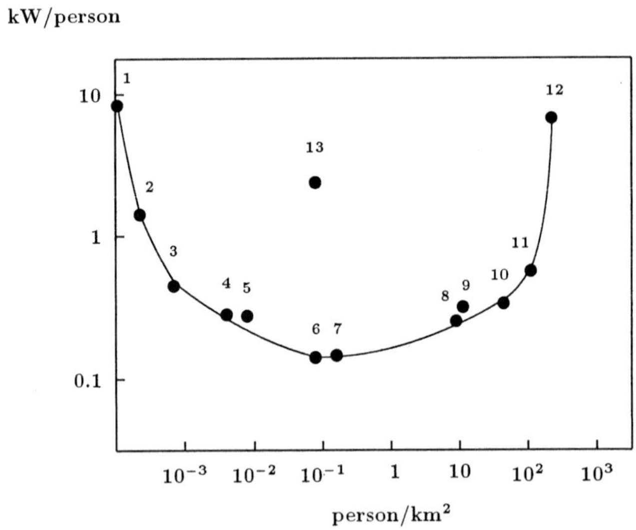

Figure 12.1. Anthropogenic *per capita* energy consumption in relation to population density. Figures stand for different energetic modes of human existence on a normal diet (i.e. not starving). 1 – motorised hunting; 2 – hunting on horseback; 3 – modern hunting by Eskimos; 4 – primitive hunting; 5 – primitive animal husbandry; 6 – primitive fishery; 7 – foraging; 8 – animal husbandry on pastures; 9 – primitive slash-and-burn agriculture; 10 – modern slash-and-burn agriculture (Tsembago); 11 – traditional agriculture (Russia, year 1897); 12 – modern world as a whole; 13 – motorised ocean fishery (Gorshkov, 1980; Gorshkov, 1995). Note that per capita energy consumption is rigidly correlated with population density. At a given population density the per capita energy consumption cannot be reduced. It can be only increased, as, e.g. in the case of motorised fishery – the most inefficient way of food consumption in the human history. Any species (humans are not an exception) is adapted to its natural ecological niche. Stable existence of a species outside its natural ecological niche requires additional energy expenditures. The greater the deviation of environmental conditions from the natural ecological niche of the species, the greater energy expenditures are needed to support life of the species individuals. Hence, hopes for reduced energy consumption of the humanity at sustained current global population number, are unjustified.

consumption and anthropogenic environmental impact is caused by the unevenness of the technological progress. Some countries develop quickly, others more slowly, but growth of consumption is common to all of them.

In the meantime, the observed degradation of the environment on both local and global scales is not due to the global economic and population *growth*. The already existing world human population goes beyond all possible ecological limits. It is insufficient simply to stop population growth. To solve the ecological problems of humanity, a significant reduction of the global population number is absolutely indispensable. Whatever is contemporary people's attitude towards negative population growth, this problem will have to be solved on the basis of the cumulative scientific potential of the humanity involving knowledge of ecological, genetic and social properties of *Homo sapiens*.

In this book we quantified the threshold value for the world population number beyond which global changes started (Figure 6.5). It is of the order of half a billion people. On a local scale, the threshold population density was exceeded long before the cumulative world population reached that value. There is no doubt that the information about the ecologically permissible species population density in humans, as in all other species, is encoded into the species genome and determines peculiarities of human behaviour in overpopulated areas. By all appearances, the critical population density of humans was reached when massive bloody wars were first registered in the history of humanity, a phenomenon never observed in any other biological species under natural (!) environmental conditions. In no natural biological species do adult individuals kill each other in masses, as humans do (Tinbergen, 1968). This indicates that humans have been living locally in distorted environmental conditions of overpopulation for at least several thousand years.

During thousands of years different countries have been competing for the territorial and material resources of the biosphere using military wars as the major way of competitive interaction. The bravest soldiers who killed the largest number of enemies enjoyed high social prestige. The greatest countries were those having the largest number of colonies. Wars undermined the power of the belligerent countries, but practically did not affect the rate at which the biosphere was being destroyed.

At the present time, colonies and large-scale military wars have become a thing of the past. They have been replaced by no less cruel economic wars which are aimed, as ever, at gaining control over biospheric resources. The highest social prestige is now associated with businessmen who have been able to use the largest amount of resources and gain the highest profit. Economic wars are not accompanied by visible loss of human lives, so they prove to be the most efficient way of resource use and cause the quickest possible degradation of the biosphere. The unconscious aggressive desire of *Homo sapiens* individuals to struggle with overpopulation by killing their conspecifics, that had been catered for during the historical period of global wars, is now satisfied in virtual reality, which is the only explanation of the unprecedented popularity of violence in the videos and computer games of the modern world.

The traditional global process of plundering the biosphere may be stopped only when people realise that the biosphere is not a resource to be used to increase personal or national wellbeing. In very much the same manner as modern ethics prohibits economic growth based on the sale of slaves, drugs and human organs, the future ethics of humanity is sure to prohibit extensive cultivation of the biosphere. Civilisation can and must develop, but not at the expense of the biosphere, which is the only guarantee for long-term global environmental stability. The development of civilisation in the direction of further cultivation of the biosphere should be stopped.

Such limitation of civilisation's development does not represent a constraint of human rights or individual freedom of human beings. Humanity has long ago got used to *physical constraints* imposed on human activities by the laws of nature. Nobody's personal freedom is violated by the fact that it is impossible to move at a speed higher than that of light, that it is impossible to construct a perpetuum mobile or by the fact that any individual life has a start and an end. Apart from these, there is a great number of *juridical constraints* on human activities that are written in human laws and aimed at maintaining stability of the modern, overpopulated world. The majority of these laws are never treated as a threat to human rights or personal freedom. Similarly, *ecological constraints* on human activities, when properly and clearly explained to the population, will not present any threat to human rights. These ecological constraints will govern the long-term stability of the existence of humanity on our planet. It is necessary to incorporate the knowledge of ecological constraints into the education system of modern humanity. Were people to observe the ecological constraints, the necessity of many contemporary constraints will be relaxed. In modern society, there are no laws that prohibit jumping from skyscrapers, because such constraint is already inherent to the genetically-encoded normal behaviour of humans. Similarly, there will be no necessity for laws prohibiting wars in a world with normal population density. Under conditions of their natural ecological niche, humans should be as incapable of killing their conspecifics as all other natural biological species.

The contemporary ecological problems of humanity are due to the fact that scientific investigation of the ecological constraints on civilisation development began very late in human history. People have been living too long in the imagined comfortable world of the unlimited possibilities of exploring natural resources of the Earth and space. People have been too long convinced of the principal inevitability of non-stationary development of the civilisation following the non-stationary development of the Universe itself (Shklovskii, 1987). The most important ecological problem of today is how to reduce the current level of the anthropogenic perturbation of the biosphere down to the threshold level compatible with long-term environmental stability. The problem is so non-trivial and unconventional that people are often unconsciously unwilling to admit the very existence of ecological constraints on economic and population growth.

The world human population has been exponentially growing since the middle of this millennium. Such demographic explosion became possible as a result of the economic growth based on scientific and technological progress. Since many generations of humans have already been existing under conditions of exponential economic growth accompanied by anthropogenic transformation of the biosphere, the very phenomenon of growth and non-stationarity is perceived by people as a natural state of affairs. Exponential economic growth is manifested in positive annual increments in industrial production, bank deposits and values of securities, which is regarded as an indispensable condition of people's existence.[1]

Industrial production is based on consumption of certain types of energy and matter known as natural resources. Non-renewable resources are those not used by the biota and not involved in the biochemical cycles, i.e. mineral ores and fossil fuels. The so-called renewable resources are solar energy and production of the biota itself.

The existence of limits to economic growth related to complete extinction of non-renewable resources and approaching the natural limits of consumption of renewable resources were convincingly demonstrated to the world community in a series of papers of the Rome club (Meadows *et al.*, 1972; 1974; 1992; Mesarovic and Pestel, 1974; Tinbergen *et al.*, 1977; see also Gore, 1992).

However, in all these papers the biosphere, i.e. the global biota and its environment, was treated from the economic point of view as but one of the available types of resources. It was assumed that the anthropogenic pollution of the environment was to be counteracted by adequately compensating human activities, which, as well as all the other types of human activities, were to be based on consumption of natural resources. The question of whether it is at all possible to keep the global environment stable by technological means did not even arise. The question of natural limits to the biospheric stability and the existence of biotic regulation of the environment was completely ignored as well.

Until the present time the problem of human population growth has been almost exclusively related to the task of feeding the global population on the basis of exploitation of renewable resources (Bongaarts, 1994; Daily *et al.*, 1998). The so-called optimistic forecasts are based on the fact that at present humans are only consuming about 10% of the net primary production (a renewable resource) of the biosphere (Figure 1.1). It is then concluded that, on the basis of the law of matter conservation, the global consumption and, hence, world population can be increased by an order of magnitude. However, renewable biospheric resources only remain renewable until the anthropogenic share of their consumption does not go beyond the threshold of biospheric stability. When the threshold is passed, renewable resources become non-renewable. On the other hand, were the consumption of renewable resources to remain below the ecological threshold, the very notion of resource loses its meaning (see below and Section 5.7).

Economic growth of free market economies is based on the possibility that a businessman finds an area of capital investment and production of goods which will be in demand and bring profit. The profit is then used for the personal enrichment of the businessman and the workers, on the one hand, and further expansion of the business, on the other. As a result, the industrial production, as well as the businessman's welfare, grows exponentially and may take a much shorter time than the human lifespan. The welfare of the businessman determines his social prestige in the human society, i.e. increases his competitive capacity. To increase the individual competitive capacity is an inherent desire of any living being. Stability of life is solely maintained because of this unique property of living matter (see Chapter

[1] A constant relative increment of any variable indicates its exponential growth.

2). Thus, prestige determines all the actions of the businessman and promotes economic growth. Socialism and centralised economic planning has lost the competition with capitalism and free market economy because this basic genetically-programmed instinct of humans was not paid sufficient attention within the socialist system. Those who worked hard and those who worked little were equally paid and, hence, were equally competitive in the society.

Production of anything is in its essence an ordered process maintained at the expense of external energy. Goods are ultimately produced from raw materials regarded as natural resources. Thus economic growth is inevitably followed by extensive cultivation of the biosphere. A free market makes it possible to effectively balance the offer and the demand, determine relative prices of all goods and introduce their money equivalent. Economic growth is the cornerstone of modern economic science, which completely loses any sense were the notion of growth eliminated from it (Leontieff, 1977). Economic growth is manifested in increasing rates of production of goods and, consequently, their money equivalent. The latter value is considered all over the world to be the major indicator of living standards. Thus, politicians in all countries tend to enhance economic growth by all available means, economic stagnation being the worst of all possible evils (Gore, 1992).

Any living organism in the biosphere is genetically adapted to its natural ecological niche. By definition, all organisms feel most comfortable under environmental conditions that are as close to the natural ones as possible. For the overwhelming majority of humans, such desired existence is only possible during leave when they are freed from work for a sufficiently long period of time. In most countries, yearly leave does not exceed 2–4 weeks, which is no more than 10% of the average lifetime. Thus, an average 70-year-old person has spent only seven years under natural environmental conditions. However, with the increasing degree of anthropogenic transformation of the biosphere fewer and fewer environments can be considered as close to the natural ecological niche of humans. The remaining natural areas begin to acquire high market value available to rich people only. One of the most expensive is the possibility of spending some time under conditions of low population density. Only the richest people are able to spend their free time on nearly uninhabited tropical islands. The overwhelming majority of humans have to spend most of their lives under unnatural environmental conditions distorted by civilisation. In a completely transformed biosphere it will be impossible to buy normal conditions of existence for any sums of money.

Hence, intensive work of modern people is ultimately aimed at transformation of the natural environment into a comfortable prison cell with the highest possible rate of per capita consumption of all possible goods. Prison, however comfortable, cannot substitute for the natural environment and cannot be associated with high living standard. Neither can such work constitute the meaning of human life.

It is often assumed that the question of the meaning of life is very complex, possibly unscientific and ambiguous. However, all the time people return to this question, which in itself is a scientific fact that calls for explanation. The meaning of human life includes necessary work aimed at reproduction and genetic stabilisation of the *Homo sapiens* species. That was evident to humans at the very moment of the origin of *Homo sapiens*. Were that not the case, there would be no humans and no such question. During the last several thousand years it has become evident that the meaning of human life also includes necessary work on stabilisation (i.e. preventing from decay) of culture and civilisation. Finally, during the last several decades it becomes evident that the meaning of human life also includes, as one of the major tasks, necessary work on stabilisation of the biosphere, i.e. the environment and the natural biota. Human culture cannot develop independently from this work and should be completely subordinate to the latter. Otherwise, as well as in the case of stopping reproduction and genetic stabilisation of the species, there will be no humans and no question of the meaning of life at all.

The meaning of life of individuals of all natural biological species is not reduced to existence in a given environment, adaptation to it and consumption of its resources, as is usually thought. Among all possible biological species that are capable of existence on Earth, the biota has selected only those that are also capable of performing strictly specified work on the stabilisation of their own environment in close cooperation with other species of their ecological community. Mechanisms of such selection are described in Chapter 5.

In locomotive animals, the strategy of necessary actions aimed at environmental stabilisation is genetically programmed as the system of positive and negative emotions. The meaning of life of any individual is to follow positive emotions and avoid negative ones. The tactic diversity of possible actions determined by the genetically-programmed strategy is dictated by information of individual memory which accumulates during the whole lifespan of the individual. In humans, the natural tactic realisation of the strategic programme of positive and negative emotions is considerably masked owing to the presence of artificial cultural information and requires a detailed scientific analysis. All the available evidence testifies for the statement that the meaning of human life cannot be principally different from that of all living beings of the rest of the natural biota.

As discussed in Section 6.7, the natural ecological function of large animals, including *Homo sapiens*, is to regularly introduce perturbations into the ecological community in order to create environments suitable for existence of so-called species-repairers, i.e. those that restore the environment after large-scale environmental perturbations. One of the modern manifestations of such destructive character of genetically-programmed natural human behaviour is the fact that the majority of people enjoy scenes of explosions, fires, catastrophes, which is extensively used by film producers. Had it not been the case, films showing catastrophes of all kinds could have not dominated the modern video market, as they now do.

However, in natural communities the population density of large animals is kept at a low level, so that their destructive activity (which is useful for the community in small doses) does not go beyond permissible ecological thresholds (Figure 3.3). Development of cultural heritage has provided humans with powerful tools of transformation of the biosphere. However, the basic strategy of human behaviour has remained unaffected by culture. As a result, with increasing population density of humans, the destructive anthropogenic forcing went far beyond the ecological threshold and gradually transformed into the global process of degradation of the biosphere which threatens the existence of life as a whole. It is time to remember that, apart from genetic information governing behaviour of individuals of all biological species, humans are in possess of cultural information that both can and must impose limits on instinctive destructive behavioural strategies of humans, stop the degradation of the biosphere and return humanity to a state of global stability.

While people certainly need to spend some time in the artificial environment of their civilisation, the total area affected by civilisation should constitute a small part of the area occupied by the natural non-perturbed biota. Most part of the work performed by humans should be aimed not at extensive development of civilisation, but at conservation of the natural biota in its non-perturbed state below the limit threatening the stability of the biosphere as a whole.

Functioning of natural biota is not based on consumption of environmental resources. The environment itself is prepared by the biota and maintained in the optimal for life state. Destruction of whatsoever substance in the biosphere is compensated by its production. All ordered processes in the biota work on a cyclic basis. Thus, there is no such notion as resource or limiting resource for the natural biota in the biosphere (Section 5.7).

Free competitive interaction of conspecific individuals in the natural biota serves as a tool of elimination from the population of decay individuals with impaired genetic programmes that destroy the existing optimal environment. In a natural ecological community there is no place for species-gangsters with all their individuals disrupting the correlated community structure and preventing it from performing regulation of the environment. Individuals of such species-gangsters do use the biota and the environment as a resource, not caring about the consequences. The consequences are disintegration of the community that gives shelter to the species-gangster, violation of biotic regulation and degradation of the local environment. As a result, the community loses competitiveness and is forced out from the biosphere (i.e. becomes extinct together with the species-gangster) by normal communities resistant to species-gangsters.

Hence, the notion of resource has no meaning for the self-regulatory system *community – its environment*. As soon as there are no resources, it is impossible to compete for them or privatise them. Similarly, it is impossible to divide conspecific individuals into the rich, i.e. those having property and governing the resource use, and the poor, i.e. those deprived of property and incapable of using resources. Under natural environmental conditions, decay individuals prevented from reproduction or suppressed in any other way constitute a negligible part of the population. The overwhelming majority of the population is represented by normal, equally-competitive individuals capable of performing equal amounts of work on the stabilisation of the environment and consuming equal amounts of life-necessary goods from the latter. Environmental conditions of the natural ecological niche of any species provide for 'democracy' and equal rights for the overwhelming majority of individuals in the population, these individuals working themselves on stabilisation of such optimal environment.

Under distorted environmental conditions, competitive interaction among conspecific individuals is weakened or completely switched off, so that decay individuals may accumulate and form the majority (Chapter 9). Decay individuals are incapable of performing the necessary work on stabilisation of the environment. To return the population and, hence, the environment, to the initial state, the remaining minority of normal individuals should nevertheless be able to govern the population. In such a situation, 'democracy' that equals normal and decay individuals is absolutely inadmissible. It would only work to further increase the number of decay individuals and, by doing so, facilitate loss of genetic information of the species and increase the probability of the species' extinction.

Being meaningless for the stable biosphere, the notion of resource acquires a meaning for any degrading system during the finite time of its degradation, e.g. when a species-gangster is introduced into an ecological community. Species-gangsters do not contain genetic information about the necessity of ensuring stability of the community and its environment. When several species-gangsters are introduced into the community, they may begin to compete for one and the same resource. Such rivalry of gangsters reveals the most competitive ones, i.e. those capable of using the resource at the highest speed and, hence, destroying the system in the shortest time. Such rivalry has nothing to do with the natural competitive interaction of conspecific individuals aimed at elimination of decay individuals destabilising the community and its environment.

Legal economy of the modern world considers the biosphere as one of the major resources of humanity and cannot avoid destroying the natural biota and the global environment. In comparison, criminal structures envisage the legal economy and civilisation itself as their resource. The legalised economic methods of plundering of the biosphere do not differ in their essence from legally prohibited actions of economic criminals. Competition of legal businessmen for the most efficient use of natural resources has much in common with rivalry of criminal structures dividing their zones of interests in using the resources of civilisation. Criminal structures destroy civilisation. Legal economy destroys the biosphere. The object of plundering is different, the methods are nearly the same. In the meantime, degradation of the biosphere would mean disappearance of the natural ecological niche of humans, which cannot be compensated by whatsoever high *per capita* rate of production of goods.

Humanity has learnt long ago that in order to keep society healthy and competitive, one needs to keep activity of criminal structures below a certain critical level. Consumption of criminal structures should not exceed a certain small share of the gross economic production of civilisation compatible with stable existence of the latter. In very much the same manner, stable existence of the biosphere and, ultimately, of the civilisation itself, is only possible when activity of civilisation does not go beyond an ecological threshold.

As shown in this book, stable existence of civilisation is only possible when the anthropogenic share of consumption does not exceed 1% of the net primary production of the biosphere. This estimate has been obtained by several independent ways using

(1) data on the global carbon cycle (Sections 6.4 and 6.5);
(2) data on the distribution of consumption over heterotrophic organisms of different body size in natural ecosystems (Section 3.7);
(3) data on the rate of diffusion of the excreta of large animals under natural environmental conditions (Section 4.6).

The present-day direct anthropogenic consumption of the net primary production on land is of the order of 10% exceeding the permissible ecological threshold by an order of magnitude (Sections 6.4 and 6.5). The cultivated part of land perturbed by the modern civilisation in one way or another constitutes more than 60% of the total land surface (Gorshkov, 1980; Vitousek *et al.*, 1986, 1997; World Resources, 1988).

A balanced reduction of the anthropogenic consumption of the biospheric production down to the ecologically-permissible level can be only performed as a result of reduction of the world human population approximately to the value observed in the beginning of the 18th century (Section 6.4 and Figure 6.5). Such reduction can be performed over a time period not exceeding a century following the one-family-one-child pattern in family planning. Were such reduction to be performed, the restrictions on the living standards of people related to over-consumption can be relaxed. Gradual reduction of the world population can thus be accompanied by continuously increasing living standards in all countries. It does not violate human rights and commonly accepted ethical norms. To return to the state of stable civilisation in a stable environment one only needs to overcome the non-scientific conservatism of the traditional attitude towards depopulation common to professional politicians, demographers, economists and the mass media. People should finally learn to treat the future generations with true (instead of declarative) respect and begin to really care for their interests, as all natural biological species do by maintaining stable environmental conditions in their ecological communities.

The global strategy towards depopulation can be accompanied by tactic measures causing more immediate improvement of the global environment. Modern forest exploitation represents one of the most serious threats to the environmental stability of today. Forests represent one of the most productive types of ecosystems encountered in the biosphere. When non-perturbed, they perform the largest amount of work on biotic regulation on land. Paradoxical as it is, the regulatory mechanism of forests that is vitally important for the whole of humanity is now being destroyed by a few per cent of the world's working population. Forest industry as a whole contributes not more than a few per cent to the Gross World Product (World Bank, 1997). Thus, complete abandonment of forest use could not significantly disturb the world economic system.

In the meantime, modern exploitation of forests leading to complete degradation of natural forest communities on the remaining large forest-covered territories, completely undermines the biotic regulation of the environment on all continents. Natural forests cannot be considered as a resource belonging to individual businessmen or even nations. The soonest possible prohibition of forest use and subsequent protection of forests by the united world community may give hope for the sustainable development of civilisation in the future. Immediate improvements in the water cycle, and many other components of the environment on a regional scale after restoration of natural forests, will benefit those countries where the forests are situated. The subsequent global improvement of the environment (e.g. with respect to the carbon cycle, Section 6.5) could compensate economically those relatively small economic losses related to stopping forest use. Those countries that are deprived of forests but benefit from the global environmental improvement will be able to compensate economic losses to the countries that will have stopped forest use. It is high time that the world community takes coordinated measures in this respect.

As shown in Section 6.5, the modern carbon cycle change and accumulation of CO_2 in the atmosphere can be stopped even if the modern rate of fossil fuel burning remains unchanged if the modern forest exploitation were reduced by approximately 40% on a global scale and natural forests were restored on previously-exploited territories. Note that only natural forest communities are able to perform biotic regulation of the environment. Plantations and managed forests are deprived of such ability and, on the contrary, contribute to the anthropogenic perturbation of the environment (Section 6.5). We speak therefore about restoration of the natural regulatory potential of the terrestrial biota achieved by restoration of natural forests, which has absolutely nothing to do with fixation of a certain amount of carbon in industrial wood, as it is sometimes proposed.

Were forest use reduced by the above value, further gradual reduction of the global world population can be prolonged over time and performed without imposing any undesirable effects on the world population. In other words, natural forests will give the people time to return the planet to the state of maximum possible environmental stability and restore the maximum efficiency of the biotic regulation on a global scale. Hence, humanity does have a real opportunity to overcome the global ecological crisis and restore the stability of favourable (for humans and life as a whole) environment on our planet.

References

Ahti T. and Oksanen J. (1990) Epigenic lichen communities of taiga and tundra regions. *Vegetatio*, **86**, 39–70.

Ajtay G.L., Ketner P. and Duvigneaud P. (1979) Terrestrial primary production and phytomass. In: Bolin B. (Ed.) *The Global Carbon Cycle*, SCOPE, 13. Wiley, Chichester, 129–182.

Alexeyev V.A. (1975) *Forest light regime*. Nauka, Leningrad (in Russian).

Alfaro R.I. and Singh P. (1997) Forest health management: a changing perspective. In: Vol 1. *Forest and tree resources*. Proc. XI World Forestry Congress, 13–22 October 1997, Antalia, Turkey, 157–166.

Allen C.W. (1955) *Astrophysical Quantities*. Athlone Press, London.

Altukhov Yu.P. (1991) *Population Genetics, Diversity and Stability*. Harwood Academic, New York.

Alwarez W. and Asaro F. (1990) An extraterrestrial impact. *Sci. Am.*, **263**, 78.

Amabile-Cuevas C.F. and Chicurel M.E. (1992) Bacterial Plasmids and Gene Flux. *Cell*, **70**, 189–199.

Andelman S.J. (1987) Evolution of concealed ovulation in vervet monkeys *Cercopithecus aethiops*. *Am. Nat.*, **129**, 785–799.

Andrijanov A.P. (1954) *Fishes of North Seas of USSR*. Acad., Moscow (in Russian).

Atkins A.R. and Nicholson J.D. (1963) An accurate constant-work-rate ergometer. *J. Appl. Physiol.*, **18**, 205.

Avise J.C. and Aquadro C.F. (1982) A comparative summary of genetic distances in vertebrates. *Evol. Biol.*, **15**, 151–185.

Avise J.C., Walker D. and Johns G.C. (1998) Speciation durations and Pleistocene effects on vertebrate phylogeography. *Proc. R. Soc. Lond. B.*, **265**, 1707–1712.

Ayala F.J. (1999) Molecular clock mirages. *BioEssays*, **21**, 71–75.

Ayala F.J. and Fitch W.M. (1997) Genetics and the origin of species: an introduction. *Proc. Nat. Acad. Sci. USA*, **94**, 7691–7697.

Ayala F.J. and Kiger J.A. (1984) *Modern Genetics*. Benjamin, London.

Baerlocher F. (1990) The Gaia hypothesis: A fruitful fallacy? *Experimentia*, **46**, 232–238.

Bannikova L.V. and Zubareva L.A. (1995) Genetic structure of some native and commercial breeds of cattle (*Bos taurus*) from Eurasia. *Genetika*, **31**, 697–708.

Barnola J.M., Pimienta P., Raynaud D. and Korotkevich Y.S. (1991) CO_2 climate relationship as deduced from Vostok ice core: a reexamination based on new measurements and on re-evolution of the air dating. *Tellus*, **43B**, 83–90.

Barry R.G. and Chorley R.J. (1987) *Atmosphere, Weather and Climate*. Methuen, New York.

Bauer J.E., Druffel E.R.M., Williams P.M., Wolgast D.M. and Griffin S. (1998) Temporal variability in dissolved organic carbon and radiocarbon in the eastern North Pacific Ocean. *J. Geophys. Res.*, **103C2**, 2867–2881.

Baumgartner H. and Reichel E. (1975) *The World Water Balance: Mean Annual Global Continental and Maritime Precipitation, Evaporation and Runoff*. Elsevier, Amsterdam.

Bazzaz F.A. (1975) Plant species diversity in old field successional ecosystem in Southern Illinois. *Ecology*, **56**, 485–488.

Begon M., Harper J.L. and Townsend C.R. (1996) *Ecology*. 3rd Ed. Blackwell, Oxford.

Begon M., Harper J.L. and Townsend C.R. (1986) *Ecology. Individuals, Populations and Communities*. Blackwell Science Publishers, London.

Bell G. (1982) *The Masterpiece of Nature: The Evolution and Genetic of Sexuality*. Univ. California, San Francisco.

Belov V.S. (1976) The effect of winds on forest. *Forest science, forest cultures and soil science*, **5**, 103–108 (in Russian).

Bergeron Y., Prairie Y.T, Richard P.J.H. and Duchesneau R. (1996) Holocene fire frequency and forest composition in the Canadian southeastern boreal forest. In: Bergeron Y., Frisque J. (Eds.) *Workshop Abstracts of Second International workshop on disturbance dynamics in Boreal Forests. Rouyn-Noranda, Québec, Canada, August 1996*. Université du Québec, Rouyn-Noranda, 11–14.

Berggren W.A. and Van Couvering J.A. (Eds.) (1984) *Catastrophes and Earth History: The New Uniformitarianism*. Princeton University, New York.

Berman E.R. (1975) *Geothermal Energy*. Neyes Data Co, London.

Bisazza A., Rogers L.J. and Vallortigara G. (1998) The origins of cerebral asymmetry: a review of evidence of behavioural and brain lateralization in fishes, reptiles and amphibians. *Neurosci. Biobehav. Rev.*, **22**, 411–426.

Bishop J.A. and Cook L.M. (1975) Moths, melanism and clean air. *Sci. Am.*, **232**, 90–99.

Blondel J., Pradel R. and Lebreton J.-D. (1992) Low fecundity insular blue tits do not survive better as adults than high fecundity mainland ones. *J. Animal. Ecol.*, **61**, 205–213.

Bolin B., Björkström A., Holmén K. and Moore B. (1983) The simultaneous use of tracers for ocean circulation studies. Tellus, **35B**, 206–236.

Bongaarts J. (1994) Can the growing human population feed itself? *Sci. Am.*, **270**, 36–42.

Bonnell M.L. and Selander R.K. (1974) Elephant seals: genetic variation and near extinction. *Science*, **184**, 908–909.

Borisenkov E.P. and Pasetsky V.M. (1988) *Thousandth chronicle of extraordinary natural phenomena.* Mysl, Moscow (in Russian).

Bormann B.T. and Sidle R.C. (1990) Changes in productivity and distribution of nutrients in a chronosequence at Glacier Bay National Park, Alaska. *Journal of Ecology*, **78**, 561–578.

Bormann F.N. and Likens G.E. (1979) *Pattern and process in a forested ecosystem.* Springer, New York.

Bormann F.N. and Likens G.E. (1994) *Pattern and process in a forested ecosystem. Disturbance, Development and the Steady State Based on the Habbard Brook Ecosystem Study.* 2nd Ed. Springer, Berlin.

Borovikov G.A. (1912) *Brief description of West Trans Angara vegetation.* Materials of the soil-botabical expedition on colonisation regions of the Asian Russia. Part II. Botanical research of the 1909. St.-Petersburg (in Russian).

Bouwman A.F. (1989) *The role of soil and land use in the greenhouse effect.* International Soil Reference and Information Center (ISRIC), Wageningen, The Netherlands.

Bowling A.T. and Ryder O.A. (1987) Genetic studies of blood markers in Przewalski's horse. *J. Hered.*, **78**, 75–80.

Bradbury J.W. and Andersson M.B. (1987) *Sexual Selection Testing the Alternatives.* Wiley, New York.

Brain K. and McNab B.K. (1980) Food habits, energetics, and population biology of mammals. *Am. Nat.*, **116**, 106–124.

Brais S., Camiré C., Bergeron Y. and Paré D. (1995) Changes in nutrient availability and forest floor characteristics in relation to stand age and forest composition in the southern part of North-Western Quebec. *Forestry Ecol. Management*, **76**, 181–189.

Brillouin L. (1956) *Science and Information Theory.* Academic Press, New York.

Brody S. (1945) *Bioenergetics and Growth.* Reinhold, New York.

Broecker W.S. (1982) Ocean chemistry during glacial time. *Geochim. Cosmochim. Acta*, **46**, 1689–1705.

Broecker W.S. and Peng T.-H. (1974) Gas exchange rates between air and sea. Tellus, **26**, 21–24.

Broecker W.S., Peng T.-H., Ostlund G. and Stuiver M. (1985) The distribution of bomb radiocarbon in the ocean. *J. Geophys. Res.*, **90**, 6953–6970.

Broecker W.S., Takahashi T., Simpson H.J. and Peng T.-H. (1979) Rate of fossil fuel carbon dioxide and the global carbon budget. *Science*, **206**, 409–418.

Brutsaert W. (1982) *Evaporation into the Atmosphere: Theory, History and Applications.* D. Riedel Publ., Dorrdecht, Holland.

Bryant D., Nielsen D. and Tangley L. (1997) *The Last Frontier Forests: Ecosystems and Economies on the Edge.* World Resources Institute, Washington.

Buckingham R.H. and Grosjean H. (1986) The accuracy of mRNA-tRNA recognition. In: Galas D.J., Kirkwood T.B.L. and Rosenberger R.F. (Eds.) *Accuracy in Molecular Processes: Its Control and Relevance to Living Systems.* Chapman and Hall, London.

Budyko M.I., Ronov A.B. and Yanshin A.L. (1987) *History of the Earth's Atmosphere*. Springer, Berlin.

Burt A., Bell G. and Harvey P.H. (1991) Sex differences in recombination. *J. Evol. Biol.*, **4**, 259–277.

Calder W.A. (1974) Consequences of body size for avian energetics. In: Paynter R. A. (Ed.) *Avian Energetics*. Publ. Nuttal Ornithological Club, Cambridge, No.15, 41–68.

Campbell S. and Liegel L. (1996) *Disturbance and Forest Health in Oregon and Washington*. USAD For. Serv. Pac. Northwest Sta. Gen. Tech. Rep. PNV-GTR-381.

Cano M.I. and Santos J.L. (1990) Chiasma frequencies and distributions in gomphocerine grasshoppers: a comparative study between sexes. *Heredity*, **64**, 17–23.

Cavagna, G.A. and Kaneko, M. (1977) Mechanical work and efficiency in level walking and running. *J. Physiol.*, **268**, 467–481.

Chahine M.T. (1992) The hydrological cycle and its influence on climate. *Nature*, **359**, 373–380.

Charlesworth B., Coyne J.A. and Barton N.H. (1987) The relative rates of evolution of sex chromosomes and autosomes. *Am. Nat.*, **130**, 113–146.

Chazdon R.L. and Field N. (1984) Light environments of tropical forests. In: Medina E. *et al.* (Eds.) *Physiological ecology of plants of the wet tropics*. W. Junk, The Hague.

Chen C.-T. A., Gong G.-C., Wang S.-L. and Bychkov A.S. (1996) Redfield ratios and regeneration rates of particulate matter in the Sea of Japan as a model of closed system. *Geophysical Res. Let.*, **23**, 1785–1790.

Chislenko L.L. (1981) *Structure of Flora and Fauna as Related to Body Size of Organisms*. Moscow Univ., Moscow (in Russian).

Ciais P., Tans P.P., Trolier M., White J.W.C. and Francey R.J. (1995) A large Northern Hemisphere terrestrial CO_2 sink indicated by the $^{13}C/^{12}C$ ratio of atmospheric CO_2. *Science*, **269**, 1098–1102.

Cloud P.E. (1972) A working model of the primitive earth. *Am. J. Sci.*, **240**, 537–548.

Clutton-Brock T.H. (1982) Sons and daughters. *Nature*, **298**, 11–12.

Cohn J.P. (1986) Surprising cheetah genetics. *Bioscience*, **36**, 358–361.

Courtillot V.E. (1990) Volcanic eruption. *Sci. Am.*, **263**, 85.

Covington W.W. (1976) Forest floor organic matter and nutrient content and leaf fall during secondary succession in northern hardwoods. *Ph.D. Thesis*, Yale University, New Haven, CT.

Covington W.W. (1981) Changes in forest floor organic matter and nutrient content following clear cutting in northern hardwoods. *Ecology*, **62**, 41–48.

Coyne J.A. and Orr H.A. (1989) Patterns of speciation in Drosophila. *Evolution*, **43**, 362–381.

Crocker R.L. and Major J. (1955) Soil development in relation to vegetation and surface age at Glacier Bay, Alaska. *Journal of Ecology*, **43**, 427–448.

Curio E. (1989) Some aspects of avian mortality patterns. *Mitteilungen Zoologisches Museum Berlin* (*Annalen für Ornithologie*), **65**, Suppl. 13, 47–70.

Daily G., Dasgupta P., Bolin B., Crosson P., du Guerny J., Ehrlich P., Folke C., Jansson A.M., Jansson B.-O., Kautsky N., Kinzig A., Levin S., Mler K.-G., Pinstrup-Andersen P., Siniscalco D. and Walker B. (1998) Food production, population growth and environment. *Science*, **281**, 1291–1292.

Damuth J. (1981a) Population density and body size in mammals. *Nature*, **290**, 699–700.

Damuth J. (1981b) Home range, home range overlap and species energy use among herbivorous mammals. *Biol. J. Linn. Soc.*, **15**, 185–193.

De Grandpré L., Gagnon D. and Bergeron Y. (1993) Changes in the understory of Canadian southern boreal forests after fire. *Journal of Vegetation Science*, **4**, 803–810.

De Voogs C.G.N. (1979) Primary production in aquatic environments. In: Bolin B., Degens E.T., Kempe S. and Ketner P. (Eds.) *The Global Carbon Cycle*. Wiley, New York, 259–292.

Degens E.T., Berendt M., Gotthardt B. and Reppmann E. (1968) Metabolic fractionation of carbon isotopes in marine plankton. *Deep-Sea Res.*, **15**, 11–20.

Degens E.T., Kempe S. and Spitzy A. (1984) Carbon dioxide: a biogeochemical portrait. In: Hutziger O. (Ed.) *The Handbook of Environmental Chemistry*, vol. 1, part C. Springer, Berlin, 127–215.

Dickinson E. (Ed.) (1987) *The geophysiology of Amazonia: Vegetation and climate interaction*. Wiley, New York.

Dinerstein E. and McCracken G.F. (1990) Endangered greater one-horned rhinoceros carry high levels of genetic variation. *Conserv. Biol.*, **4**, 417–422.

Dobzhansky T. (1951) *Genetics and the Origin of Species*. 3rd ed. Columbia Univ., New York.

Dokuchayev V.V. (1954) *Selected Works*. AN SSSR, Moscow (in Russian).

Domingo E., Escarmis C., Sevilla N., Moya A., Elena S.F., Quer J., Novella I.S. and Holland J.J. (1996) Basic concepts in RNA virus evolution. *FASEB J.*, **10**, 859–864.

Doran J.W., Elliot E.T. and Paustian K. (1998) Soil microbial activity, nitrogen cycling, and long-term changes in organic carbon pools as related to fallow tillage management. *Soil and Tillage Research*, **49**, 3–18.

Dorit R.L., Akashi H. and Gilbert W. (1995) Absence of polymorphism at the ZFY locus on the human Y chromosome. *Science*, **268**, 1183–1185.

Drake J.W. (1969) Comparative rates of spontaneous mutation. *Nature*, **221**, 1128–1132.

Drake J.W. (1974) The role of mutation in bacteria evolution. *Symp. Soc. Gen. Microbiol.*, **24**, 41–58.

Drake J.W. (1991) A constant rate of spontaneous mutation in DNA-based microbes. *Proc. Natl. Acad. Sci. USA*, **88**, 7160–7164.

Drake J.W. (1993) Rates of spontaneous mutations among RNA viruses. *Proc. Natl. Acad. Sci. USA*, **90**, 4171–4175.

Druffel E.R.M. and Benavides L.M. (1986) Input of excess CO_2 to surface ocean based on $^{13}C/^{12}C$ ratios in a banded Jamaican sclerosponge. *Nature*, **321**, 58–61.

Druffel E.R.M. and Williams P.M. (1990) Identification of a deep marine source of particulate organic carbon using bomb ^{14}C. *Nature*, **347**, 172–174.

Druffel E.R.M., Williams P.M. and Suzuki Y. (1989) Concentrations and radiocarbon signatures of dissolved organic matter in the Pacific ocean. *Geophys. Res. Lett.*, **15**, 991–994.

Dugdale R.C. and Goering J.J. (1967) Uptake of new and regenerated forms of nitrogen in primary productivity. *Limnol. Oceanogr.*, **12**, 196–207.

Duvigneaud P. (1974) *La Synthese Ecologue*. Doin, Paris.

Dyrenkov S.A. (1984) *Structure and dynamics of spruce forest of taiga zone*. Nauka LO, Leningrad (in Russian).

Efroimson V.P. (1932) On some problems of accumulation and action of lethals. *Biological Journal*, **1**, 87–101 (in Russian).

Ehrhardt J.P. and Seguin G. (1978) *Le plancton, composition, ecologie, pollution*. Borgas, Paris.

Eigen M. (1971) *Selforganization of Matter and the Evolution of Biological Macromolecules*. Springer, Berlin.

Eisenberg J. (1981) *The Mammalian Radiations: an Analysis of Trends in Evolution, Adaptation and Behavior*. Athlone Press, London.

Emlen S.T. and Oring L.W. (1977) Ecology, sexual selection, and the evolution of mating systems. *Science*, **197**, 215–223.

Engelmark O. (1987) Fire history correlations to forest type and topography in northern Sweden. *Annales Botanici Fennici*, **24**, 317–324.

Falkowski P.G. and Wilson C. (1992) Phytoplankton productivity in the North Pacific Ocean since 1900 and implications for absorption of anthropogenic CO_2. *Nature*, **358**, 741–743.

Falkowski P.G. and Woodhead A.D. (Eds.) (1992) *Primary Productivity and Biogeochemical Cycles in the Sea*. Plenum, New York.

FAO (1995) *Forest fire statistics, 1992–94*. Timber Bulletin, UN Publication, **48**(4).

Farrar J.F. (1976) The lichen as an ecosystem: observation and experiment. In: Brown D.H., Hawksworth D.L. and Bayley R.H. (Eds.) *Lichenology: Progress and Problems*. Academic, New York, 385–406.

Fedorov V.D. and Gilmanov T.G. (1980) *Ecology*. Moscow Univ., Moscow (in Russian).

Fisher R.A. (1930) *The Genetical Theory of Natural Selection*. Clarendon Press, Oxford.

Fogg G.E. (1975) Biochemical pathways in unicellular plants. In: Cooper J.P. (Ed.) *Photosynthesis and Productivity in Different Environments*. Cambridge.

Fonda R.W. (1974) Forest succession in relation to river terrace development in Olympic National Park, Washington. *Ecology*, **55**, 927–942.

Forest Fund of Russia (1995) *Statistical collection*. ARIC, Moscow (in Russian).

Foster T.J. (1983) Plasmid-determined resistance to antimicrobial drugs and toxic metal ions in bacteria. *Microbiol. Rev.*, **47**, 361–409.

Friedli H., Lotscher H., Oeschger H., Siegenthaler U. and Stauffer B. (1986) Ice core record of the $^{13}C/^{12}C$ ratio in atmospheric CO_2 in the past two centuries. *Nature*, **324**, 237–238.

Fuerst P.A. and Ferrell R.E. (1983) The analysis of hidden electrophoretic variation: interspecific electrophoretic differentiation and amino acid divergence. *J. Mol. Evol.*, **19**, 449–454.

Furyaev V.V. (1996) *The Role of Fires in Forest Formation Process.* Nauka SO, Novosibirsk (in Russian).

Furyaev V.V. and Kireyev D.M. (1979) *Investigation of Post-fire Dynamics of Forests on the Landscape Basis.* Nauka SO, Novosibirsk (in Russian).

Gagarina E.I. (1996) Soils and soil cover of plateau-form heights in the north-west of the Russian plain. *Vestnik (Herald) of St.-Petersburg University.* Ser. 3, **1**, 62–74 (in Russian).

Galasyeva T.V. (1977) Localization of hotbeds of tree pests in forest, burnt-out in 1972 in the Moscow region. *PhD Thesis*, Leningrad (in Russian).

Galenko E.P. (1983) *Phytoclimate and Energy Factors of Productivity of North European Coniferous Forest.* Nauka LO, Leningrad (in Russian).

Gammon R.H., Komhyr W.D. and Peterson J.T. (1986) The global atmospheric CO_2 distribution 1968–1983: Interpretation of the results of the NOAA/GMCC measurement program. In: Trabalka J.R. and Reichle D.F. (Eds.) *The Changing Carbon Cycle: A Global Analysis.* Springer, Berlin.

Gavrilov V.M. (1994) General regularities of temperature impact on energetics of endothermic animals. *Doklady (Proceedings) of the Russian Academy of Sciences*, **334**, 121–126 (in Russian).

Geodakyan V.A. (1987) *Evolutionary logic of sex differentiation in phylogenesis and ontogenesis.* Doctoral Thesis. Koltsov Inst. of Developmental Biology, Moscow (in Russian).

Gessaman J.A. (Ed.) (1973) *Ecological Energetics of Homeotherms.* Utah State Univ., Logan, Utah.

Giller P.S. (1984) *Community Structure and the Niche.* Chapman and Hall, London.

Ginatulin A.A. (1984) *Structure, Organisation and Evolution of Genome of Vertebrates.* Nauka, Moscow (in Russian).

Ginzburg V.L. (1985) *Physics and Astrophysics.* Pergamon Press.

Goldammer J. and Furyaev V.V. (Eds.) (1996) *Fire in Ecosystem of Boreal Eurasia.* Kluwer.

Golley F.B. (1973) Impact of small mammals on primary production. In: Gessaman J.A. (Ed.) *Ecological Energetics of Homeotherms.* Utah State Univ., Logan, Utah, 147–142

Goodman M.F. and Fygenson K.D. (1998) DNA polymerase fidelity: from genetics toward a biochemical understanding. *Genetics*, **148**, 1475–1482.

Gore A. (1992) *Earth in the Balance: Ecology and the Human Spirit.* Houghton Mifflin, New York.

Gorshkov V.G. (1979) On the role of the terrestrial and marine biota in the global carbon budget. *Preprint 534.* Leningrad Nuclear Physics Inst., Leningrad.

Gorshkov V.G. (1980) Structure of biospheric energy flows. *Botanical J.*, **65**, 1579–1590 (in Russian).

Gorshkov V.G. (1981) The distribution of energy flows among the organisms of different dimensions. *J. General Biology*, **42**, 417–429 (in Russian).

Gorshkov V.G. (1982a) Energetics of locomotive animals. *Ecologija*, **1**, 1–14 (in Russian).

Gorshkov V.G. (1982b) *Energetics of the Biosphere*. Leningrad Polytechnical Inst., Leningrad (in Russian).

Gorshkov V.G. (1982c) The possible global budget of carbon dioxide. *Nuovo Cimento*, **5C**, 209–222.

Gorshkov V.G. (1983a) Power and rate of locomotion in animals of different sizes. *J. General Biology*, **44**, 661–678 (in Russian).

Gorshkov V.G. (1983b) The influence of the oceanic biota on atmospheric carbon dioxide. *Preprint 831*. Leningrad Nuclear Physics Inst., Leningrad.

Gorshkov V.G. (1984a) Energetical efficiency of flight and swmming. *J. General Biology*, **45**, 779–794 (in Russian).

Gorshkov V.G. (1984b) Possible role of the oceanic biota in the global carbon balance. *Oceanologija*, **24**, 453–459 (in Russian).

Gorshkov V.G. (1985a) Natural selection of communities and the stability of biogeochemical cycles. In: Mlikovsky J. and Novak V.J.A. (Eds.) *Evolution and Morphogenesis*. Academia, Prague, 787–797.

Gorshkov V.G. (1985b) Stability of the biogeochemical matter cycle. *Ecologija*, **2**, 3–12 (in Russian).

Gorshkov V.G. (1995) *Physical and Biological Bases of Life Stability: Man, Biota, Environment*. Springer, Berlin.

Gorshkov V.G. and Dolnik V.R. (1980) Energetics of the biosphere. *Uspekhi Fizicheskikh Nauk*, **131**, 441 478. English transl.: *Sov. Phys. Usp.*, **23**, 386, American Institute of Physics.

Gorshkov V.G. and Makarieva A.M. (1997) Heterozygosity dependence on body mass in mammals. *Doklady (Proceedings) of the Russian Academy of Sciences*. **355**, 418–421 (in Russian).

Gorshkov V.G. and Makarieva A.M. (1998). Impact of terrestrial and oceanic biota on the modern carbon and oxygen cycles. *Ecological Chemistry*, **7**, 129–137.

Gorshkov V.G. and Makarieva A.M. (1999) Haldane's rule and somatic mutations. *Russian Journal of Genetics*, **35**, 611–617. (Translated from *Genetika*, **35**, 1999, 725–732)

Gorshkov V.V. (1993) Characteristics of forest ecosystem recovery after fires. *Doklady (Proceedings) of the Russian Academy of Sciences*, **333**, 111–114 (in Russian).

Gorshkov V.V. (1994) Change in species diversity of ground lichens under impact of pollution with respect to the postfire age of the forest community. *Doklady (Proceedings) of the Russian Academy of Sciences*, **334**, 665–668 (in Russian).

Gorshkov V.V. (1998) Boreal forest community: its internal environment, recovery after disturbances and stationary state. In: Gorshkov V.G., Makarieva A.M., Kharkina T. (Eds.) Proc. Int. Sem. *Role of virgin terrestrial biota in the modern processes of global change: Biotic regulation of the environment*. 12–16 October 1998, Petrozavodsk, Russia. PNPI, Gatchina, 138–197.

Gorshkov V.V. and Bakkal I.Yu. (1996) Species richness and structure variations of Scots pine forest communities during the period from 5 to 210 years after fire. *Silva Fennica*, **30**, 329–340.

Gorshkov V.V. and Gorshkov V.G. (1992) *Recovery characteristics for ecosystems. Preprint 1850*. Petersburg Nuclear Physics Inst., St. Petersburg.

Gorshkov V.V., Bakkal I.Yu. and Stavrova N.I. (1996) Postfire recovery of forest litter in Scots pine forests in two different regions of boreal zone. *Silva Fennica*, **30**, 209–219.

Gorshkov V.V., Bakkal I.Yu. and Stavrova N.I. (1997) An estimate of possible impact of global warming on relaxation of characteristics in northern forest communities. *Doklady (Proceedings) of the Russian Academy of Sciences*, **355**, 414–417 (in Russian).

Goryshina T.K. (1979) *Plant Ecology*. Vysshaja shkola, Moscow (in Russian).

Gould S.J. and Eldridge N. (1993) Punctuated equilibrium comes of age. *Nature*, **366**, 223–227.

Govindjee O. (Ed.) (1982) *Photosynthesis*. Academic Press, New York.

Grant V. (1977) *Organismic Evolution*. Freeman Co, San Francisco.

Green N.P.O., Stout G.W., Taylor D.J. and Soper R. (1989) *Biological Science*. CUP, Cambridge.

Gromtsev A.N. (1993) Fire regime in spontaneous forests in the landscapes of NW taiga. *Ecologija*, **3**, 22–26 (in Russian).

Grosjean H., Cedergren R.J. and McKay W. (1982) Structure in tRNA data. *Biochimie*, **64**, 387–397.

Gustavson M.R. (1979) Limits to wind power utilization. *Science*, **204**, 13–17.

Haldane J.B.S. (1922) Sex ratio and unisexual sterility in hybrid animals. *J. Genet.*, **12**, 101–109.

Haldane J.B.S. (1954) The stasis of evolution. In: Huxley J., Hardy A.C., Ford E.B. (Eds.) *Evolution as a Process*. George Allen and Unwin Ltd., London, 109–121.

Haldane J.B.S. (1957) The cost of natural selection. *J. Genet.*, **55**, 511–524.

Hamilton W.D., Axelrod R. and Tanese R. (1990) Sexual reproduction as an adaptation to resist parasites: a review. *Proc. Natl. Acad. Sci. USA*, **87**, 3566–3573.

Hansen J.E. and Lacis A.A. (1990) Sun and dust versus greenhouse gases: an assessment of their relative roles in global climate change. *Nature*, **346**, 713–719.

Hardy Y., Mainville M. and Schmitt D.M. (1986) *An atlas of spruce budworm defoliation in Eastern North America, 1938–1980*. United States Dept. of Agriculture, Forest Service, Miscellaneous publication.

Harestad A.S. and Bunnell F.L. (1979) Home range and body weight – a reevaluation. *Ecology*, **60**, 389–402.

Hawksworth D.L. and Rose F. (1970) Qualitative scale for estimation for sulphur dioxide air pollution in England and Wales using epiphytic lichens. *Nature*, **5254**, 145–148.

Hayes J.M. (1996) The earliest memories of life on Earth. *Nature*, **384**, 21–22.

Hays G.C., Harris R.P., Head R.N. (1997) The vertical nitrogen flux caused by zooplankton diel vertical migration. *Mar. Ecol. Prog. Ser.*, **160**, 57–62.

Heglund N.C., Cavagna G.A., Fedak M.A. and Taylor C.R. (1979) Muscle efficiency during locomotion: how does it vary with body size and speed? (Museum of Comp. Zool. Harvard Univ. Cambridge) MA 02138 and Univ. of Milan (FASEB Abstract speed).

Heimann M. and Maier-Reimer E. (1996) On the relations between the oceanic uptake of carbon dioxide and its carbon isotopes. *Global Biogeochemical Cycles*, **10**, 89–110.

Heinselman M.L. (1981) Fire intensity and frequency as factors in the distribution and structure of northern ecosystems. In: *Fire Regimes and Ecosystem Properties*. Proceedings of the conference December 11–15, 1978, Honolulu, Hawaii. United States Department of Agriculture, Forest Service, General Technical Report WO-26, 7–57.

Helli P. and Niemi G. (1996) Bird community dynamics in boreal forest. In: DeGraaf R.M. and Miller R.I. (Eds.) *Conservation of Faunal Diversity in Forested Landscapes*. Chapman and Hall, London.

Hemmingsen A.M. (1960) Energy metabolism as related to body size and respiratory surfaces, its evolution. *Rept. Steno Mem. Hosp. Nord. Insulinlab.*, **9**, 7–110.

Henderson-Sellers A. and Cogley J.G. (1982) The Earth's early hydrosphere. *Nature*, **298**, 832–833.

Hesin R.V. (1984) *Nonconstancy of Genom*. Nauka, Moscow (in Russian).

Hill A.V. (1960) Production and absorption of work by muscle. *Science*, **131**, 897.

Hofker M.H., Scraastad M.I., Bergen A.A.B., Wapenaar M.C., Bakker E., Millington-Ward A., van Ommen G.J.B. and Pearson P.L. (1986) The X chromosome shows less genetic variation at restriction sites than the autosomes. *Am. J. Hum. Genet.*, **39**, 438–451.

Holing K. (Ed.) (1978) *Adaptive Environmental*. John Wiley & Sons, Chichester.

Holland J., Spindler K., Horodyski F., Grabau E., Nichol S. and VandePol S. (1982) Rapid evolution of RNA genomes. *Science*, **215**, 1577–1585.

Holligan P.M. and de Boois H. (Eds.) (1993) *Land-Ocean Interactions in the Coastal Zone*. Global Change, IGBP report No.25, Stockholm.

Houghton R.A. (1989) The long-term flux of carbon to the atmosphere from changes in land use. *Extended Abstracts of Papers Presented at the Third Internat. Conf. on Analysis and Evolution of Atmospheric CO_2*, Data. W.M.O. Univ. Heidelberg, 80–85.

Houghton R.A., Hobbie J.E., Melillo J.M. *et al.* (1983) Changes in the content of terrestrial biota and soils between 1860 and 1980: a net release of CO_2 to the atmosphere. *Ecological Monographs*, **53**, 235–262.

Hubbert M.K. (1971) The energy resources of the Earth. *Sci. Am.*, **225**, 61–70.

Hutchinson G.E. and MacArthur R.H. (1959) A theoretical ecological model of size distributions among species of animals. *Am. Nat.*, **43**, 117–125.

IPCC (1994) *Climate Change 1994: Radiative Forcing of Climate Change and an Evaluation of the IPCC IS92 Emission Scenarios*. Houghton J.T., Meira-Filho

L.G., Bruce J. Lee Hoesung, Callander B.A., Haites E., Harris N. and Maskell K. (Eds.). CUP, Cambridge.

IPCC (1996) *Climate Change 1995: The Science of Climate Change*. Houghton J.T., Meira-Filho L.G., Callander B.A., Harris N., Kattenberg A. and Maskell K. (Eds.). CUP, Cambridge.

Isayev A.S. and Petrenko E.S. (1968) Biogeocenotic peculiarities of tree pests quantity dynamics. *Forest Science* (*Lesovedenie*), **3**, 1–12 (in Russian).

Isayev A.S. and Utkin A.I. (1963) Ground fires in larch forests in East Siberia and tree pests significance in postfire stand state. In: *Protection of Siberian Forests against Insect Pests*. AN SSSR, Moscow (in Russian).

Isayev A.S., Khlebopros R.G., Nedoresov L.V., Kondakov Ju.P. and Kiselev V.V. (1984) *Population Dynamics of the Forest Insects*. SO Nauka, Novosibirsk (in Russian).

Ivanoff A. (1972), (1975) *Introduction a L'Océanographie*. vols 1, 2. Librarie Vuibert, Paris.

Ivanter E.V. (1975) *Population Ecology of Small Mammals in Taiga Zone in the North-West of USSR*. Nauka LO, Leningrad (in Russian).

Ivashkevich B.A. (1929) Virgin forest, its structure and development peculiarities. *Forest Husbandry and Forest Industry* (*Lesnoe khozjajstvo i lesnaja promishlennost*), **10**, 36–44; **11**, 40–47, **12**, 41–46 (in Russian).

Jackson I.J. (1971) Problems of throughfall and interception under tropical forests. *J. Hydrology*, **12**, 234–254.

Jackson J.B.C. (1990) Evolutionary significance of morphospecies: a test with cheilostome Bryozoa. *Science*, **248**, 579–583.

Jackson J.B.C. (1994) Constancy and change of life in the sea. *Phil. Trans. R. Soc. Lond. B*, **344**, 55–60.

Jakubiczka S. Arnemann J., Cooke H.J., Krawczak M. and Schmidtke J. (1989) A search for restriction fragment length polymorphism on the human Y chromosome. *Hum. Genet.*, **84**, 86–88.

Jarvis P.G. (Ed.) (1989) *Forests, Weather and Climate*. A discussion organized by Jarvis P.G. *et al.*, London.

Jemisson G.H. (1934) The significance of the effect of stand density upon the weather beneath the canopy. *Journal of Forestry*, **32**, 416–446.

Johnson E.A. (1979) Fire recurrence in the Subarctic and its implications for vegetation composition. *Canadian Journal of Botany*, **57**, 1374–1379.

Johnson E.A. (1996) *Fire and Vegetation Dynamics. Studies from North American Boreal Forests*. CUP, Cambridge.

Johnston D.W. and Odum E.P. (1956) Breeding bird population in relation to plant succession on the piedmont of Georgia. *Ecology*, **37**, 50–62.

Jones D. (1990) Time twisters. *Nature*, **347**, 130.

Jordan R.A. and Brosemer R.W. (1974) Characterization of DNA from three bee species. *J. Insect. Physiol.*, **20**, 2513–2520.

Jukes T.H. (1987) Transversions and molecular evolutionary clock. *J. Mol. Evol.*, **26**, 87–98.

Kamenir Yu.G. (1991) Comparison of size spectra of living matter of three aquatic ecosystems. *Synt. Anal. Model. Stimul.*, **8**, 72–75.

Kamenir Yu.G. and Khailov K.M. (1987) The metabolism parameters and the world oceanic living matter integral spectra: comparison of size spectra. *Okeanologija*, **27**, 656–661 (in Russian).

Kanwisher J.W. and Ridgway S.H. (1983). The physiological ecology of whales and porpoises. *Sci. Am.*, **248**, 110–121.

Karpov V.G. (1969) *Experimental Phytocenology of Dark-coniferous Taiga*. Nauka LO, Leningrad (in Russian).

Kasting J.F. (1987) Theoretical constraints on oxygen and carbon dioxide concentrations in the Precambrian atmosphere. *Precambrian Research*, **34**, 205–229.

Kasting J.F., Toon O.B. and Pollack J.B. (1988). How climate evolved on the terrestrial planets. *Sci. Amer.*, **258**, 90–97.

Keeling C.D. (1973) The carbon dioxide cycle. In: Rasool S. (Ed.) *Chemistry of the Lower Atmosphere*. Plenum Press, New York, 251–329.

Keeling R.F. and Shertz R. (1992) Seasonal and interannual variations in atmospheric oxygen and implications for the global carbon cycle. *Nature*, **358**, 723–726.

Keeling, R.F., Piper, S.C. and Heimann M. (1996) Global and hemispheric CO_2-sinks deduced from changes in atmospheric O_2 concentration, *Nature*, **381**, 218–221.

Kellman M. and Ruolet N. (1990) Nutrient flux and retention in a tropical sand-dune succession. *Journal of Ecology*, **78**, 664–676.

Kellogg E.A. and Appels R. (1995) Intraspecific and interspecific variation in 5S RNA genes are decoupled in diploid wheat relatives. *Genetics*, **140**, 325–343.

Kellogg W.W. and Schneider S.H. (1974) Climate stabilisation, for better of for worse? *Science*, **186**, 1163–1172.

Kendeigh S.C. (1974) *Ecology with Special References to Animals and Man*. Prentice-Hall, New York.

Kershaw K.A. (1975) Studies of lichen-dominated systems. XV. The temperature and humidity profiles in a Cladina alpestris mat. *Canadian Journal of Botany*, **53**, 2614–2620.

Kerzina M.N. (1956) The influence of clear-cutting and burned-out forests on forest fauna formation. In: *Role of Animals in the Forest Life*. AN SSSR, Moscow (in Russian).

Kettlewell H.B.D. (1973) *The Evolution of Melanism*. Clarendon Press, Oxford.

Khailov K.M. (1971) *Ecological Metabolism in the Sea*. Naukova Dumka, Kiev (in Russian).

Kimura M. (1968) Evolutionary rate at the molecular level. *Nature*, **217**, 624–625.

Kimura M. (1983) *The Neutral Theory of Molecular Evolution*. CUP, Cambridge.

Kimura M. (1987) Molecular evolutionary clock and neutral theory. *J. Mol. Evol.*, **26**, 24–33.

Kimura M. (1989) The neutral theory of molecular evolution and the world view of neutralists. *Genome*, **31**, 24–31.

King T. (1974) Seasonal allocation of time and energy resources in birds. In: Paynter R.A. (Ed.) *Avian Energetics*. Nuttall Ornithological Club (Publ. No. 15), Boston, 4–28.

Kitredge D. (1951) *The Influence of Forest on Climate, Soil and Water Regime*. Foreign Literature (Inostrannaja literatura), Moscow (Russian edition).

Kleiber M. (1961) *The Fire of Life*. Wiley, New York.

Knowlton N., Weigt L.A., Solœrzano L.A., Mills D.K. and Bermingham E. (1993) Divergence in proteins, mitochondrial DNA, and reproductive compatibility across the Isthmus of Panama. *Science*, **260**, 1629–1632.

Kohlmaier G.H., Brohl H., Size E.O., Plochl M. and Ravelle R. (1987) Modelling stimulation of plants and ecosystem response to present levels of excess atmospheric CO_2. *Tellus*, **39B**, 155–170.

Kolomiets N.G. (1961) Cold-resistance of Siberian silkworm caterpillars and temperature regime in their hibernation habitat. *Izvestija SO AN SSSR*, **1**, 113–120 (in Russian).

Kondakov Yu.P. (1974) Regularities of massive outbreaks of the Siberian silkworm. In: *Population ecology of forest animals*. Nauka SO, Novosibirsk (in Russian), 206–265.

Kondrashov A.S. (1982) Selection against harmful mutations in large sexual and asexual populations. *Genet. Res.*, **40**, 325–332.

Kondrashov A.S. (1988) Deleterious mutations and the evolution of sexual reproduction. *Nature*, **336**, 435–440.

Kondratyev K.Ya. (1999) *Climatic Effects of Aerosols and Clouds*. Praxis Publishing Ltd., Chichester, UK.

Kordum V.A. (1982) *Evolution and Biosphere*. Naukova Dumka, Kiev (in Russian).

Korovin G.N. (1996) Analysis of the distribution of forest fires in Russia. In: Goldammer J.G. and Furyaev V.V. (Eds.) *Fire in Ecosystem of Boreal Eurasia*. Kluwer, 112–128.

Kovda V.A. and Rozanov B.G. (1988) *Soil Science*. Vysshaja shkola, Moscow (in Russian).

Krylov P.N. (1919) *Brief Description of Siberian vegetation*. Tomsk (in Russian).

Kurbatsky N.P. (1964) The problem of forest fires. In: *Appearance of Forest Fires*. Nauka, Moscow (in Russian).

Kurkhinen Yu.P. (1983) Change of environment, distribution character and quantity of herbivorous mammals in relation to industrial forestry. In: *Fauna and Ecology of Birds and Mammals of the North-West of the USSR*. Karelian filial AN SSSR, Petrozavodsk (in Russian).

Kuznetsov N.I. (1915) *Brief description of vegetation in Narymsk area of Tomsk region. Materials of the soil-botabical expedition on colonisation regions of the Asian Russia. Part II. Botanical research of the 1911*. St.-Petersburg (in Russian).

la Riviere J.W.M. and Marton-Lefevre J. (1992) United Nation conference on environment and development and global change science. *Global Change Newsletter*, **11**, 1–3.

Landau L.D., Akhiezer A.I., Lifshitz E.M. (1965) *General Physics. Mechanics and Molecular Physics*. Nauka, Moscow (in Russian).

Langman V.A., Roberts T.J., Black J., Maloiy G.M.O., Heglund N.C., Weber J.-M., Kram R. and Taylor C.R. (1995) Moving cheaply: Energetics of walking in the African elephant. *J. Exp. Biol.*, **198**, 629–632.

Larcher, W. (1980) *Ökologie der Pflanzen*. UTB Ulmer, Stuttgart.

Larson A. (1984) Neontological inferences of evolutionary pattern and process in the salamander family Plethodontidae. *Evol. Biol.*, **17**, 119–217.

Laurie C.C. (1997) The weaker sex is heterogametic: 75 years of Haldane's rule. *Genetics*, **147**, 937–951.

Lavrenko E.M. and Sochava V.B. (Eds.) (1956) *Vegetation Cover of USSR*. Explanatory text to 'Geobotanical map of USSR' (1 : 4 000 000). AN SSSR, Moscow-Leningrad (in Russian).

Lebedev V.L., Aizatullin T.A. and Khailov K.M. (1974) *Ocean as a Dynamic System*. Gidrometeoizdat, Leningrad (in Russian).

Leontieff W. (1977) *The Future of World Economy*. OUP, Oxford.

Leuenberger M., Siegenthaler U. and Langway C.C. (1992) Carbon isotope composition of atmospheric CO_2 during the last ice age from Antarctic ice core. *Nature*, **357**, 488–490.

Levitus S. (1982) *Climatological Atlas of the World Ocean*. NOAA Professional Paper, **13**, Geophysical Fluid Dynamics Lab., Princeton, New Jersey.

Lewin B. (1987) *Genes*. Wiley, New York.

Lewis M.P., Harrison W.G., Oakey N.S., Hebert D. and Platt T. (1986) Vertical nitrate fluxes in the oligotrophic ocean. *Science*, **234**, 870–873.

Li Z.-X., Le Treut H. and Ghil M. (1997) Atmospheric radiative equilibria in a simple column model, *Climate Dynamics*, **13**, 429–440.

Likens G.E., Bormann F.H., Pierce R.S. and Reiners W.A. (1978) Recovery of a deforested ecosystem. *Science*, **199**, 492–496.

Linnaeus, K. (1789–96) *Systema naturae*. V. 1–3, 13 ed. Lugdini.

Liski J. and Westman C. J. (1995) Density of organic carbon in soil at coniferous forest sites in southern Finland. *Biogeochemistry*, **29**, 183–197.

Listov A.A. (1986) *Scots Pine Forests of Lichen Site Type*. Agropromizdat, Moscow (in Russian).

Loeb L.A. and Kunkel T.A. (1982) Fidelity of DNA synthesis. *Ann. Rev. Biochem.*, **52**, 429–447.

Loftus G.R. and Bell S.M. (1975) Two types of information in picture memory. *J. Exp. Psychol.* [*Human Learning and Memory*], **104**, 103–113.

Lorius C. and Oeschger H. (1994) Paleo-perspectives: reducing uncertainties in global change? *Ambio*, **3**, 30–36.

Lorius C., Jouzel J., Raynaud D., Hansen J. and Le Treut H. (1990) The ice-core record: climate sensitivity and future greenhouse warming. *Nature*, **347**, 139–145.

Lotka A.J. (1925) *Elements of Physical Biology*. Williams Wilking Co., Baltimore.

Loudon A.S.I. (1985) Lactation and neonatal survival of mammals. *Symp. Zool. Soc. Lond.*, **54**, 183–207.

Lovelock, J.E. (1988) *The Ages of Gaia, a Biography of Our Living Earth*. Bantam Books: New York.

Ludwig W., Amiotte-Suchet P. and Probst J.-L. (1996) River discharges of carbon to the world's oceans: determining local inputs of alkalinity and of dissolved and particulate organic carbon. *C. R. Acad. Sci. Paris*, **323**, série IIa, 1007–1014.

Lurin B., Rasool S.I., Cramer W., Moore B. III. (1994) Global terrestrial primary productivity. *Global Change Newsletter*, **19**, pp. 6–8.

Lvovitch M.I. (1974) *World Water Resources and their Future*. Mysl, Moscow (in Russian).

Maak R.K. (1859) *Travel Along the Ussury River*. Tom 2. St.-Petersburg (in Russian).

MacFadden B.J. (1993) *Fossil Horses. Systematics, Paleobiology, and Evolution of the Family Equidae*. CUP, Cambridge.

MacFadden B.J. and Hubbert R.C. Jr. (1988) Explosive speciation at the base of the adaptive radiation of Miocene grazing horses. *Nature*, **336**, 466–468.

Maksimowski L.F. (1988) On the role of non-transcribed chromatin in the cellular processes of morphogenesis. *Ontogenes*, **19**, 461–467 (in Russian).

Malaspina P., Persichetti F., Noveletto A., Iodice C., Terrenato L., Wolfe J., Ferraro M. and Prantera G. (1990) The human Y chromosome shows a low level of DNA polymorphism. *Ann. Hum. Genet.*, **54**, 297–305.

Margulis L. (1975) Symbiosis in cell evolution life and its environment on the yearly Earth. In: Jennings D.H. and Lee D.L. (Eds.) *Symbiosis*. CUP, Cambridge, 21–38.

Marienfeld J.R., Unseld M., Brandt P. and Brennicke A. (1997) Viral nucleic acid transfer between fungi and plants. *Trends in Genetics*, **13**, 260–261.

Martin J.H. and Fitzwater S.E. (1992) Dissolved organic carbon in Atlantic, Southern and Pacific oceans. *Nature*, **256**, 699–700.

Martin J.H., Gordon R.M. and Fitzwater S.E. (1990) Iron in Antarctic waters. *Nature*, **345**, 156–158.

Martin J.P. and Fridovich I. (1981) Evidence for a natural gene transfer from the ponyfish to its bioluminiscent bacterial symbiont *Photobacter beioguanthi*. *J. Biol. Chem.*, **256**, 6080–6089.

Martin W., Brinkmann H., Savonna C. and Cerff R. (1993) Evidence for a chimeric nature of nuclear genomes: Eubacterial origin of eukaryotic glyceraldehide-3-phosphate dehydrogenase genes, *Proc. Natl. Acad. Sci. USA*, **90**, 8692–8696.

Maslov A.D., Vedernikov N.M. and Andreyeva G.I. (1988) *Forest Protection from Pests and Illnesses: Reference Book*. Agropromizdat, Moscow (in Russian).

May R.M. (1990) How many species? *Phyl. Trans. Roy. Soc. London*, **330B**, 293–304.

Maynard Smith J. (1964) Group selection and kin selection. *Nature*, **201**, 1145–1147.

Maynard Smith J. (1978) *The Evolution of Sex*. CUP, Cambridge.

Mayr E. (1963) *Animal Species and Evolution*. Harvard Univ., Cambridge, MA.

McFarland D. (1985) *Animal Behaviour: Phychobiology, Ethology and Evolution*. Pitmann Publishing Ltd., London.

McNab B.K. (1983) Energetics, body size, and the limits to endothermy. *J. Zool. (Lond.)*, **199**, 1–29.

Meadows D.H., Meadows D.L. and Randers J. (1992) *Beyond the Limits*. Chelsea Green Publ. Co, Post Milles, VT.

Meadows D.H., Meadows D.L., Randers J. and Behrens W.W. III (1972) *The Limits to Growth*. Potomac, New York.

Meadows D.L., Behrens W.W. III, Meadows D.H., Naill R.F., Randers J. and Zahn E.K.O. (1974) *Dynamics of Growth in a Finite World*. Wright Allen Press, New York.

Melekhov I.S. (1971) On regularities and periodicity of forest fires. In: *Miscellaneous Papers on Forest Husbandry and Forest Chemistry*. Severo-Zapadnoe knizhnoe izdatelstvo, Arkhangelsk (in Russian).

Melillo J.M., Prentice I.C., Farquhar G.D., Schulze E.-D. and O.E. Sala (1996) Terrestrial biotic responses to environmental change and feedbacks to climate. In: Houghton J.T., Meira Filho L.G., Callander B.A., Harris N., Kattenberg A., Maskell K. (eds.) *Climate Change 1995. The Science of Climate Change*. CUP, Cambridge, 445–516.

Merola M. (1994) A reassessment of homozygosity and the case for inbreeding depression in the cheetah, *acinonix jubatus*: implications for conservation. *Conserv. Biol.*, **8**, 961–971.

Mesarovic M. and Pestel E. (1974) *Mankind at the Turning Point*. Dutton, New York.

Michod R.E. and Lewin B.R. (1988) *The Evolution of Sex: An Examination of Current Ideas*. Sinauer, Sunderland.

Middendorf A.F. (1867) *Travel to the North-East of Siberia*. Part 1. North and East of the Siberia from natural and historical point of view. Chapter 4. Siberian vegetation. St-Petersburg (in Russian).

Miller D.H. (1976) *Water at the Surface of the Earth – An Introduction to Ecosystem Hydrodynamic*. Academic Press, New York.

Mitchell J. (1989) The 'greenhouse' effect and climate change. *Rev. Geophys.*, **27**, 115–139.

Molchanov A.A. (1960) *Hydrological Role of Forest*. AN SSSR, Moscow (in Russian).

Molchanov A.A. (1961) *Forest and Climate*. AN SSSR, Moscow (in Russian).

Molchanov A.A. (1971) *Productivity of Organic Phytomass in Forests of Different Geographical Zones*. Nauka, Moscow (in Russian).

Molchanov A.A. and Preobrazhensky I.F. (1957). *Forests and Forest Management in Arkhangelsk Region*. AN SSSR, Moscow (in Russian).

Mopper K. and Degens E.T. (1979) Organic carbon in the ocean: nature and cycling. In: Bolin B., Degens E.T., Kempe S. and Ketner P. (Eds.) *The Global Carbon Cycle*. Wiley, New York, 293–316.

Morin H., Laprise D. and Bergeron Y. (1993) Chronology of spruce budworms outbreak in the Lace Duparquet region, Abitibi, Québec. *Canadian Journal of Forest Research* **23**, 1497–1506.

Morneau C. and Payette S. (1989) Postfire lichen—spruce woodland recovery at the limit of the boreal forest in northern Québec. *Canadian Journal of Botany*, **67**, 2770–2782.

Moses Ph.B. and Chua N.-H. (1988) Light switches for plant genes. *Sci. Am.*, **258**, 64.

Mozolevskaja E.G., Lebedeva G.S. and Galasyeva T.V. (1979) Estimation and dynamics of secondary insect quantity in clear-cutting forests. *Forest Science (Lesovedenie)*, **2**, 91–98 (in Russian).

Nakagoshi N., Nehira K. and Takahashi F. (1987) The role of fire in pine forests of Japan. In: *The Role of Fire in Ecological Systems.* SPB Academic Publishing, The Hague.

Nei M. (1975) *Molecular Population Genetics and Evolution.* North-Holland, Amsterdam.

Nei M. (1984) Genetic polymorphism and neomutationism. *Lecture Notes in Biomathematics*, **53**, 214–241.

Nei M. and Graur D. (1984) Extent of protein polymorphism and the neutral mutation theory. *Evol. Biol.*, **17**, 73–118.

Neshyba S. (1987) *Oceanography Perspectives on a Fluid Earth.* John Wiley, New York.

Nevo E., Beiles A. and Ben-Shlomo R. (1984) The evolutionary significance of genetic diversity: ecological, demographic and life history correlates. *Lecture Notes in Biomathematics*, **53**, 13–213.

Newbold G. (1990) The path of drug resistance. *Nature*, **345**, 202–203.

Newkirk G. (1980) Solar variability of time scales of 10^5 years to $10^{9.6}$ years. In: Pepin R.O. and Eddy J.A. (Eds.) *Proc. Conf. Ancient Sun USA*, 293–320.

Nicolis J.S. (1986) *Dynamics of Hierarchical Systems: An Evolutionary Approach.* Springer-Verlag, Berlin.

Nikolayev D.S. (1932) *Forest Fires in the Ural Region and Fire-fighting*. Uraltekhizdat, Sverdlovsk (in Russian).

Nikonov V.V. (1987) *Soil Formation on the North Limit of Scots Pine Biogeocenoses.* Nauka, Leningrad (in Russian).

Ninio J. (1998) Acquisition of shape information in working memory, as a function of viewing time and number of consecutive images: evidence for a succession of discrete storage classes. *Brain. Res. Cogn. Brain. Res.*, **7**, 57–69.

North G.H., Cahalan R.F., Cookley J.A. (1981) Energy balance climate models. *Rev. Geophys. and Space Phys.*, **19**, 91–121.

O'Brien S.J. (1994) A role for molecular genetics in biological conservation. *Proc. Natl. Acad. Sci. USA*, **91**, 5748–5755.

O'Brien S.J., Gail M.H. and Levin D.L. (1980) Correlative genetic variation in natural populations of cats, mice and men. *Nature*, **288**, 580–583.

Ochman H. and Wilson A.C. (1987) Evolution in bacteria: evidence for a universal substitution rate in cellular genomes. *J. Mol. Evol.*, **26**, 74–86.

Odum E.P. (1971). *Fundamentals of Ecology*. 3rd Ed. W.B. Sounders, Philadelphia.

Odum E.P. (1983) *Basic Ecology*. Saunders College Pub., New York.

Oeschger H. and Stauffer B. (1986) Review of the history of atmospheric CO_2 recorded in ice cores. In: Trabalka J.R. and Reichle D.F. (Eds.) *The Changing of the Carbon Cycle: A Global Analysis*. Springer-Verlag, Berlin.

Ogawa H. and Ogura N. (1992) Comparison of two methods for measuring dissolved organic carbon in sea water. *Nature*, **256**, 696–698.

Ohta T. (1987) Very slightly deleterious mutations and the molecular clock. *J. Mol. Evol.*, **26**, 1–6.

Olson J.S. (1958) Rates of succession and soil changes on southern lake Michigan sand dunes. *Botanical Gazette*, **119**, 125–170.

Orgel L.E. (1992) Molecular replication. *Nature*, **358**, 203–209.

Orlov V.N. and Bulatova N.Sh. (1983) *Comparative Cytogenetics and Caryosystematics of Mammals*. Moscow, Nauka.

Orr H.A. (1995) Somatic mutation favors the evolution of diploidy. *Genetics*, **139**, 1441–1447.

Palmen E. and Newton C.W. (1969) *Atmospheric Circulation Systems*. Academic, New York.

Panteleyev P.A. (1983) *Bioenergetics of Small Mammals*. Nauka, Moscow (in Russian).

Parker G.A. and Maynard Smith J. (1990) Optimality theory in evolutionary biology. *Nature*, **348**, 27–33.

Partridge L. and Harvey P. (1986) Contentious issues in sexual selection. *Nature*, **323**, 580–581.

Paulyukyavichyus G.B. (1989) *The Role of Forest in Ecological Landscape Stabilization*. Nauka, Moscow (in Russian).

Paune R.B. and Westneat D.F. (1988) A genetic and behavioural analysis of mate choice and song neighborhoods in indigo buntings. *Evolution*, **42**, 935–947.

Payovski V.A. (1985) *Bird Demography*. Leningrad, Nauka (in Russian).

Peixoto J.P. and Oort A.H. (1984) Physics of climate. *Rev. Modern Phys.*, **56**, 365–378.

Penafiel S.R. and Terrado C.B. (1977) The 1970–1973 forest fires in the Upper Agno River Basin: a preliminary analysis. *Sylvatrop*, **2**, 59–63.

Peng T.-H. and Broecker W.S. (1991) Dynamic limitations on the Antarctic iron fertilisation strategy. *Nature*, **349**, 227–229.

Peng T.-H., Takahachi T., Broecker W.S. and Olafsson J. (1987) Seasonal variability of carbon dioxide, nutrients and oxygen in the northern North Atlantic surface water: observations and a model. *Tellus*, **39B**, 439–458.

Peters R.H. (1983) *The Ecological Implications of Body Size*. CUP, Cambridge.

Pierce B.A. and Mitton J.B. (1980) The relationship between genome size and genetic variation. *Am. Nat.*, **116**, 850–861.

Platt T. and Rao D.V.S. (1975) Primary production of marine microphytes. In: Cooper J.P. (Ed.) *Photosynthesis and Productivity in Different Environments*. CUP, Cambridge.

Platt T., Harrison W.G., Lewis M.R. *et al.* (1989) Biological production of the oceans: the case for a consensus. *Mar. Ecol. Prog. Ser.*, **52**, 77–88.

Pollack J.B. (1979). Climate change on terrestrial planets. *Icarus*, **37**, 479–553.

Powell J.R. (1975) Protein variation in natural populations of animals. *Evol. Biol.*, **8**, 79–119.

Prentice K.C. and Fung I.Y. (1990) The sensitivity of terrestrial carbon storage to climate change. *Nature*, **346**, 48–51.

Prinn R.G. (1982) Origin and evolution of planetary atmospheres: an introduction to the problem. *Planet. Space Sci.*, **30**, 741–753.

Prinn R.G. (1994) The interactive atmosphere: global atmospheric biospheric chemistry. *Ambio*, **23**, 50–61.

Protopopov V.V. (1975) *The Role of Dark-coniferous Forest in the Environment Formation*. Nauka SO, Novosibirsk (in Russian).

Quay P.D., Tilbrook B. and Wong C.S. (1992) Oceanic uptake of fossil fuel CO_2: carbon-13 evidence. *Science*, **256**, 74–79.

Quine C., Coutts M., Gardiner B. and Pyatt G. (1995) Forests and wind: management to minimize damage. *Forestry Commission Bulletin 114*. London.

Rabotnov T.A. (1983) *Phytocenology*. Moscow State University, Moscow (in Russian).

Radman M. and Wagner R. (1988) The high fidelity of DNA duplication. *Sci. Am.*, **259**, 40–46.

Rakhmanov V.V. (1981) *Forest Hydrology*. **3**. VINITI, Moscow (in Russian).

Ramanathan V. (1987) The role of Earth radiation budget studies in climate and general circulation research. *J. Geophys. Res.*, **92D**, 4075–4087.

Ramanathan V., Callis L., Cess R., Hansen J., Isaksen I., Kuhn W., Lacis A., Luther F., Mahlman J., Reck R., Schlesinger M. (1987) Climate–chemical interactions and effects of changing atmospheric trace gases. *Rev. Geophys.*, **25**: 1441–1482.

Randi E., Tosi G., Toso G., Lorenzini R. and Fusco G. (1990) Genetic variability and conservation problems in Alpine ibex, domestic and feral goat populations (genus *Capra*). *Z. Säugetierkunde*, **55**, 413–420.

Rasch E.M., Cassidy J.D. and King R.C. (1977) Evidence for dosage compensation in parthenogenetic *Hymenoptera. Chromosoma*, **59**, 323–340.

Raup D.M. and Sepkoski J.J.Jr. (1982) Mass extinction in the marine fossil fuel record. *Science*, **215**, 1501–1503.

Raval A. and Ramanathan V. (1989) Observational determination of the greenhouse effect. *Nature*, **342**, 758–761.

Raven P.H. and Johnson G.B. (1988) *Understanding Biology*. Times Mirror, Mosby Coll. Pub., St. Louis.

Raynaud D., Jouzel J., Barnola J.M., Chappellaz J., Delmas D.J. and Lorius C. (1993) The ice record of greenhouse gases. *Science*, **259**, 926–934.

Redfield A. (1958) The biological control of chemical factors in the environment. *Amer. Nat.*, **46**, 205–221.

Redfield A.C., Ketchum B.H. and Richards P.A. (1967) The influence of organisms on the composition of sea water. In: *The Sea*, vol. 2. Wiley, New York, 26–77.

Reimers N.F. (1972) *Ecological Successions and Fur-bearing Animals*. TSNIIL Glavokhoty RSFSR, Moscow (in Russian).

Repnevsky V.V. (1961) Sources of seeding in clear cutting areas. In: Melekhov I.S. (Ed.) *Forests of the Kola Peninsular and their Recovery*. AN SSSR, Moscow, 71–97 (in Russian).

Rowe J.S. and Scotter G.W. (1973) Fire in the boreal forest. *Journal of Quaternary Research*, **3**, 444–464.

Rozhkov A.S. and Massel G.I. (1982) *Resinous substances of conifers and xylophagous insects*. Nauka, Novosibirsk (in Russian).

Ryapolov V.Ya. (1987) Landscape position of Siberian silkworm hotbeds. In: Vladishevsky D.V. and Petrenko E.S. (Eds.) *Ecological Estimation of Habitat Condition of Forest Animals*. Nauka, Novosibirsk, 76–89 (in Russian).

Ryttman N., Tegelstrøm H. and Jansson H. (1979) G- and C- banding in four related Larus species (*Aves*). *Hereditas*, **91**, 143–148.

Sager R. (1972) *Cytoplasmic Genes and Organelles*. Academic, New York.

Sannikov S.N. and Sannikova N.S. (1985) *Ecology of Scots Pine Self-regeneration under Forest Canopy*. Nauka, Moscow (in Russian).

Sannikov S.N. (1992) *Ecology and Geography of Natural Regeneration of Scots Pine*. Nauka, Moscow (in Russian).

Sarmiento J.L., Toggweiler J.R. and Najjar R. (1988) Ocean carbon-cycle dynamics and atmospheric pCO_2. *Trans. R. Soc. London*, **A325**, 3–21.

Savin S. (1977) The history of the Earth s surface temperature during the past 160 million years. *Ann. Rev. Earth Planetary Sci.*, **5**, 319–355.

Schlesinger W.H. (1986) Changes in soil carbon storage and associated properties with disturbance and recovery. In: Trabalka J.R. and Reichle D.E. (Eds.) *The Changing Carbon Cycle. A Global Analysis*. Springer-Verlag, Berlin, 194–220.

Schlesinger W.H. (1990) Evidence from chronosequence studies for a low carbon-storage potential of soils. *Nature*, **348**, 232–234.

Schmidt-Nielsen K. (1972) *How Animals Work*. CUP, Cambridge.

Schmidt-Nielsen K. (1984) *Scaling: Why is Animal Size so Important?* CUP, Cambridge.

Schneider S.H. (1989) The greenhouse effect: science and policy. *Science*, **243**, 771–781.

Scholes B. (1999) Will the terrestrial carbon sink saturate soon? *Global Change Newsletter*, **37**, 2–3.

Schrödinger E. (1945) *What is Life? The Physical Aspect of the Living Cell*. Dublin Trinity College, Inst. for Advanced Studies, Dublin.

Schwartzman D.W. and Volk T. (1989) Biotic enhancement of weathering and habitability of Earth. *Nature*, **340**, 457–460.

Scientific and Applied Reference Book on the Climate of the USSR. (1988–1990) Ser. 3. Many years' data. Parts 1–6. Gidrometeoizdat, Leningrad (in Russian).

Selander R.K. and Johnson W.E. (1973) Genetic variation among verterbrate species. *Ann. Rev. Ecol. Syst.*, **4**, 75–91.

Session S.K. and Larson A. (1987) Developmental correlates of genome size in plethodontid salamanders and their implications for genome evolution. *Evolution*, **41**, 1239–1251.

Sheppard P.M. (1975). *Natural Selection and Heredity*. 4th Ed. Hutchinson University Library, London.

Sher A.V. (1990) Actualism and disconformism in studies of ecology of pleistocene mammals. *J. General Biology*, **51**, 163–177 (in Russian).

Shklovskii I.S. (1987) *Universe. Life. Mind.* Nauka, Moskow (in Russian).

Shukla J. and Mintz Y. (1982) Influence of land-surface evapotranspiration on the Earth's climate. *Science*, **215**, 1498–1500.

Shvidenko A., Nilsson. S. and Roshkov V. (1997) Possibilities for increased carbon sequestration through the implementation of rational forest management in Russia. *Water, Air and Soil Pollution*, **94**, 137–162.

Sieburth J.M. (1976) Bacterial substrates and productivity in marine ecosystems. *Ann. Rev. Ecol. Syst.*, **7**, 259–285.

Siegenthaler U. and Oeschger H. (1987) Biospheric CO_2 emission during the past 200 years reconstructed by deconvolution of ice core data. *Tellus*, **39B**, 140–154.

Siegenthaler U. and Sarmiento J. L. (1993) Atmospheric carbon dioxide and the ocean, *Nature*, **365**, 119–125.

Simpson G.G. (1944) *Tempo and Mode in Evolution.* Columbia University Press, New York.

Siren G. (1955) The development of spruce forest on raw humus sites in northern Finland and its ecology. *Acta Forestalia Fennica*, **62**, 1–363.

Skinner B.J. (1986) *Earth Resources.* Prentice-Hall Inc., New Jersey.

Skvortsova E.B., Ulanova N.G. and Basevich N.G. (1983) *Ecological Role of Wind Treefalls.* Lesnaja promishlennost, Moscow (in Russian).

Soberón J.M. (1990) Deforestation and the greenhouse effect. *Trends in Ecology and Evolution*, **5**, 63–64.

Southwood T.R.E., Brown W.K. and Reader M.E. (1978) The relationships of plant and insect diversities in succession. *Biological J.Linneian Society.*, **12**, 327–348.

Spurr S.H. and Barnes B.V. (1980) *Forest Ecology.* 3rd Ed. Wiley, New York.

Staffelbach T., Stauffer B., Sigg A. and Oeschger H. (1991) CO_2 measurements from polar ice cores: more data from different sites. *Tellus*, **43B**, 91–96.

Starke L. (Ed.) (1987), (1990) *The State of the World 1987, 1990. A Worldwatch Inst. Report on Progress toward a Sustainable Society.* Norton and Co, New York.

Starr C. (1971) Energy and power. *Sci. Am.*, **225**, 37–49.

Sugimura Y. and Suzuki Y. (1988) A high temperature catalytic oxidation method for determination of non-volatile dissolved organic carbon in seawater by direct injection of liquid samples. *Mar. Chem.*, **24**, 105–131.

Suzuki J. (1993) On the measurement of DOC and DON in seawater. *Marine Chemistry*, **41**, 287–288.

Syrjaenen K., Kalliola R., Puolasmaa A. and Mattsson J. (1994) Biodiversity in the fennoscandian forests: natural variation and its management. *Ann. Zool. Fenn*, **31**, 19–34.

Syvanen M. (1984) Conserved regions in mammalian β-globins: Could they arise by cross-species gene exchange? *J. Theor. Biol.*, **107**, 685–696.

Syvanen M. (1987) Molecular clocks and evolutionary relationships: Possible distortions due to horizontal gene flow. *J. Mol. Evol.*, **26**, 16–23.

Syvanen M. (1989) Classical plant taxonomic ambiguities extend to the molecular level. *J. Mol. Evol.*, **28**, 536–544.

Szathmary E. and Maynard Smith J. (1995) The major evolutionary transitions. *Nature*, **374**, 227–232.

Takahashi T., Broecker W.S. and Bainbridge A.E. (1981) Supplement to the alkalinity and total carbon dioxide concentration in the world oceans. In: Bolin B. (Ed.) *Carbon Cycle Modelling*, SCOPE, **16**, Wiley, Chichester, 159–199.

Takahashi T., Broecker W.S. and Langer S. (1985) Redfield ratio based on chemical data from isopycnal surface. *J. Geophys. Res.*, **90**, 6907–6924.

Tang B. and Weaver A.J. (1995) Climate stability as deduced from an idealized coupled atmosphere-ocean model. *Climate Dynamics*, **11**, 141–150.

Tans P.P., Fung I.Y. and Takahashi T. (1990) Observational constraints on the global atmospheric CO_2 budget. *Science*, **247**, 1431–1438.

Thomas C.D. (1990) Fewer species. *Nature*, **347**, 237.

Thompson K., Bakker J. and Bakker R. (1996) *The Soil Seed Banks of North West Europe*. CUP, Cambridge.

Thribus M. and McIrvine E.C. (1971) Energy and information. *Sci. Am.*, **224**, 179–188.

Timmermans K.R., Gledhill M., Nolting R.F., Veldhuis M.J.W., de Baar H.J.W. and van den Berg C.M.G. (1998) Responses of marine phytoplankton in iron enrichment experiements in the Northern North Sea and northeast Atlantic Ocean. *Marine Chemistry*, **61**, 229–242.

Tinbergen J., Dolman A. and von Ettinger J. (Eds.) (1977) *Reshaping the International Order. A report to the Club of Rome*. Times Mirror, New York.

Tinbergen N. (1953) *The Herring Gull's World*. Collins, London.

Tinbergen N. (1968) On war and peace in animal and man. *Science*, **160**, 1411–1418.

Tkachenko M.E. (1911) Forests of North. In: *Forestry Investigations in the Arkhangelsk region. Part 1. Studies of Experimental Forestry in Russia*. Vipusk 25, St.-Petersburg, pp. 1-91 (in Russian).

Toggweiler J.R. (1990) Diving into the organic soup. *Nature*, **345**, 203–204.

Tolonen M. (1978) Palaeoecology of annually laminated sediments in Lake Ahvenainen, S. Finland. I. Pollen and charcoal analyses and their relation to human impact. *Annales Botanici Fennici*, **15**, 177–208.

Townshend J.R.G. (Ed.) (1992) Improved global data for land application: A proposal for new high resolution data set. *Report of Land Cover Working Group of IGBP – DIS*. Global Change. Rep. 20, Stockholm.

Townshend J.R.G., Justice C.O., Li W., Gurney C. and McManus J. (1991) Global land cover classification by remote sensing: present capabilities and future possibilities. *Remote Sensing of Environment*, **35**, 243–256.

Trivett N.B.A. (1989) A comparison of seasonal cycles and trends in atmospheric CO_2 concentration as determined from robust and classical regression techniques. *Extended Abstracts of Papers Presented at the Third Internat. Conf. on Analysis and Evolution of Atmospheric* CO_2, Data. W.M.O., Univ. Heidelberg, 299–304.

Turelli M. and Begun D.J. (1997) Haldane s rule and X-chromosome size in *Drosophila*. *Genetics*, **147**, 1799–1815.

Turkov V.G. (1979) Wind uprooting (treefall) in virgin forest as a biogeocenotic phenomenon (by example of mountain fir-spruce forests in the Central Ural). In: *Dark Coniferous Forests of the Middle Ural*. Works of Animal and Plant Ecology Institute, Vipusk 128, Sverdlovsk, 121–140 (in Russian).

Vasilevich V.I. (1983) *Essays of Theoretical Phytocenology*. Nauka, Leningrad (in Russian).

Vasilevich V.I. (1993) Some new trends in the vegetation dynamics study. *Botanical Journal*, **78**, 1–15 (in Russian).

Vetrova V.P. (1987) The influence of stand structure on character of consortium connection of big black capricorn beetle with Siberian fir trees. In: Vladishevsky D.V. and Petrenko E.S. (Eds.) *Ecological Estimation of Habitat Condition of Forest Animals*. Nauka, Novosibirsk, 106–116 (in Russian).

Vinogradov M.E. and Shushkina E.A. (1988) Quantitative characteristics of the Pacific pelagic community. Production regions and estimates of photosynthetic primary production. *Oceanologija*, **28**, 819–829 (in Russian).

Vitousek P.M., Ehrlich P.R., Ehrlich A.H.E. and Matson P.A. (1986) Human appropriation of the products of photosynthesis. *Bioscience*, **36**, 368–375.

Vitousek P.M., Mooney H.A., Lubchenko J. and Mellilo J.M. (1997) Human domination of earth ecosystems. *Science*, **277**, 494–499.

Vladyshevsky D.V. and Petrenko E.S. (Eds.) (1987). *Ecological Estimation of Habitat Condition of Forest Animals*. Nauka, Novosibirsk (in Russian).

Vogel F. and Motulsky A.G. (1979), (1982), (1986) *Human Genetics*. Springer-Verlag, Berlin.

Vogel F. and Rathenberg R. (1975) Spontaneous mutation in man. *Adv. Hum. Genetics*, **5**, 223–318.

Volkov A.D. (1998) Natural properties of primary forests in the taiga zone and problem of their conservation. In: Gorshkov V.G., Makarieva A.M., Kharkina T. (Eds.) Proc. Int. Sem. *Role of Virgin Terrestrial Biota in the Modern Processes of Global Change: Biotic Regulation of the Environment*. 12–16 October 1998, Petrozavodsk, Russia. PNPI, Gatchina, 240–248.

Vorisek P., Votypka J., Zvara K. and Svobodova M. (1998) Heteroxenous coccidia increase the predation risk of parasitized rodents. *Parasitology*, **117**, 521–524.

Vorontsov A.I. (1963) *Biological Fundamentals of Forest Protection*. 2nd Ed. Vysshaja shkola, Moscow (in Russian).

Wang S., Foote W.C. and Bunch T.D. (1990) Genetic variability in domesticated and wild sheep based on blood protein characters. *Comp. Biochem. Physiol. B*, **96**, 201–207.

Watts J.A. (1982) The carbon dioxide question: Data sampler. In: Clark W.C. (Ed.) *Carbon Dioxide Review*. Clarendon Press, New York.

White M.J.D. (1973) *Animal Cytology and Evolution*. CUP, Cambridge.

White R. and Lalouel J.-M. (1988) Chromosome mapping with DNA markers. *Sci. Am.*, **258**, 22.

Whitmore T. (1991) *An Introduction to Tropical Rain Forest*. Clarendon Press, Oxford.

Whittaker R.H. (1975) *Communities and Ecosystems*. Macmillan, New York.

Whittaker R.H. and Likens G.E. (1975) The biosphere and man. In: Lieth H. and Whittaker R. (Eds.) *Primary Productivity of the Biosphere*. Springer-Verlag, Berlin, 305–328.

Whittaker R.H. and Woodwell G.M. (1968) Dimension and production relations of trees and shrubs in the Brookhaven Forest, New York. *Journal of Ecology*, **56**, 1–25.

Whittaker R.H. and Woodwell G.M. (1969) Structure production and diversity of the oak-pine forest at Brookhaven, New York. *Journal of Ecology*, **57**, 157–176.

Wiebinga C.J. and de Baar H.J.W. (1998) Determination of the distribution of dissolved organic carbon in the Indian sector of the Southern Ocean. *Marine Chemistry*, **61**, 185–201.

Williams G.C. (1975) *Sex and Evolution*. Princeton University, New York.

Willson R.C. (1984) Measurements of solar total irradiance and its variability. *Space Science Reviews*, **38**, 203–242.

Wilson D.S. (1980) *The Natural Selection of Populations and Communities*. Benjamin Cummings, Menlo Park, California, USA.

Wilson D.S. (1997) Biological communities as functionally organised units. *Ecology*, **78**, 2018–2024.

Wilson E.O. (1975) *Sociobiology*. Harvard University Press, Cambridge MA.

Wilson E.O. (1989) Threats to biodiversity. *Sci. Am.*, **261**, 108.

Wilson E.O. (Ed.) (1988) *Biodiversity*. Natl. Acad. Press, Washington.

Winberg G.G. (1976) Dependence of energetic metabolism on body mass in water ectothermic animals. *J. General Biology*, **37**, 56–64 (in Russian).

Winberg G.G. (Ed.) (1979) *Fundamentals of Aquatic Ecosystems Research*. Nauka, Leningrad (in Russian).

Wofsy S.C., Goulden M.L., Munger J.W., Fan S.-M., Bakwin P.S., Daube B.C., Bassow S.L. and Bazzaz F.A. (1993) Net exchange of CO_2 in a mid-latitude forest. *Science*, **260**, 1314–1317.

World Bank (1997) *Russia. Forest Policy During Transition*. The World Bank, Washington, DC, USA.

World Resources 1988-1989, (1988) **16**. *Land Cover Settlements*. Basic Books. Inc., New York, 64–265.

Wright S. (1932) The roles of mutation, inbreeding, crossbreeding, and selection in evolution. Proc. 6th Int. Congr. *Genet.*, **1**, 356–366.

Wright S. (1988) Surfaces of selective value revisited. *Am. Nat.*, **131**, 115–123.

Wynne-Edwards V.C. (1986) *Evolution Through Group Selection*. Blackwell, Oxford.

Yurgenson P.B. (1973) *Ecological Fundamentals of Hunting Economy in Forests*. (Forest Industry) Lesnaja promishlennost, Moscow (in Russian).

Zaboyeva I.V. (1975). *Soils and Land Resources in Komi ASSR*. Komi knizhnoe izdatelstvo, Syktyvkar (in Russian).

Zackrisson O. (1977) Influence of forest fires on the North Swedish boreal forests. *Oikos*, **29**, 22–32.

Zjabchenko S.S. (1984). *Pine Forests of the European North*. Nauka LO, Leningrad (in Russian).

Zonn S.V. (Ed.) (1991) *Degradation and Recovery of Forest Soils*. Nauka, Moscow (in Russian).

Zuckerkandl E. (1987) On the molecular evolutionary clock. *J. Mol. Evol.*, **26**, 34–46.

Zuckerkandl E. and Pauling L. (1965) Evolutionary divergence and convergence in proteins. In: Bryson V. and Vogel H.J. (Eds.) *Evolving Genes and Proteins*. Academic, New York, 97–165.

Index